JN434622

Aviation Technical English

항공기술영어

ASD-STE100

이상희 · 오경훈 · 한상경 · 하영태 편찬

한국항공우주기술협회 경문사

Copyright statement

Copyright

The information in this document is the property of ASD (previously AECMA). Transmittal, receipt, or possession of the information does not express license or imply any rights to use, sell, or manufacture from this information and no reproduction or publication of it, in whole or in part, shall be made without the written authority of an officer of ASD.

Copyright in the whole and every part of this document is fully owned by:

ASD, AeroSpace and Defence Industrise Association of Europe
Rue du Trône 100
1050 Brussels, Belgium
www.asd-europe.org
© ASD, 2005, 2007, 2010, 2013, 2017, 2021
(previously AECMA, 1986, 1987, 1988, 1989, 1995, 1998, 2001, 2004)

Special usage rights

Irrevocable permission to use, sell, or manufacture from the information contained in this document or any subsequent modification or revision thereof, and the right to reproduce or publish this document or any subsequent modification or revision thereof, in whole or in part, free of charge, is hereby given to the following:

1. National Associations who are members of ASD, and all their member companies.
2. Members of Aerospace Industries Association of America (AIA) and Canada (AIAC).
3. Members of International Coordinating Council of Aerospace Industries Associations (ICCAIA) not included in categories 1 thru 4.
4. Airlines that are customers of companies included in categories 1 thru 3.
5. Ministries of Defense of the member countries of ASD.
6. The Department of Defense of the USA.
7. Airlines for America (A4A), formerly known as Air Transport Association of America (ATA).
8. Airworthiness Authorities.

Statement of liability

No legal significance or implications should be attached to the rules and terms in this document, the purpose of such terms being to provide definitions which are of practical value for technical communication. No legal significance or implications should be attached to the contents of this document when it is used outside the intended technical context.

추천사

항공우주 및 방위산업분야의 기술지침서는 모두 영어로 작성되어 있으나 실제 이 지침서를 활용하는 기술자의 구성은 비영어권이 많은 실정입니다. 통일된 간편기술영어의 국제적인 사용으로 영어가 모국어가 아닌 항공분야 종사 기술자에게 크게 도움이 될 것입니다.

조선대학교 항공우주공학과 명예교수 공창덕

STE-100(항공기술영어)은 항공기에 관련된 기술도서를 작성할 때 단어와 문장 작성방법을 국제표준으로 규정한 것으로, 비영어권에 속한 우리 항공산업계(제작자, 사용자, 운용자 등)가 항공기술영어를 필수적인 기술영어 지식으로 정착시킴으로써 항공안전의 토대를 다지는 수단이 될 것입니다.

전 국토교통부 항공철도사고조사위원회 사무국장 김관연

우리 항공분야에 종사하는 기술자는 여러 항공기 제작사, 엔진 및 부품 제작사 등의 기술도서를 접하는데 문장구조, 단어 조합 등이 서로 달라 이해하는 데 애로가 많았습니다. 이러한 문제점을 해결하기 위하여 항공기술도서가 통일된 기준으로 작성되도록 국제적인 지침서가 규정되어, 누구나 보다 쉽고 확실하게 이해할 수 있는 길잡이 역할을 함으로써 항공기의 안전운항에 크게 기여할 것입니다. 항공기술인을 양성하는 초도 교육기관에서부터 적극적으로 항공기술영어(STE-100)를 필수 기본과목으로 운영할 것을 권유합니다.

전 대한항공 부사장 강영식

발간사

<Aviation Technical English(Simplified Technical English)의 탄생 배경>

영어가 모든 산업분야에서 주된 공용 언어로 광범위하게 사용되는 국제적인 추세에서 특히 항공우주 및 방위산업분야의 기술지침서는 모두 영어로 작성되고 있으나 이 지침서를 활용하는 전 세계의 항공기술분야의 인력 분포를 보면 무려 80%의 사용자가 비영어권에 속해 있다. 외국어로서의 한정된 영어지식을 가진 항공분야의 기술자들이 직면하는 애로사항은 다양한 문장구조도 문제지만 의미가 비슷한 단어(동의어)를 제작사별로 무분별하게 적용하는 수많은 단어의 조합도 기술문서를 이해하는 데 있어 커다란 장애요소가 되어 왔다. 항공분야에 종사하는 기술자는 조종사와 더불어 승객의 생명과 안전을 보장하기 위해서 제작사의 지침서에 따라 정확하게 점검 및 정비 등의 적절한 조치가 필수적인데 해당 기술도서와 작동지침을 완벽하게 이해하지 못한다면 이것은 매우 심각한 문제라 할 것이다.

이에 대한 근본적인 대책으로 항공관련 기술문서와 도서를 보다 간편하고 명확하게 해득할 수 있도록 유럽항공우주산업협회(ASD)와 미국의 항공우주산업협회(AIA)가 공동으로 국제적으로 적용할 항공기술영어 작성규정을 제정하였다. 이렇게 탄생한 규정이 바로 "ASD-STE100"이며 항공기 제작사를 포함한 모든 항공산업계(부품제작자, 사용자, 운영자 등)가 항공기술 문서와 지침서를 작성할 때 **일반단어의 사용범위를 제한하고 문장의 구조도 간결하게 구성시키는 항공기술영어(STE)가 탄생하게 되었다.**

<항공기술영어 도서 발간의 의의>

항공기술이 빠르게 고도로 발전하는 환경에서 신기술에 대한 완벽한 적응력을 갖추기 위해서는 영어가 선택사항(optional)이 아니라 필수사항(essential)이 될 정도로 항공분야의 종사자에게 기술능력 못지 않게 영어능력이 중요한 요소가 되었다. 더욱이 항공산업이 국제적인 규모로 성장하면서 항공사만이 아니라 정비지원이나 지상조업을 하는 서비스업체도 모든 작업문서를 **영어**로 작성하는 것이 기본으로 되었고 또한 국내와 국외에 관계 없이 항공기운항을 지원하는 자(조종, 정비, 운항, 조업 등)와의 기본적인 의사소통을 위한 공용어로 **영어**의 사용범위가 보편화되고 있다.

이러한 시대적인 변화에 따라 대부분의 항공산업체에서 항공종사자에게 영어능력을 갖추도록 요구하고 있는 추세이나, 항공기술업무와는 거리가 먼 일반적인 사회영어의 지식평가방식이 획일적으로 채택되고 운용되어 왔다.

항공우주기술협회는 이러한 비효율적이고 실용성도 없는 단순평가방식을 개선하는 보완책으로 국제적으로 권고하는 항공기술영어를 항공기술인의 양성단계에서 필수교육으로 시행할 수 있도록 기술영어도서를 발간하기로 하였다.

항공기술영어를 익히는 것은 단순하게 학문으로서의 영어지식을 높이기 위한 목적이 아니라, 우리의 항공산업계가 보다 신뢰성과 효율성을 보장하고 국제적인 위상을 높이며 궁극적으로는 비행안전의 토대를 다지기 위한 필수적인 수단이라 하겠다.

나아가서 모든 항공산업계가 항공기술영어를 필수적인 기술영어 지식으로 정착시켜 항공안전을 위한 기본적인 도구로 활용하기를 강력히 권고하고자 한다.

본 도서가 영어지식 전달의 범위를 넘어 항공산업계에서 실무적으로 적용할 수 있는 내용으로 구성되도록 각고의 노력을 기울인 집필위원과 감수위원의 노고를 치하하며, 한편으로 항공기술영어의 필요성과 적용성에 대한 현업에서의 평가를 위하여 파일럿과정을 운영해 준 대한항공에게도 진심으로 감사를 드린다

부연해서 ASD-STE100도 살아있는 언어이기 때문에 계속해서 기술의 변화를 반영하여 3~4년 주기로 개정이 이루어지며 STEMG web site(stemg@asd-ste100.org)를 통해서 최신판을 확인할 수 있고 무료로 다운로드받을 수 있다. 참고로 본 도서는 가장 최신판인 2021년에 발행한 ASD-STE100-issue-8.pdf를 기준으로 편집하였다. 또한 본 규정의 국내 정착을 위하여 당 협회가 추진하는 출판 의지에 뜨거운 격려를 보내준 유럽항공우주산업협회(ASD) 당국자인 Orlando 씨에게도 이 자리를 빌어 감사를 드린다.

Note: ASD Simplified Technical English, ASD-STE100 is a Copyright and a Trade Mark of ASD, Brussels, Belgium.

2022.

항공우주기술협회장 이상희

GENERAL INTRODUCTION

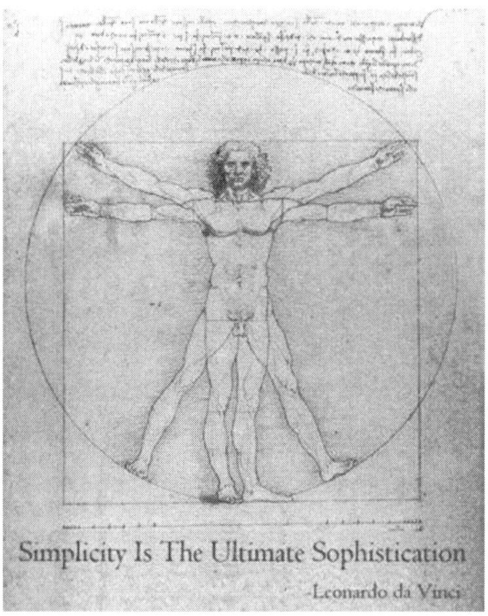

What is ASD Simplified Technical English?

ASD Simplified Technical English (STE) is an international specification for the preparation of technical documentation in a controlled language.

STE has two parts: a set of writing rules (PART 1) and a controlled dictionary (PART 3). The writing rules are about grammar and style. The dictionary gives the general words that a writer can use.

When and why was STE developed?

English is the international language of science, technology, and human relations. It is also the language of the aerospace and defense industry. However, it is not often the native language of the readers of technical documentation. Many readers have a limited knowledge of English. Complex sentence structure and the large number of meanings and synonyms that many English words have can cause confusion.

In aviation, it is mandatory to correctly understand maintenance and operation documentation to make sure that systems operate safely and correctly and to protect human lives.

In the late 1970s, the Association of European Airlines (AEA) asked the European Association of Aerospace Industries (AECMA, now ASD www.asd-europe.org) to investigate the readability of maintenance documentation in the civil aviation industry and find a solution to simplify the language used to write such documentation.

AECMA asked the Aerospace Industries Association (AIA) of America to help with this project and two project groups from AECMA and AIA were formed. These two groups explored the existing controlled languages and researched texts in several maintenance manuals. The results confirmed that a simplified language was necessary.

On June 30, 1983, in Amsterdam, the AECMA Simplified English Working Group was founded and the AECMA Simplified English project started.

The product of this effort was the AECMA Simplified English Guide (first release in 1986) which, in 2005, became the ASD Simplified Technical English Specification, ASD-STE100. In 1987, the Air Transport Association of America (ATA) included the requirement to write in STE in its ATA100 specification for technical publications. Thus, the name ASD-STE 100 was assigned as a tribute to the name of the ATA 100 specification.

Today, the success of STE is such that other industries use it beyond its original intended purpose of aerospace maintenance documentation. Interest in STE has also increased dramatically in the areas of language services, professional translation and interpreting, as well as in the academic world.

What is the purpose of STE?

The purpose of STE is to tell technical writers how to write technical texts in a clear, simple and unambiguous manner that readers throughout the world will find easy to understand.

The STE specification is not a treatise on the English language or technical writing. Thus, linguistic matters are not included in this specification because they are explained in other language reference books and guidelines.

Which words are available to the writer?

STE has a controlled dictionary that gives the words that are most frequently used in technical writing.

The approved words were selected because they were simple and easy to recognize. In general, each word has only one meaning and is approved with only one part of speech. For example, "to fall" has the approved meaning of "to move down by the force of gravity," and not "to decrease".

When there are several words in English for the same thing (synonyms), STE permits one of these synonyms to the exclusion of the others. For example, STE uses “start” instead of “begin”, “commence”, “initiate”, or “originate”. STE approved meanings and spelling are based on American English (Merriam-Webster’s dictionary).

In addition to its dictionary, STE permits the use of company-specific or project-specific technical words (referred to in STE as technical names and technical verbs). STE gives you the rules and the categories for these technical words.

Basically, writers can use the approved words in the dictionary as a core vocabulary. But they can also use terms that are usual in their companies or industries and applicable to their projects and products.

Does STE give rules for abbreviations of nomenclature?

No. Industries, companies, and projects can use different abbreviations and nomenclature. Thus, it is not practical to create such rules.

Does STE regulate text formatting?

No. STE regulates the way to express the content. It does not regulate formatting (for example, typeface, numbering, and lettering). Usually, these requirements are included in the applicable specifications for technical publications, style guides, and other official directives.

Does STE regulate units of measurement?

No. There are different methods to show units of measurement in a technical text. Projects and companies must decide which one to use. Usually, these requirements are included in the applicable specifications for technical publications, style guides, and other official directives.

Can STE be used alone?

No. It is intended to be used with other applicable specifications for technical publications, style guides, and official directives. A high standard of professionalism is necessary to use the STE specification correctly.

Can STE be used for oral communication?

No. STE was developed for technical documentation only. But the rules and a simpler vocabulary can help communication in meetings and presentations.

Can STE be used to teach a writer English?

No. STE is not an English course book. This specification will help the writer give complex information in an easy form. Writing clearly is a complex task and it is necessary to have a high level of fluency in English to write STE correctly.

Can STE help technical translation?

Yes. STE can be a good aid for technical translation. When STE was created, one of the primary objectives was to make translation easier. If the vocabulary, meanings of words, and the types of sentence constructions in a text are controlled, the variation between texts will be minimal.

Thus, it is easier for translators or translation machines to translate text written in STE into the target language.

Can the writer get training in STE and find supporting software products?

There are organizations, companies, and individuals that market and give training courses in the use of STE, and there are producers of software products or authoring tools that support STE.

Neither ASD, the Simplified Technical English Maintenance Group (STEMG), nor any organization or company associated with the production of STE intend or imply any warranty or endorsement to these organizations, companies, and individuals that give training or produce supporting software products.

For training only, the above statement is not applicable to organizations, companies, and individuals who, through official agreements, are accredited by ASD to give authorized training. For more information, refer to the STEMG website at www.asd-ste100.org.

ASD-STE100 SIMPLIFIED TECHNICAL ENGLISH

Table of Contents

PART 1 - WRITING RULES

George Orwell "Politics and the English Language"

Rules for Writers:

1. *Never use a metaphor, simile, or other figure of speech which you are used to seeing in print.*

2. *Never use a long word where a short one will do.*

3. *If it is possible to cut a word out, always cut it out.*

4. *Never use the passive where you can use the active.*

5. *Never use a foreign phrase, a scientific word, or a jargon word if you can think of an everyday English equivalent.*

6. *Break any of these rules sooner than say anything outright barbarous.*

"If you simplify your English, you are freed from the worst follies of orthodoxy... ...And when you make a stupid remark, its stupidity will be obvius, even to yourself."
George Orwell (1946) - Politice and the English Language.

Blank Page

Section 1 - Words

Summary of the rules

Which words can you use?

Rule 1.1	You can use words that are: - Approved in the dictionary - Technical names - Technical verbs.

Part of speech

Rule 1.2	Use approved words from the dictionary only as the part of speech given.

Approved meaning

Rule 1.3	Use approved words only with their approved meanings.

Forms of verbs and adjectives

Rule 1.4	Use only the approved forms of verbs and adjectives.

Technical names

Rule 1.5	You can use words that you can include in a technical name category.
Rule 1.6	Use a word that is unapproved in the dictionary only when it is a technical name or part of a technical name.
Rule 1.7	Do not use words that are technical names as verbs.
Rule 1.8	Use technical names that agree with approved nomenclature.
Rule 1.9	When you must select a technical name, use one which is short and easy to understand.
Rule 1.10	Do not use slang or jargon words as technical names.
Rule 1.11	Do not use different technical names for the same item.

Technical verbs

Rule 1.12	You can use verbs that you can include in a technical verb category.
Rule 1.13	Do not use technical verbs as nouns.

Spelling

Rule 1.14	Use American English spelling (unless other official directives tell you differently.)

Which words can you use?

Rule 1.1	**You can use words that are:** **- Approved in the dictionary** **- Technical names** **- Technical verbs.**

Simplified Technical English (STE) has a controlled dictionary (part 2) that gives you the words most frequently used in technical writing. You can also use words that are not in the dictionary if you can include them in the specified categories of technical names and technical verbs.

Examples:

The word "use" is an approved verb in the dictionary.
The word "engine" is a technical name.
The word "ream" is a technical verb.

The dictionary also gives a selection of unapproved words, with examples that show how to use alternative words.

Part of speech

Rule 1.2	**Use approved words from the dictionary only as the part of speech given.**

In the dictionary, each approved word has a specified part of speech. When you use an approved word, make sure that you only use it as that part of speech.

Examples:

The word "test" is an approved noun, but not an approved verb.

STE: Test B is an alternative to test A.

Non-STE: *Test the system for leaks.*
STE: Do the leak test of the system.
or
STE: Do a test for leaks in the system.

The word "dim" is an approved adjective, but not an approved verb.

STE: A dim light comes on.

Non-STE: *Dim the lights.*
STE: Set the lights to the dim position.

There are some words (not many) that are approved as more than one part of speech. For example, the word "clean" is an approved verb but also an approved adjective. The position of the word in the sentence will tell you its function (and its meaning,) because verbs and adjectives have different positions.

STE: Clean the inner surface of the container.
("Clean" is a verb here.)

STE: Make sure that the area is clean.
("Clean" is an adjective here.)

The word "acceptable" is an unapproved adjective. The dictionary gives an approved alternative that has the same part of speech. You can use that word to replace the unapproved word in the sentence (word-for-word replacement.)

Non-STE: A value of 2 mm is acceptable.
STE: A value of 2 mm is permitted.

The word "operable" is an unapproved adjective. The dictionary gives an approved alternative that has a different part of speech ("operate" as a verb.) Then, you must use a different sentence construction.

Non-STE: Make sure that the valve is operable.
STE: Make sure that the valve can operate.

If a word that you want to use is not in the dictionary, search for that word in an English dictionary and find which is the best synonym listed in the STE dictionary. Then, use the approved STE word.

When you replace a word, always make sure that the alternative you select does not change the meaning of the sentence. If the meaning changes, use a different construction.

Approved meaning

Rule 1.3	**Use approved words only with their approved meanings.**

Each approved word in the dictionary has a specified approved meaning which can be more restricted than its meaning in standard English. Always use these words with their approved meanings.

Example:

The approved meaning of the word "follow" is "come after, go after" and not "obey."

Non-STE: Follow the safety instructions.
STE: Obey the safety instructions.

But you can write:

STE: Do the instructions that follow:

STE: Follow the green lights to the nearest staircase.

Forms of verbs and adjectives

Rule 1.4 Use only the approved forms of verbs and adjectives.

Each approved verb in the dictionary is given with its approved forms. Approved adjectives are given in their basic form, with their comparative and superlative forms in parentheses where applicable.

Examples:

REMOVE (v),
REMOVES,
REMOVED,
REMOVED

This word tells you that you can use the approved verb "remove" as follows:

Infinitive / Imperative	Present tense	Simple past tense	Past participle (as an adjective)
(To) Remove / Remove	Remove(s)	Removed	Removed

The past participle of the verb is often the same as the simple past tense. This is why it is given two times in the dictionary.

SLOW (adj)
(SLOWER, SLOWEST)

This word tells you that you can use the approved adjective "slow" as follows:

Basic form	Comparative form	Superlative form
Slow	Slower	Slowest

Adjectives that make their comparative and superlative forms with "more" and "most" do not have these forms listed in the dictionary, because "more" and "most" are also listed as approved words.

Technical names

Rule 1.5	**You can use words that you can include in a technical name category.**

Technical names are words related to categories that are specified in this rule. The dictionary does not include technical names as approved words, because there are too many, and each manufacturer uses different technical names. STE gives you a list of categories, with examples, to help you use technical names correctly.

The listed technical names are only examples, and this is not a complete list of all possible technical names. Words in the list have capital letters only when it is necessary. For example, units of measurement, titles, official names, or quoted text.

Words are technical names if you can include them in one or more of these 20 categories.

1. **Names in the official parts information**
   (for example, Illustrated Parts Catalog/List or engineering drawing):
   bolt, cable, clip, conductor, contact, engine, ferry tank, filter, hatch, indicator, light, logo, oil seal, prelubricated seal, pipe, propeller, retractor link, screw, switch, transceiver
2. **Names of vehicles or machines, and locations on them:**
   aircraft, aircraft carrier, airframe, airplane, bicycle, cabin, car, cargo compartment, cargo hold, cockpit, deck, engine room, fuselage, helicopter, galley, lifeboat, overhead panel, ship, submarine, tank, train, truck, wing, wing root
3. **Names of tools and support equipment, their parts, and locations on them:**
   access ladder, blade, brush, cap, chock, clamp, cover, display, file, gauge (gage), handle, jack, label, rigging pin, roller, rope, rung, shaft, stand, tag, test rig, torque wrench, trestle
4. **Names of materials, consumables, and unwanted material:**
   acid, adhesive, aluminum alloy, ammunition, compound, copper, debris, detergent, dirt, disinfectant, dust, foam, foreign object, fuel, grease, oil, paint, penetrant spray, plastic, primer, sealant, sealing, soap, stainless steel, tape, waste, water, wire
5. **Names of facilities, infrastructure, and logistic procedures:**
   airport, apron, base, building, camp, dock, engine shop floor, flight simulator, gate, handling, hangar, packaging, packing, port, service bay, shipping, shop, store, storage, transport
6. **Names of systems, components and circuits, their functions, configurations, and parts:**
   air conditioning, amplifying circuit, armament, audio, aural warning system, collapsed position, exhaust, flight management, hardware, inhibiting signal, injection, inlet, input frequency, latch, pedal, power unit, pump, reverse mode, reverse position, standby mode, upright position, vent

7. **Mathematical, scientific, engineering terms, and formulas:**
acceleration, allowance, astronomy, atom, average, biochemistry, biology, biome, burr, capacitance, carbon, category, cavitation, center, circle, coefficient, configuration, conversion, count, critical temperature, curve, cycle, defect, degree, deceleration, density, diameter, displacement, duty cycle, elapsed time, electricity, energy, exponent, failure, ferry flight, flutter, force, fumes, genetics. geology. geophysics, graph, gravity, hardness, heat treatment, idle speed, ignition, incidence, inhibition, instrumentation, interference, issue, light, line replaceable unit, load, loss, measurement, modification, momentum, motoring, overhaul, oversized hole, oxygen, performance, phase, polarity, power, pressure, process, radius, rating, ratio, reduction, resistance, scan, shutdown, signal, specific gravity, stall, standard, steam, stiffness, strength, suction, temperature, tension, thread, tightness, torque, toxic property, vapor, voltage, water vapor, "C = (A - B) - 0.063 mm"

8. **Navigation and geographic terms:**
air, altitude, attitude, axis, bank, clearance, climb, coordinates, critical approach, datum, delay, deviation, drag, east, France, glideslope, gradient, heading, landing, leeway, Lima, north, pitch, roll, skid, south, west

9. **Numbers, units of measurement and time (and their symbols):**
92, 303, ½, ¼, ampere (A), degree (°), first, half, hour (h), kilogram (kg), knot, liter (L or l), meter (m), mile, minute ('), month, ohm (Ω), one, one-quarter, second ("), second (s), second, square inch (sq.in.), spring, third, three, year, winter, zero

10. **Quoted text** (for example, that on placards, labels, signs, markings, and display units):
abort button, EXIT sign, INOP system, OXYGEN pushbutton switch, ON position, NEXT button, FAULT legend, NO STEP marking, FASTEN SAFETY BELT sign, WEAR PROTECTIVE CLOTHING sign

11. **Names of persons, groups, or organizations:**
air traffic control, captain, commander, copilot, crew, crew chief, European Aviation Safety Agency (EASA), Federal Aviation Administration (FAA), manufacturer, operator, Transport Canada Civil Aviation (TCCA)

12. **Parts of the body:**
blood, digestive system, ear, eyes, hair, hand, head, lung, mouth, respiratory tract, skin, stomach

13. **Common personal effects, food, and drink:**
beans, bread, cigarette lighter, clothing, coffee, flour, footwear, high-heeled shoes, jewelry, lipstick, matches, milk, mineral water, nail scissors, perfume, pizza, shampoo, wine

14. **Medical terms:**
allergy, aspirin, asthma, blood poisoning, breathing, circulation, dermatitis, diabetes, dizziness, female, headache, heart rate, irritation, male, medication, nausea, pneumonia, pregnancy, pulse, skin irritation, virus

15. **Names of official documents and parts of documentation**
(this includes manuals, technical records, standards, specifications, and regulations):
Acceptance Test, Allowable Damage, attention, caution, chapter, Checklist, Class, Cleaning, Compass Correction Card, danger, data module, Description and Operation, diagram, engine logbook, Federal Aviation Regulations, Fault Isolation, figure, flow chart, font, Functional Test, Ice and Rain Protection, Inspection/Check, issue, letter, maintenance planning, maintenance practice, maintenance records, Normal Braking, note, notice, packaging, page, paragraph, parentheses, post-flight report, post-mod, pre-mod, preservation, reference, Removal/Installation, Repair Scheme, revision, section, Service Bulletin, Standard Practices Manual, storage, Structural Repair Manual, table, test procedure, Transportation, valid welding certificate, warning

16. **Environmental and operational conditions:**
atmosphere, cloud, day, daylight, ice, hail, humidity, lightning, moisture, night, rain, sand, snow, storm, turbulence, volcanic ash, wind

17. **Colors:**
beige, black, cyan blue, dark brown, gray, green, magenta, light green, orange, red, white, yellow
NOTE: Colors are adjectives, but – conventionally – STE defines them as technical names. Comparative and superlative forms of colors (for example, blacker, the reddest) are not permitted in STE.

18. **Damage terms:**
buckle, chafing, corrosion, crack, crack propagation, deformation, dent, discoloration, distortion, erosion, fracture, fraying, galling, kink, nick, score, scratch, stain, spurious fault message

19. **Information technology and telephony terms:**
add-in, add-on, arrow, authentication, backup, backup file, bookmark, content, cursor, database, dialog check box, e-mail, field, file, firewall, HTML, icon, interface, internet, laptop, local operation, memory, menu, mouse, network, operating system, phone, plug-in, pre-loaded software, preset value, remote operation, screen, smartphone, status bar, store, tablet, toolbar, touchscreen, tweet, update, voice mail, XML

20. **Military terms:**
armed forces, assault, bomb, bullet, checkpoint, combat plan, deployment, echelon, ejection seat, evacuation, formation, general, ground zero, gun, lieutenant, machine gun, mission, patrol, rank

Rule 1.6 Use a word that is unapproved in the dictionary only when it is a technical name or part of a technical name.

The dictionary includes some unapproved words that can be technical names if you can put them in the applicable technical name category.

Examples:

"Base" is an unapproved word in the dictionary. But you can use this word as a technical name.

STE: The base of the triangle is 5 cm.
("Base" is a technical name, category 7, mathematical, scientific, engineering terms, and formulas.)

Non-STE: Make sure that the two spigots at the base of the unit engage.
("Base" is an unapproved word that is related to a surface.)
STE: Make sure that the two spigots at the bottom of the unit engage.

The same word "base" can go into different technical name categories when it is used with different meanings in different contexts.

Example:

STE: Access to the base is permitted between 9 a.m. and 6 p.m.
("Base" is a technical name, category 5, names of facilities, infrastructure, and logistic procedures.)

"Backup" is an unapproved word in the dictionary. But you can use this word as a technical name.

Example:

STE: Do the backup of the computer at regular intervals.
("Backup" is a technical name, category 19, information technology and telephony terms.)

"Backup" is a one-word technical name. But you can also write "backup file," a two-word technical name which contains the unapproved word "backup."

Examples:

STE: Keep the backup file in a safe area.
("Backup file" is a technical name, category 19, information technology and telephony terms.)

Non-STE: For this procedure, make sure that one person is available as backup.
("Backup" is unapproved here and you cannot use it.)
STE: Two persons are necessary to do this procedure.

"Main" is an unapproved word and its approved alternative is "primary." But you can use this word as part of a technical name.

Examples:

Non-STE: *The laptop has these main parts:*
("Main part" is not a technical name, and it is correct to replace "main" with "primary.")
STE: The laptop has these primary parts:

STE: Retract the main landing gear.
("Main landing gear" is a technical name. Do not replace "main" with "primary" here, because "primary landing gear" is not the official name given, for example, in company nomenclature.)

Rule 1.7 Do not use words that are technical names as verbs.

Use a technical name only as a noun or as an adjective that is part of a technical name. Do not use the same word as a verb.

Examples:

"Oil" is a technical name (category 4, names of materials, consumables, and unwanted material.) Do not use "oil" as a verb. Use a different construction which lets you use it as a noun.

Non-STE: *Oil the steel surfaces.*
STE: Apply oil to the steel surfaces.

"Snow" is a technical name (category 16, environmental and operational conditions.) Do not use "snow" as a verb. Use a different construction which lets you use it as a noun.

Non-STE: *If you think it will snow, make sure that the vehicle is in the applicable configuration.*
STE: If you think that snow will fall, make sure that the vehicle is in the applicable configuration.

Words that can be technical names and also technical verbs

The same word can be a technical name and a technical verb when you can put this word in a technical name category and also in a technical verb category (rule 1.12.)

Examples in STE:

Remove the rivets from the flange.
("Rivet" is a technical name, rule 1.12, category 1, names in the official parts information.)

Rivet the panel in its position.
("Rivet" is a technical verb, category 1 c), manufacturing processes, attach material.)

Rule 1.8	**Use technical names that agree with approved nomenclature.**

If there is a designated technical name for a system, component, part, or process, use that technical name. Usually, those technical names are included in official parts data and in company documentation.

Example:

STE: The front panel of the phone has a touchscreen and a home button.
("Touchscreen" and "home button" are technical names that are included in company nomenclature.)

Technical names from approved nomenclature can be long groups of words (noun clusters.) When possible, you can make these technical names shorter (refer to section 2.)

Rule 1.9	**When you must select a technical name, use one which is short and easy to understand.**

When there is no technical name in approved nomenclature, select one that is short and easy to understand. Always make sure that the technical names that you select are well-known words.

Example:

Non-STE: Remove the four stainless steel pan head machine screws (10) that attach the metallic machined flange (15) to the front housing cover (20).
STE: Remove the four screws (10) that attach the flange (15) to the cover (20).

In this example, it is sufficient to use the words "screws," "flange," and "cover," because these parts have index numbers and are clearly identified in the related illustration.

Rule 1.10 Do not use slang or jargon words as technical names.

Some words are only used in a specific geographical area. These words will be very difficult to understand for people from outside of this area. Always use the word that most people will know. This is also applicable to technical jargon. If a word is only understood by a very small number of people in a specific technical area, it will cause confusion.

Example:

Non-STE: *Make a sandwich with two washers and the spacer.*
STE: Install the spacer between the two washers.

Non-STE: *Remove your gear from the work area.*
STE: Remove your tools and equipment from the work area.

Rule 1.11 Do not use different technical names for the same item.

When you select a technical name, do not use a different name in other parts of your text to refer to the same item. For example, if you use “actuator” as a technical name, always use that name. Do not use “servo control unit” or other technical names.

Examples:

Non-STE: *1. Make sure that the servo control unit is in the open position.*
2. Do the operational test of the actuator.
3. Disconnect the control unit from the test rig.
STE: 1. Make sure that the actuator is in the open position.
2. Do the operational test of the actuator.
3. Disconnect the actuator from the test rig.

In the non-STE example, “servo control unit,” “actuator,” and “control unit” refer to the same item. Refer to your company nomenclature and use the word that occurs there. If, as in the example, the word is "actuator," then use this word consistently in your procedure (and throughout your documentation.)

Technical verbs

Rule 1.12 You can use verbs that you can include in a technical verb category.

Technical verbs are words that give instructions and information in specified technical and operational contexts. The dictionary does not include technical verbs because there are too many, and each manufacturer uses different words to describe the same action. STE gives you a list of categories, with examples, to help you use technical verbs correctly.

The listed technical verbs are only examples, and this is not a complete list of all possible technical verbs. Technical verbs must obey the same rules as other approved verbs in STE (refer to section 2.)

Words are technical verbs if you can include them in one or more of these four categories.

1. **Manufacturing processes**

   Technical verbs in this category give instructions to:

   a) **Remove material:**
   drill, grind, mill, ream

   b) **Add material:**
   flame, insulate, remetal, retread

   c) **Attach material:**
   braze, crimp, rivet, solder, weld

   d) **Change the mechanical strength, the structure, or physical properties of a material:**
   anneal, cure, decay, freeze, heat-treat, magnetize, normalize, vaporize

   e) **Change the surface finish of a material:**
   buff, burnish, dress, passivate, plate, polish

   f) **Change the shape of a material:**
   blend, cast, extrude, spin, stamp

2. **Computer processes and applications**

   This category of technical verbs is for:

   a) **Input/output processes**:
   click, digitize, enter, press, print, swipe, tap, type

   b) **User interface and application processes**:
   clear, close, copy, cut, delete, deselect, disable, drag, drag and drop, enable, encrypt, erase, filter, highlight, maximize, minimize, navigate, open, paste, save, scroll, sort, store, tweet, zoom in, zoom out

   c) **System operations**:
   abort, boot, communicate, debug, download, format, install, load, manage, process, reboot, update, upgrade, upload

3. **Descriptions**

This category of technical verbs is only for descriptive texts. For example, texts applicable to general information, system description and operation, descriptive parts of service bulletins, technical reports, technical and legal papers. Do not use these verbs when you write procedures. These verbs refer to:

a) **Mathematical, scientific, and engineering processes**
bisect, compensate for, convert, detect, emit, modulate, radiate, transform

b) **Military processes**
aim, arm, detect, disable, enable, explode, fire, intercept, load, lock on, parachute, unload

c) **Regulatory language**
waive (for inspection and requirements), comply with, conform to, supersede, meet (a requirement)

4. **Operational language**

This category of technical verbs is only for texts that refer to an operational context.

Operational texts tell users how to operate and use something correctly. For example, a manual on how to use a phone, a tablet, a medical device, or a television set is an operational text. Aircrew manuals and land or sea information sets are also operational texts.

Examples:

airdrop, alert, approach, authorize, brief, call, contact, crank, descend, deviate, disembark, drift, dry-motor, enable, evacuate, fasten, ferry, fly, hover, inform, inhibit, land, load, maintain, navigate, observe, provide, reach, respond, retard, retrim, return, rotate, serve, sanitize, shut down, sideslip, sit, sleep, sterilize, switch off, switch on, take off, take over, taxi, tie, trigger, trim, unfasten, unlatch, unload, verify, wet-motor

If there is an approved verb in the dictionary that accurately gives the instruction and information, use it. Do not use a technical verb if it is possible to write the same sentence with the words that are approved in the dictionary.

Examples:

Non-STE: *If you detect broken wires, repair them.*
STE: If you find broken wires, repair them.

But you can write:

STE: The security scanner detects metallic objects.

If you must use technical verbs, use only specific verbs. Do not use verbs that are general and vague.

Example:

Non-STE: *Machine the hole until it has a diameter of 8.00 +/- 0.003 mm.*
STE: Ream the hole until it has a diameter of 8.00 +/- 0.003 mm.

Do not create a technical verb if it is not necessary. If possible, use a simple verb that is approved in the dictionary and an applicable technical name.

Examples:

"Clamp" is a technical name (category 1, names in the official parts information.) Do not use "clamp" as a technical verb.

Non-STE: *Clamp the cable in position.*
STE: Put clamps on the cable to hold it in position.

"Grease" is a technical name (category 4, names of materials, consumables, and unwanted material.) Do not use "grease" as a technical verb.

Non-STE: *Grease the fasteners.*
STE: Apply grease to the fasteners.

"Wire" is a technical name (category 1, names in the official parts information.) Do not use "wire" as a technical verb.

Non-STE: *Wire the cable to the structure.*
STE: Attach the cable to the structure with wire.

<u>Unapproved words that can be technical verbs</u>

The dictionary includes some unapproved words that can be technical verbs if you can put them in the specified categories.

Examples:

STE: Enter your password.
("Enter" is a technical verb, category 2 a), computer processes and applications, input/output processes.)

Non-STE: *Do not enter the engine test area without approval.*
("Enter" is an unapproved word that is related to different contexts.)
STE: Do not go into the engine test area without approval.

STE: If the tower does not respond, use a different channel.
("Respond" is a technical verb, category 4, operational language.)

Non-STE: *If the instrument fails to respond, do a test.*
("Respond" is an unapproved word that is related to different contexts.)
STE: If the instrument does not operate correctly, do a test.

Rule 1.13 Do not use technical verbs as nouns.

In English, words that look the same do not always have the same function in a sentence. Only use technical verbs as verbs, not as nouns.

Example:

Non-STE: *Give the hole 0.20-inch over-ream.*
STE: Ream the hole 0.20 inch larger than the standard.

But you can use the past participle of technical verbs as adjectives. In the same way that you can use the participle form of an approved verb to make an adjective, you can also make adjectives from technical verbs.

Example:

STE: Lubricate the reamed hole.
(The adjective "reamed" is the past participle of the technical verb "ream". "Reamed hole" is a technical name.)

<u>Words that can be technical verbs and also technical names</u>

The same word can be a technical verb and a technical name when you can put this word in a technical verb category and also in a technical name category.

Examples in STE:

Make sure that the plate is not damaged.
("Plate" is a technical name, category 1, names in the official parts information.)

There are two methods to plate the ring nut (2).
("Plate" is a technical verb, category 1 c), manufacturing processes, attach material.)

Spelling

Rule 1.14 Use American English spelling (unless other official directives tell you differently.)

Use the spelling specified in the STE dictionary (American English spelling). Use a different spelling only if other technical publication specifications, style guides, contracts, or other official directives apply.

Examples:

Non-STE: *The door is made of carbon fibre reinforced plastic.*
("Fibre" is British English spelling.)
STE: The door is made of carbon-fiber-reinforced plastic.
("Fiber" is American English spelling.)

Non-STE: Change the colour of the display.
("Colour" is British English spelling.)
STE: Change the color of the display.
("Color" is American English spelling.)

Section 2 - Noun clusters

Summary of the rules

Noun clusters

Rule 2.1	Write noun clusters of no more than three words.
Rule 2.2	When a technical name has more than three words, write it in full. Then you can simplify it as follows: - Give a shorter name or - Use hyphens (-) between words that are used as a single unit.

Articles and demonstrative adjectives

Rule 2.3	When applicable, use an article (the, a, an) or a demonstrative adjective (this, these, that, those) before a noun or a noun cluster.

Noun clusters

Rule 2.1 Write noun clusters of no more than three words.

In English, you can use one or more words to describe (or modify) a noun.

Technical texts often contain long groups of words that have the function of one part of speech in a sentence. Usually, these groups of words are made of nouns and/or adjectives and are the subject or the object in a sentence. Such word groups are known as noun clusters.

General examples:

Horizontal cylinder pivot bearing

Stainless steel corrosion protection strips

Actuator operating rod

Long noun clusters are not easy to understand because the words in the noun cluster can connect to each other in different ways. The "main noun", or "head", in the cluster is usually the last word of the noun cluster. When words can link up in different ways, ambiguity occurs. Thus, short noun clusters are easier to understand.

General examples:

Runway light connection
(This is a short noun cluster (3 words). The main noun is "connection.")

Runway light connection resistance calibration
(This is a long noun cluster (5 words). The main noun is "calibration.")

The long noun cluster in the example does not tell the reader the relation between “runway” and “calibration.” The reader must understand four modifying words to get to the main noun “calibration”.

Long noun clusters are also confusing for non-native English readers, because in some languages the main noun is the first noun in the cluster. Thus, the more words there are in a cluster, the more difficult it is to understand.

To help your reader, keep noun clusters to a maximum of three words. To do this, you can use prepositions (for example, “of,” “on,” “in,” and “for”) to explain the noun cluster. In STE, articles and prepositions do not count as words in a noun cluster.

Examples:

Non-STE: *Runway light connection resistance calibration.*
STE: Calibration of the resistance of the runway light connection.

Non-STE: *Install the forward turbine overheat thermocouple terminal tags.*
STE: Install the terminal tags on the forward overheat thermocouple of the turbine.

Non-STE: *Remove the engine transmission housing attachment bolts.*
STE: Remove the bolts that attach the transmission housing to the engine.

Non-STE: *Adjust to obtain door operating rod alignment with the attachment point.*
STE: Adjust the door operating rod until it aligns with the attachment point.

Rule 2.2 When a technical name has more than three words, write it in full. Then you can simplify it as follows:
- Give a shorter name
or
- Use hyphens (-) between words that are used as a single unit.

A long noun cluster can be a technical name on its own or it can be a combination of shorter technical names. Frequently, it is not possible to divide them into smaller parts (because they are part of company nomenclature) and you must write them as they are.

Shorter names

If a long technical name comes from an official document (for example, an engineering drawing or an illustrated parts catalog,) write it in full the first time that it occurs in the text. Then, if possible, in the remaining text of your document, use a shorter name or its approved abbreviation.

Examples in STE:

Before you do this procedure, engage the ramp service door safety connector pin (the pin that holds the ramp service door, referred to in this procedure as the “safety connector pin”).

> The Main Fuel Metering Unit (MFMU) is an aluminum alloy unit that includes a Main Engine Control Unit (MECU) and a Distribution Block (DB). The MFMU is installed in the engine bypass duct and operates in the engine fuel system. The function of the MFMU is to meter and distribute the fuel from the Main Engine Fuel Pump (MEFP) to the fuel manifolds and the starter jets. The Digital Engine Control Unit (DECU) sends electrical signals to operate the MFMU.

In the first example, you write "ramp service door safety connector pin" in full. Then, after an explanation, you give a shorter name: "safety connector pin." This shorter name has three words and obeys rule 2.1.

In the second example, the explanation is not necessary because the text gives all the necessary information about the unit. All official technical names that have more than three nouns are written in full the first time that they occur. Then, in the remaining parts of the text, the related approved abbreviations are used.

Hyphens (-)

A hyphen is a punctuation mark that connects words or parts of words. You can use hyphens between words to show how related words function as one unit.

Examples in STE:

> Make sure that the cutoff-switch power connection is safe.
>
> Inspection of the lavatory rapid-decompression device.

Make sure that you do not connect words which are not related, because this will change the meaning of the noun cluster. If you are not sure, only explain the noun cluster. Then, use a shorter version, or an official approved abbreviation.

If a technical name from company nomenclature includes hyphens, do not change it. If it is too long, write it in full the first time it occurs and then use the recommended method (shorter names) specified in this rule.

Do not hyphenate groups of more than three words. It does not help your reader to understand the noun cluster if you hyphenate all of the words in it:

Example:

Non-STE: *Main-gear-door-retraction-winch handle.*
STE: Main-gear-door retraction-winch handle.

Articles and demonstrative adjectives

Rule 2.3	**When applicable, use an article (the, a, an) or a demonstrative adjective (this, these, that, those) before a noun or a noun cluster.**

Articles and demonstrative adjectives show where nouns and noun clusters are. Use articles and demonstrative adjectives correctly and do not omit them to make the text shorter.

Examples:

Non-STE: *Turn shaft assembly.*
STE: Turn the shaft assembly.

Non-STE: *Data module tells you how to operate unit.*
STE: This data module tells you how to operate the unit.

It is not always correct English to put an article before a noun. Do not use articles in general statements:

Example:

STE: Solvents can cause damage to paint.

In short sentences, it can be clearer to use articles before all nouns.

Example:

STE: Install the nuts (2) and the bolts (3).

But sentences that contain a long series of items are clearer when you do not repeat the articles.

Example:

STE: Discard the O-rings (3), gaskets (4), seals (7), and washers (9).

Also, a definite article is incorrect before a noun when an alphanumeric identifier comes after it.

Example:

Incorrect: *Tag the circuit breaker 36L7.*
CORRECT: Tag circuit breaker 36L7.

Section 3 - Verbs

Summary of the rules

Forms and tenses of verbs

Rule 3.1 Use only those forms of the verb that are given in the dictionary.

Rule 3.2 Use the approved forms of the verb to make only:
- The infinitive
- The imperative (command form)
- The simple present tense
- The simple past tense
- The past participle (as an adjective)
- The future tense.

Rule 3.3 Use the past participle as an adjective.

Rule 3.4 Do not use helping verbs to make complex verb structures.

Rule 3.5 Use the "-ing" form of a verb only as a technical name or as a modifier in a technical name.

Active voice

Rule 3.6 Use only the active voice in procedural writing. Use the active voice as much as possible in descriptive writing.

How to describe an action

Rule 3.7 Use an approved verb to describe an action (not a noun or other parts of speech.)

Forms and tenses of verbs

Rule 3.1 Use only those forms of the verb that are given in the dictionary.

The STE dictionary gives you the forms that you can use for each approved verb.

Examples:

Infinitive / Imperative	Simple present tense	Simple past tense	Past participle (as an adjective)
To decrease / Decrease	Decrease(s)	Decreased	Decreased
To give / Give	Give(s)	Gave	Given

Rule 3.2 Use the approved forms of the verb to make only:
- The infinitive
- The imperative (command form)
- The simple present tense
- The simple past tense
- The past participle (as an adjective)
- The future tense.

Use only the verb tenses that are approved.

Examples:

Infinitive	Imperative (command form)	Simple present tense	Simple past tense	Past participle (as an adjective)	Simple future tense
(To) Adjust	Adjust + object	You/we/they adjust It adjusts	You/we/they adjusted It adjusted	The adjusted linkage	You/we/they will adjust It will adjust

Do not use other verb tenses that are not approved, for example:
- The present perfect (have/has adjusted)
- The past perfect (had adjusted)
- The present/past progressive (is/was adjusting)
- And all other complex verb forms.

Rule 3.3 Use the past participle as an adjective.

You can use the past participle of a verb as an adjective, as follows:

- Before a noun
- After a form of the verb “to be” or “to become.”

When you use the past participle as an adjective, it describes the condition of something. This is not passive voice.

Do not use the past participle if it is not in the dictionary. The past participle is not permitted for some irregular verbs, for example, “been,” the past participle of “to be.” The participle “been” is never an adjective. It is always part of a complex verb form.

Examples:

STE: Connect the disconnected wires.
(“Disconnected” is an adjective before the noun “wires.”)

STE: The wires are disconnected.
("Disconnected" is an adjective after the verb "to be" that shows the condition of the wires.)

There are also approved adjectives in the dictionary that are the past participle of unapproved verbs. For example, "specified," "permitted," "completed," and "damaged." Their approved part of speech in the dictionary is "(adj)" and thus you can use these words.

Examples:

STE: These areas have the specified damage limits.
STE: Do not put more than the permitted weight on the trolley.
STE: Send the completed report to the supervisor for approval.
STE: Make sure that the mating surfaces are not damaged.

Rule 3.4 Do not use helping verbs to make complex verb structures.

Do not use the past participle with the auxiliary verb "to have." This will make a tense that is not approved.

Example:

Non-STE: The operator has adjusted the linkage.
(The present perfect is an unapproved tense.)
STE: The operator adjusted the linkage.
(The simple past tense is approved.)

Some complex verb structures include auxiliary or modal verbs ("can," "must," "will," or "to be,") the infinitive of "to be," and the past participle. Such sentences are complex passive sentences.

Examples:

Non-STE: The volume control can be adjusted.
STE: You can adjust the volume control.

Non-STE: The temperature must be adjusted.
STE: Adjust the temperature.

Non-STE: The sleeve will be adjusted.
STE: You will adjust the sleeve.

Non-STE: The seat is to be installed before you install the cushion.
STE: Before you install the cushion, install the seat.

Rule 3.5	**Use the "-ing" form of a verb only as a technical name or as a modifier in a technical name.**

In English, words that end in "-ing" can have different functions in a sentence (different parts of speech).

They can be part of a verb to describe an action in the present.

General example:

Be careful while the door is opening.

They can also be adjectives.

General example:

An opening door can be dangerous.

They can be nouns or parts of noun phrases.

General example:

Opening a door can be dangerous.

They can be used to make long strings of modifiers, noun phrases, and dependent clauses.

General example:

A mechanic opening a door without obeying the specified safety precautions can easily cause injury to persons standing near the door.

Because these different functions in a sentence can frequently cause ambiguity or lead to long, complex sentences, it is usually not permitted in STE to use words that end in "-ing."

Examples:

Non-STE: *When you are doing this procedure, obey all the safety precautions.*

STE: When you do this procedure, obey all the safety precautions.

Non-STE: *Mechanics wearing insufficient protective clothing and opening containers containing hazardous materials in areas where there is a lack of ventilation, using inappropriate tools without observing the manufacturer's instructions, are in danger of coming into contact with these materials and thus suffering from skin irritation and breathing problems.*

STE: Before you use dangerous materials, obey these precautions:

(1) Read the manufacturer's instructions.
(2) Make sure that there is sufficient airflow in the work area.
(3) Put on a face mask and protective clothing.
(4) Get the correct tools to open the containers for these materials.

If you do not obey these precautions, injury to your skin and lungs can occur.

Words with an "-ing" ending that are technical names or parts of technical names

You can use a word with an "-ing" ending as a technical name (for example, in procedural titles.)

Examples:

Cleaning, Testing and Fault Isolation, Handling, Packaging, Shipping, Troubleshooting

You can also use the "-ing" form of a verb as a modifier in a technical name. This "-ing" form word is an adjective which describes the function of a system, component, part, tool, or equipment.

Examples:

Air-conditioning system, degreasing agent, grinding wheel, polishing disc, sanding machine, switching relay, welding torch

Approved words with an "-ing" ending

Only a very few approved words in the dictionary have an "-ing" ending. They are:
- Nouns (lighting, opening, routing, and servicing)
- Adjectives (mating, missing, and remaining)
- A pronoun (something)
- A preposition (during).

Active voice

Rule 3.6	**Use only the active voice in procedural writing. Use the active voice as much as possible in descriptive writing.**

Technical texts consist of procedural writing and descriptive writing. When you write in STE, use only the active voice in procedural writing and, as much as possible, in descriptive writing.

What is "active" or "passive" voice?

In the active voice, the subject of the sentence does the action of the sentence ("A" does "B"). Thus, the grammatical subject (A) is also the logical subject (agent).

In the passive voice, the subject of the sentence receives the action ("B" is done by "A"). Here, the grammatical subject is B, and the logical subject, or agent, is A.

General examples:

Active: The manufacturer supplies the safety procedures.
Passive: The safety procedures are supplied by the manufacturer.

Active: The side stay holds the main gear leg.
Passive: The main gear leg is held by the side stay.

How do you know if a sentence is written in the passive voice?

The best test for the passive voice is to ask the question "by whom or by what?" (the agent). If your text gives you an answer to this question, then it is written in the passive voice. When a sentence contains the preposition "by," it is a good indication that the sentence can be passive. The object of the preposition "by" is then the agent and you can use it as the subject of a sentence in the active voice. In the examples that follow, the agent is underlined.

In each of the passive examples, you can ask the question "by whom or by what?"

The safety procedures are given by the manufacturer.

The main gear leg is held by the side stay.

But a passive construction does not always contain an agent.

The dimensions are given in the table.

The main gear leg is held in its position.

Usually, you can write descriptive text in the active voice without problems. But when you describe how a system is made or operates, it is sometimes necessary to use the passive voice to keep the correct technical content. Use the passive voice only when it is really necessary.

How do you change a sentence that is in the passive voice to the active voice?

To change a sentence from the passive voice to the active voice, you can use one of these four methods:

Method 1

When the agent (the person or thing that does the action) is given in the sentence (usually the object of the preposition "by"), put the agent at the beginning of the sentence and use it as the subject. The subject must always be the noun that does the action in the sentence, as shown in the diagram that follows:

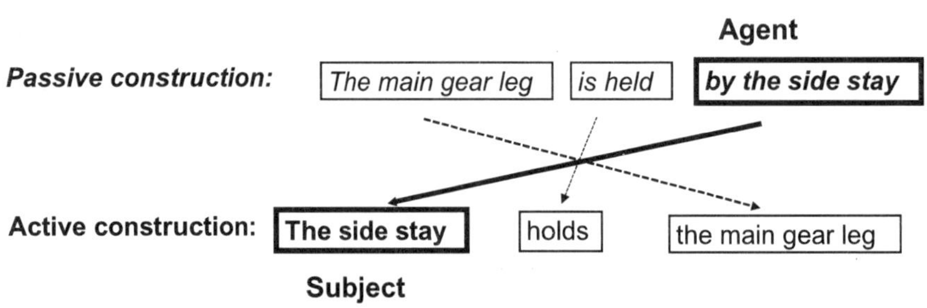

Example:

Non-STE: *The circuits are connected by a switching relay.* (Passive)

STE: A switching relay connects the circuits. (Active)

Method 2
Change an infinitive verb to an active verb.

Example:

Non-STE: *These values are used by the computer to calculate the energy consumption.* (Passive)
STE: The computer calculates the energy consumption from these values. (Active)

The construction "are used by" gives no important information here. Thus, you can use the verb "calculate" to write the sentence in the active voice.

Method 3
In procedural writing, change the verb to the imperative ("command") form.

Examples:

Non-STE: *The test can be continued by the operator.* (Passive)
STE: Continue the test. (Active)

Non-STE: *Oil and grease are to be removed with a degreasing agent.* (Passive)
STE: Remove oil and grease with a degreasing agent. (Active)

Method 4
When the agent (the person or thing that does the action) is not given in the sentence, you can use the pronouns "you" or "we" as subjects in the active form. If the agent is the reader, use "you." If the agent is your company, use "we."

Examples:

Non-STE: *On the ground, the valve can be opened with the override handle.* (Passive)
STE: On the ground, you can open the valve with the override handle. (Active)

Non-STE: *Additives are not used in this product.* (Passive)
STE: We do not use additives in this product. (Active)
or
STE: This product does not contain additives.

How to describe an action

Rule 3.7	**Use an approved verb to describe an action (not a noun or other parts of speech.)**

If there are different ways to give the same information, use an approved verb to show the action. Nouns can describe actions in an abstract way, but verbs describe actions more clearly.

Examples:

Non-STE: *The ohmmeter gives an indication of 450 ohms.*
STE: The ohmmeter shows 450 ohms.

Non-STE: *Make sure that the 600 Hz tone is audible when the GPW GND ACT switch is pushed.*
STE: (a) Push the GPW GND ACT switch.
(b) Make sure that you can hear the 600 Hz tone at the same time.

If a word is not approved as a verb in the dictionary, do not use it as a verb. Use a different construction to give the same information.

Example:

Non-STE: *Check the laptop battery.*
STE: Do a check of the laptop battery.

Section 4 - Sentences

Summary of the rules	
Short sentences and simple sentence structure	
Rule 4.1	Write short and clear sentences.
Rule 4.2	Do not omit words or use contractions to make your sentences shorter.
Vertical lists	
Rule 4.3	Use a vertical list for complex texts.
Connecting words and phrases	
Rule 4.4	Use connecting words and phrases to connect sentences that contain related topics.

Short sentences and simple sentence structure

The basic principle of STE is to make texts easy to read and understand. Thus, you must write short sentences and use simple sentence structure.

In STE, there are different rules for how to write sentences in procedures (section 5) and for how to write sentences in descriptive text (section 6.)

Rule 4.1 Write short and clear sentences.

Write short, clear sentences that give specific information.

In procedures, give short, clear instructions, and address the reader directly.

Example:

Non-STE: *To remove the cover assembly (9), first remove the four screws (10) that attach the cover (11) to the housing (12), and then, after taking the cover (11) off the housing (12), remove the preformed packing (13) and throw it away.*

STE:
1. Remove the cover assembly (9) as follows:
   A. Remove the four screws (10) that attach the cover (11) to the housing (12).
   B. Remove the cover (11) from the housing (12).
   C. Remove and discard the preformed packing (13).

In a descriptive text, make sure that each sentence has only one topic. Then, in the sentences that follow, gradually give information to develop that topic.

Example:

Non-STE: *The side stay assembly has two folding toggles hinged together and attached with hinges between the main gear strut and the side stay bracket.*

(This sentence contains two topics which are connected by the underlined conjunction "and." To make this information clearer, you can write a new sentence for each topic.)

STE: The side stay assembly has two folding toggles. The folding toggles are attached together with hinges. These toggles are also attached with hinges between the main gear strut and the side stay bracket.

For the two types of writing, always make sure that your text is not abstract and that it clearly shows how to do a task or how a system operates. Be accurate. Do not give vague, general information.

Examples:

Non-STE: *No leaks permitted.*

(This sentence is not very helpful. It does not tell the reader what to do.)

STE: Make sure that there are no leaks.

(This sentence directly tells the reader what to do.)

Non-STE: *Different temperatures will change the cure time.*

(This is an abstract sentence because it contains no information on how the cure time decreases.)

STE: When the temperature increases, the cure time will decrease.

(This sentence tells the reader the correct information on how the cure time decreases. To make your writing more accurate, give the specified temperatures and the related cure times.)

Rule 4.2 Do not omit words or use contractions to make your sentences shorter.

Each sentence must have all of its parts. Do not omit words or use contractions (for example, "don't," "isn't," "aren't") when you write sentences. If you do that, your sentence will be shorter, but it will not be easier to read. Write all words in full.

Do not omit nouns to make sentences shorter. It will be difficult for the reader to know what you mean.

Example:

Non-STE: *Can be a maximum of five inches long.*

STE: Cracks can have a maximum length of five inches.

Do not omit verbs, because the reader will not understand what the action is.

Example:

Non-STE: *Rotary switch to INPUT.*
STE: Set the rotary switch to INPUT.

Do not omit the subject, because the reader will not know what you are referring to.

Examples:

Non-STE: *If installed, remove the shims.*
STE: If shims are installed, remove them.

Non-STE: *WARNING: MAKE SURE THAT THE POTABLE WATER SYSTEM IS NOT PRESSURIZED. IF NOT, THIS CAN CAUSE INJURY TO PERSONS.*
STE: WARNING: MAKE SURE THAT THE POTABLE WATER SYSTEM IS NOT PRESSURIZED. A PRESSURIZED SYSTEM CAN CAUSE INJURY TO PERSONS.

Do not omit articles to make the sentence shorter because this can cause ambiguity.

Examples:

Non-STE: *Remove the bolt and stop.*
STE: Remove the bolt and the stop.

Do not omit parts of words to make contractions because these can be difficult to understand.

Example:

Non-STE: *If your hands are wet, don't touch the USB power adapter.*
STE: If your hands are wet, do not touch the USB power adapter.

Vertical lists

Rule 4.3	**Use a vertical list for complex text.**

When your sentence is long and you must include many different items (for example, a list of components, parts, or documents) or actions in a sentence, you can put them in a vertical list. Vertical lists make long complex sentences much easier to read and understand.

When you make a vertical list:

- Put a colon (:) at the end of first sentence, before the first item in the vertical list.
- Identify each item in the vertical list with a number, letter, punctuation mark, or symbol. For example, you can use:
    - A dash (-)
    - A bullet point (•)
    - A letter (a, b, c...)
    - A number (1, 2, 3...)

  To know which marks or symbols to use, refer to the applicable specifications for technical publications, style guides, and other official directives.
- Start each item in the vertical list with an uppercase letter.
- Where applicable, use an article before the noun that is the subject of each item in the vertical list.
- Put a full stop (period) at the end of an item in the vertical list if it is a full sentence.
- Do not put a full stop at the end of an item in the vertical list if it is not a full sentence.
- Do not put a comma or a semicolon at the end of an item in the vertical list.
- Put a full stop at the end of the last item in the vertical list.

Examples:

In the two examples that follow, the vertical lists give a list of parts and documents. Dashes and bullet points identify each of these parts and documents.

Non-STE: *The wheel assembly comprises the tire, the tube, the spokes, the spoke fittings, the valve, and the hub.*

STE: The wheel assembly has these parts:
- The tire
- The tube
- The spokes
- The spoke fittings
- The valve
- The hub.

Non-STE: *The report must include each of the following: a completed REC-1 form, a three-view drawing of the unit, a photograph of the unit, a copy of the source data.*

STE: The report must include:

- A completed REC-1 form
- A three-view drawing of the unit
- A photograph of the unit
- A copy of the source data.

In the two examples that follow, the vertical lists give a procedure and a description. Letters and numbers identify each work step. Again, to know which marks or symbols to use, refer to the applicable specifications for technical publications, style guides, and other official directives.

Non-STE: *If the RAT is retracted, remove the tag, open the isolating valves and, ensuring that there is no possibility of fouling, slowly extend the RAT to its full stroke. Close the isolating valves; tag "DO NOT OPERATE." Release the DOWN pressure.*

STE: If the Ram Air Turbine (RAT) is retracted:

(a) Remove the tag.
(b) Open the isolating valves.
(c) Slowly and fully extend the RAT and make sure that it does not touch other parts.
(d) Close the isolating valves.
(e) Tag the isolating valves with "DO NOT OPERATE."
(f) Release the DOWN pressure.

Non-STE: *During the final movement of the landing gear retraction, the door operating bar located on the leg contacts and turns the latch, withdrawing the roller from the slot and the second roller entraps the door operating bar.*

STE: When the landing gear retracts:

(1) The door operating bar on the leg touches and turns the latch.
(2) The roller moves out of the slot.
(3) The second roller keeps the door operating bar in position.

When you use a vertical list, especially in a safety instruction (for example, a warning or caution), make sure that you do not hide negative information. Use negative commands in all the applicable items in the vertical list.

Example:

Non-STE: <u>*CAUTION*</u>*: WHEN YOU GET ACCESS TO THE REAR FUSELAGE THROUGH THE ACCESS PANEL, DO NOT:*

- *PUT YOUR FEET ON THE APU LINE.*
- *USE THE APU LINE AS A HANDLE.*

STE: <u>CAUTION</u>: WHEN YOU GET ACCESS TO THE REAR FUSELAGE THROUGH THE ACCESS PANEL:

- DO NOT PUT YOUR FEET ON THE APU LINE.
- DO NOT USE THE APU LINE AS A HANDLE.

Always make sure that each item in the vertical list connects clearly and correctly to the first part of the vertical list (the sentence that is before the colon.)

Example:

Non-STE: *Do not use acetone for cleaning these parts after the repair:*
- *the service cabinet,*
- *the toilet shrouds with the supports,*
- *parts made of polycarbonate.*

STE: After the repair, do not use acetone to clean:
- The service cabinet
- The toilet shrouds
- The toilet shroud supports
- Parts made of polycarbonate.

Always make sure that the layout of your vertical list is easy to read. In the example that follows, there is a second vertical list included in the primary vertical list. If possible, use the same indenture level for all items in the vertical list.

Example:

Non-STE: *Remove these parts:*
- *The four screws (3)*
- *The four washers (4)*
- *The flange (2) that includes:*
  - *The two O-rings (6)*
  - *The seals (7)*
- *The shims (8)*
- *The spring (9).*

STE: Remove these parts:
- The four screws (3)
- The four washers (4)
- The flange (2) (that includes the two O-rings (6) and the seals (7))
- The shims (8)
- The spring (9).

Connecting words and phrases

Rule 4.4	**Use connecting words and phrases to connect sentences that contain related topics.**

Connecting words are words or phrases that connect a topic in one sentence with an idea in a sentence that follows.

In a descriptive text, they help you to structure a text logically and give information in a way that is easy to understand. Some approved connecting words are "and," "but," "then," and "thus." "As a result" and "at the same time" are examples of connecting phrases that you can use.

Examples in STE:

(The connecting words and phrases are underlined.)

The localizer course aligns with the centerline of the runway. And the glideslope path is at a constant angle to the threshold of the runway.

These safety precautions are the minimum necessary for work in the pit lane. But the local regulations can give other necessary safety precautions.

If the pressure increases, it changes the electrical currents in the transmitter. Thus, the power unit supplies currents to the indicator on the related panel.

When the hydraulic pressure is released, the sequence valve moves to the open position. As a result, the actuators are connected to the hydraulic return.

You can also use pronouns and demonstrative adjectives to connect ideas in related sentences

Examples in STE:

Identify the disassembled parts with tags. This method will help you during the subsequent assembly procedures.
(The demonstrative adjective is underlined.)

CAUTION: WHEN YOU REMOVE THE PROBES, DO NOT LET THEM TOUCH EACH OTHER. IF THE PROBES TOUCH, THEY CAN BECOME DEMAGNETIZED.
(The pronouns are underlined.)

Blank Page

Section 5 - Procedural writing

Summary of the rules

Sentences

Rule 5.1	Write short sentences. Use a maximum of 20 words in each sentence.
Rule 5.2	Write only one instruction in each sentence unless two or more actions occur at the same time.

Verbs

Rule 5.3	Write instructions in the imperative (command) form.

Descriptive statements in instructions

Rule 5.4	When you start an instruction with a descriptive statement, divide that statement from the command with a comma.

Notes

Rule 5.5	Write notes only to give information, not instructions.

Sentences

Rule 5.1 Write short sentences. Use a maximum of 20 words in each sentence.

Procedures give instructions that tell you how to do a task. Long sentences are not easy to understand.

In STE, the maximum length of a sentence for procedures is 20 words. (Refer to section 8 for the rules about word count.)

Warnings, cautions, and other safety instructions must also obey this rule.

Examples in STE:

Install the three auxiliary screws (2) in the flange of the motor assembly (9).
(This sentence has 14 words.)

CAUTION: WHEN YOU REMOVE THE SHROUD (26), BE CAREFUL NOT TO CAUSE DAMAGE TO THE SURFACE OF THE FLANGE ASSEMBLY (22).
(This sentence has 20 words.)

Non-STE: *Put preservation oil into the unit through the vent hole until the oil level is approximately 6 mm (0.24 inches) below the surface of the flange cover.* (25 words)

STE: Put preservation oil into the unit through the vent hole. (10 words) Continue until the oil level is approximately 6 mm (0.24 in) below the surface of the flange cover. (16 words)

In the STE version the long sentence is divided into two shorter sentences, which are easier to read. You cannot divide this instruction into two work steps, because the two actions occur at the same time.

Rule 5.2	**Write only one instruction in each sentence unless two or more actions occur at the same time.**

If there are too many instructions in a sentence, the sentence is not easy to understand. Put only one instruction into each sentence and clearly show (usually with numbers or letters) the sequence of work steps. You can use as many work steps as you want to complete a procedure.

Examples:

STE:
(1) De-energize the system.
(2) Make sure that all of the switches on the control panel are in their correct positions.

Non-STE: *Set the TEST switch to the middle position and release the SHORT-CIRCUIT TEST switch.*

STE:
A. Set the TEST switch to the middle position.
B. Release the SHORT-CIRCUIT TEST switch.

You can write more than one instruction in a sentence when two actions occur at the same time.

Examples in STE:

Hold the panel in its open position and install the fastener.

Slowly extend the rod fully and make sure that it does not touch other parts.

Cut and remove the wire.

Remove and discard the seal.

You can also write more than one sentence in a work step:

- When actions occur at the same time, or
- When a result occurs immediately after an action.

Examples in STE:

Make sure that the locking torque of each of the four bolts (6) is a minimum of 0.30 Nm. Then, torque each of the four bolts (6) to 4.20 Nm.

Measure the leakage from the outlet port. The leakage must not be more than 0.5 cc/minute.

Verbs

Rule 5.3	**Write instructions in the imperative (command) form**

An instruction tells the reader to do something. Write the verb in the imperative (command) form.

Examples in STE:

Set the switch to ON.

Remove the four bolts.

Increase the pressure to 60 psi.

Inflate the tires.

Install the new O-ring.

The imperative form gives the reader a clear instruction. If you use other more complicated types of sentence structure, you can cause ambiguity. Then the reader will not know:

- If it is only generally important to do a work step
- If the work step is already done
- If a different person must do the work step in the future.

Examples:

Non-STE: *The test can be continued.*
STE: Continue the test.

Non-STE: *Oil and grease are to be removed with a degreasing agent.*
STE: Remove oil and grease with a degreasing agent.

Descriptive statements in instructions

Rule 5.4	**When you start an instruction with a descriptive statement, divide that statement from the command with a comma.**

Sometimes it is only possible to do a work step under special conditions. If so, the reader must know about the condition first.

Write the condition first in the sentence, and then use a comma to show where the condition ends, and the instruction starts.

Examples:

Non-STE: *Set the switch to NORMAL when the light comes on.*
STE: When the light comes on, set the switch to NORMAL.

Non-STE: *Apply the primer when the surface is dry.*
STE: When the surface is dry, apply the primer.

Non-STE: *Disconnect the constant speed drive (CSD) from the gearbox, if it does not operate correctly.*
STE: If the constant speed drive (CSD) does not operate correctly, disconnect it from the gearbox.

The comma is important. Be careful to use it correctly because it can change the meaning of your sentence.

Examples in STE:

If the constant speed drive (CSD) does not operate correctly, disconnect it from the gearbox.

If the constant speed drive (CSD) does not operate, correctly disconnect it from the gearbox.

As you can see from the examples, these sentences are correct, but they mean two different things.

Notes

Rule 5.5 Write notes only to give information, not instructions.

Notes contain descriptive text and only give information to help the reader during a procedure.

Notes must not give instructions or requirements.

Examples in STE:

NOTE: The gyroscope will become stable after approximately 15 seconds.

NOTE: It is not necessary to remove the nameplate (33) from the flange (27).

NOTE: You can use equivalent alternatives for these items.

A note can have one or more sentences. Each sentence can have a maximum of 25 words.

Examples in STE:

NOTE: During the subsequent test, you get the cracking pressure when the fuel flow from the CROSS FEED port is more than 5 cc/minute.
(One sentence, 22 words.)

NOTE: The data collection is not completed. Thus, the statistics module can give incorrect results.
(Two sentences, 6 words and 8 words.)

Do not use the imperative form. If you do that, the note becomes an instruction for a work step.

Example:

Non-STE:	*NOTE:*	*Make sure that the avionics ventilation system continues to operate correctly.*
		(This is not a note because it contains the imperative form.)
STE:	(6)	Make sure that the avionics ventilation system continues to operate correctly.

If you include instructions in a note, it is possible that the reader will not see the information. If the information given in a note is important to prevent damage or injury, you must give such information in a safety instruction.

Examples:

Non-STE:	*NOTE:*	*When you connect the lines, do not bend them too much. If you bend the lines too much, you can cause damage to them.*
		(This is not a note. It is a safety instruction.)
STE:	CAUTION:	WHEN YOU CONNECT THE LINES, DO NOT BEND THEM TOO MUCH. IF YOU BEND THE LINES TOO MUCH, YOU CAN CAUSE DAMAGE TO THEM.

Non-STE:	*NOTE:*	*Before closing the hatch, it is mandatory to ensure that no persons are in the crew rest compartment. When the hatch is closed, there is no airflow to the compartment and therefore there is a risk of suffocation.*
		(Although the text does not contain the imperative form, it is not a note. It is a safety instruction.)
STE:	WARNING:	BEFORE YOU CLOSE THE HATCH, MAKE SURE THAT NO PERSONS ARE IN THE CREW REST COMPARTMENT. WHEN THE HATCH IS CLOSED, THERE IS NO AIRFLOW TO THE COMPARTMENT AND THERE IS A RISK OF SUFFOCATION.

When you write a procedure, and this procedure contains notes, do this test:

- Carefully read the procedure without the notes.
- Make sure that the reader can do the procedure correctly without the notes.

A satisfactory result of this test tells you that you used the notes correctly.

If important information is missing from the procedure and this information is in a note:

- Remove the information from the note.
- Write the missing information in a work step.
- Include this new work step where applicable in the procedure.
- Do the test again until you are fully sure that the reader can do the procedure without the notes.

In STE, notes are used in procedures. You can write notes in descriptions only if the notes are necessary for illustrations or tables that are parts of such descriptions.

Section 6 - Descriptive writing

Summary of the rules	
Content structure	
Rule 6.1	Give information gradually.
Rule 6.2	Use key words and phrases to organize your text logically.
Sentences	
Rule 6.3	Write short sentences. Use a maximum of 25 words in each sentence.
Paragraphs	
Rule 6.4	Use paragraphs to show related information.
Rule 6.5	Make sure that each paragraph has only one topic.
Rule 6.6	Make sure that no paragraph has more than six sentences.

Content structure

Descriptive writing gives information, not instructions. Descriptive writing can be:

- A description of an item, a product, a system, or a component, its function, how it is made and how it operates.
- A text that gives general information (for example, reports, brochures, and papers.)
- A note in a procedure.

Rule 6.1 Give information gradually.

In a descriptive text, give information gradually and make sure that each sentence contains only one topic. If you give too much information too quickly, your text will be difficult to understand, and it will be necessary for the reader to read it again.

Example:

Non-STE: *Instrument Landing System*

During the approach to the runway, deviation pointers in the course indicators give commands to fly up or down and left or right. This information comes from the VHF transceivers, which are part of the Instrument Landing System. This helps the pilot during the landing approach. When the pilot responds to the commands, the aircraft can be flown over the runway centerline (localizer) and at a fixed angle (glideslope) to the runway threshold. The localizer signals are processed by the transceiver and data are transmitted to Air Traffic Control.

The STE version that follows shows you how you can use short sentences and key words to write a text clearly.

STE: Instrument Landing System

The Instrument Landing System (the system) on the aircraft shows data that helps the pilot during the approach to the runway. This system shows the pilot the deviations from the localizer course and the glideslope path. The localizer course aligns with the centerline of the runway. And the glideslope path is at a constant angle to the threshold of the runway. During the approach to the runway, deviation pointers in the course indicators show the pilot in which direction the aircraft must go:

- Left or right (for the localizer)
- Up or down (for the glideslope).

This data about deviations from the localizer course and glideslope path comes from two VHF transceivers. These transceivers transmit this data to Air Traffic Control.

The STE version is divided into two paragraphs and has a logical structure:

- The first paragraph has five sentences
- The second paragraph has two sentences.

Rule 6.2 Use key words and phrases to organize your text logically.

Key words and phrases give structure to a text. Key words are words that often occur in a text to connect different ideas. Key phrases have the same function. Such words and phrases show how information in a text is related and give the text a logical structure.

You can also use connecting words and phrases to help the reader understand the progression of ideas in the text. They function as traffic signs and tell the reader if the information is new, or different, or a conclusion based on preceding facts. Examples of such connecting words and phrases are: “and,” “but,” “then,” “thus,” “at the same time,” and “as a result.”

The example that follows is the STE version of the example for rule 6.1. In the text, the key words and key phrases are underlined. You can see how they help to connect sentences and the ideas in them. This makes the text much easier to read.

Sentence 1	Sentence 2
The Instrument Landing System (the system) on the aircraft shows data that helps the pilot during the approach to the runway.	This system shows the pilot the deviations from the localizer course and the glideslope path.

Sentence 2 repeats the key words "system," "show," and "pilot" to add and expand on the information in sentence 1.

Sentence 2	Sentences 3 and 4
This system shows the pilot the deviations from the localizer course and the glideslope path.	The localizer course aligns with the centerline of the runway. And the glideslope path is at a constant angle to the threshold of the runway.

Sentences 3 and 4 repeat "localizer course" and "glideslope path" and add new information.

Sentence 3	Sentence 4
The localizer course aligns with the centerline of the runway.	And the glideslope path is at a constant angle to the threshold of the runway.

Sentences 3 and 4 also include the word "runway," which refers to sentence 1.

Sentences 3 and 4	Sentence 5
The localizer course aligns with the centerline of the runway. And the glideslope path is at a constant angle to the threshold of the runway.	During the approach to the runway, deviation pointers in the course indicators show the pilot in which direction the aircraft must go: - Left or right (for the localizer) - Up or down (for the glideslope).

Sentence 5 also reconnects to the previous sentences 3 and 4 with the word "runway" and gives more information about the function of indicators for the localizer and the glideslope.

Sentence 2	Sentence 5
This system shows the pilot the deviations from the localizer course and the glideslope path.	During the approach to the runway, deviation pointers in the course indicators show the pilot in which direction the aircraft must go: - Left or right (for the localizer) - Up or down (for the glideslope).

Sentences 2 and 5 are logically connected by key words and phrases.

Sentence 1	Sentence 5
The Instrument Landing System in the aircraft shows data that helps the pilot during the approach to the runway.	During the approach to the runway, deviation pointers in the course indicators show the pilot in which direction the aircraft must go: - Left or right (for the localizer) - Up or down (for the glideslope).

Sentences 1 and 5 are also connected to make a logical paragraph.

Sentence 5 also repeats the phrase "during the approach to the runway" from sentence 1 to refer to the original situation and expand on it. When you repeat the same key words and phrases, the sentences become logically connected. If you use many different words, your text will be more difficult to follow.

Also, the two paragraphs are logically connected.

Paragraph 1 (Sentences 1 thru 5): deviation, data, localizer, glideslope

Paragraph 2 (Sentences 6 and 7): This data about deviations from the localizer course and glideslope path comes from two VHF transceivers. These transceivers transmit this data to Air Traffic Control.

Sentences

Rule 6.3 Write short sentences. Use a maximum of 25 words in each sentence.

Good technical writing uses short sentences for simple and complex topics. Short sentences give a clear structure to your writing and make information easier to understand.

In descriptive writing, the maximum sentence length is 25 words. This is because descriptive text is more complex than procedural text.

Examples:

STE: During the approach to the runway, deviation pointers in the course indicators show the pilot in which direction the aircraft must go. (22 words)

Non-STE: A smartphone is a cellular telephone that has an integrated computer and many other functions, such as an operating system, internet browsing as well as the ability to run software applications. (31 words)

STE: A smartphone is a cellular telephone that has an integrated computer and many other functions. (15 words) It includes an operating system and an internet browser, and it can also operate software applications. (16 words)

Paragraphs

Rule 6.4	**Use paragraphs to show related information.**

In procedures, work steps usually have numbers and letters to show their sequence. In descriptive writing, paragraphs keep related information together and organize the text into a logical sequence.

In STE, a paragraph starts with a "topic sentence" which tells the reader what the topic of that paragraph is. Then, the sentences that follow the topic sentence explain it or give additional information to support it.

When a new paragraph starts, the reader knows that there will be new or different information in that paragraph.

Example:

STE: 1. Lightning Strike

A. General

(1) A lightning strike can cause damage to the aircraft. The damage usually occurs at the attachment points. The attachment points include:
- The nose of the fuselage
- The tail section
- The propellers
- The wing tips
- The tips of the stabilizer
- The trailing edge of the stabilizer
- The antennas
- The static dischargers.

(2) From the attachment points, the lightning strike usually goes aft through the structure of the aircraft. The high current of the lightning strike can cause damage to the mechanical systems of the aircraft. Thus, a permanent or a temporary malfunction of the electrical and electronic equipment and the systems can occur.

(3) If a lightning strike occurs, it is necessary to do an inspection of the aircraft for damage. The maintenance personnel must do this inspection before the next flight.

The example has three short paragraphs, which give the text a clear structure and make it easy for the reader to understand.

Rule 6.5 Make sure that each paragraph has only one topic.

The topic sentence is the first and most important sentence in a paragraph. The topic sentence gives new information and makes a logical connection between it and previous information. To do this, the topic sentence usually contains a key word and/or a connecting word or phrase.

From the topic sentences, the reader will understand the contents of your text and will find specific information quickly. If the reader writes down each of the topic sentences from a text, they should make a good outline of its content. The other sentences in each paragraph give additional information that supports or develops the topic of the paragraph.

If you refer to the STE version from the example for rule 6.1 and 6.2 again, you can see that the text is divided into two paragraphs:

- Paragraph 1 (sentences 1 thru 5) - "Data to help the pilot."
- Paragraph 2 (sentences 6 and 7) - "How this data is transmitted."

In the example that follows, the topic sentence at the beginning of each paragraph helps the reader know what will be explained. More information is then added gradually and the text is logically connected.

Example:

STE: Instrument Landing System

The Instrument Landing System (the system) on the aircraft shows data that helps the pilot during the approach to the runway. This system shows the pilot the deviations from the localizer course and the glideslope path. The localizer course aligns with the centerline of the runway. And the glideslope path is at a constant angle to the threshold of the runway. During the approach to the runway, deviation pointers in the course indicators show the pilot in which direction the aircraft must go:

- Left or right (for the localizer)
- Up or down (for the glideslope).

This data about deviations from the localizer course and glideslope path comes from two VHF transceivers. These transceivers transmit this data to Air Traffic Control.

Rule 6.6 Make sure that no paragraph has more than six sentences.

Paragraphs organize a text into logical units and help to hold the reader's attention. If paragraphs are too long, then they can no longer do this. Do not put different topics in the same paragraph. And if a paragraph has more than six sentences, then divide it into two smaller paragraphs. Then, your text will be easier to read.

Example:

STE: Description of the fuel manifold (refer to figure 10001)

The fuel manifold (1) has these primary parts:
- Two motorized fuel shut-off valves (referred to in this manual as "shut-off valve")
- A pressure transducer (2)
- Three fittings (10), (13) and (15). (one paragraph, 1 sentence)

The shut-off valve operates in the valve body (21). The valve body (21) has:
- An aluminum alloy ball (27)
- A retaining ring (31)
- Two seals (26)
- A spring (25). (one paragraph, 1 sentence)

The spring (25) holds the ball (27) in position to prevent internal leakage. The preformed packings (18), (2) and (30) are installed on the seals (26), the shaft (28), and in the valve body. They prevent external leakage. One of the two seals (26), on which the spring (25) operates, has the function of a relief valve. It operates when the pressure applied in the opposite direction is more than a set value. (one paragraph, 5 sentences)

The actuator assembly (4) is of a rotary type. A DC motor, which has a permanent high-speed magnet, supplies power to the actuator. Micro-switches in the actuator de-energize it at the end of travel. They also send signals to the fuel control panel to show the operation condition of the valve. A micro-relay controls the direction of shaft rotation and an electrical connector connects the actuator assembly to the aircraft electrical supply. (one paragraph, 4 sentences)

The pressure transducer (2) is installed downstream of the shut-off valve (in the RH engine line). The fitting (10) connects the pressure transducer (2) to the fuel manifold (1). A lockwire safeties the pressure transducer (2). The pressure transducer (2) has an electrical connector that connects it to the Fuel Management System of the aircraft. (one paragraph, 4 sentences)

Three fittings (10), (13), and (15) connect the two shut-off valves and the pressure transducer (2). They attach the fuel manifold (1) to the aircraft structure. (one paragraph, 2 sentences)

The example has short paragraphs, which give the text a clear structure and make it easy for the reader to understand.

Blank Page

Section 7 - Safety instructions

Summary of the rules

Definitions

How to write safety instructions

Rule 7.1	Use an applicable word (for example, "warning" or "caution") to identify the level of risk.
Rule 7.2	Start a safety instruction with a clear and simple command or condition.
Rule 7.3	Give an explanation to show the specific risk or possible result.

Definitions

Safety instructions tell the readers that procedures or steps in procedures can be dangerous or cause damage.

The words and definitions that follow agree with the specifications for technical publications that are applicable to aerospace and defense:

- A **warning** tells the reader that there is a risk of injury or death.
- A **caution** tells the reader that there is a risk of damage to objects.

It is possible that other industries use different words or categories for safety instructions.

If you use different words (for example, "danger," "attention," or "notice") or graphical symbols, always make sure that contents obey the principles of rules 7.1 thru 7.3. For more information, refer to:

- ISO 45001 - Occupational Health and Safety
- ANSI Z535 - Safety Alerting Standards (sub-standards 1 thru 6)
- ISO 3864 - International Standards for Safety Signs (parts 1 thru 4).

How to write safety instructions

Rule 7.1	**Use an applicable word (for example, "warning" or "caution") to identify the level of risk.**

Use a word – for example, "warning" or "caution" or, when applicable, a symbol, to immediately show your reader what level of risk is involved.

Examples in STE:

All examples are in uppercase letters, but STE does not give rules for formatting. Refer to the applicable specifications for technical publications, style guides, and other official directives.

WARNING: BEFORE YOU FILL THE LIQUID OXYGEN SYSTEM, PUT ON A FACE MASK AND PROTECTIVE CLOTHING. LIQUID OXYGEN CAN CAUSE IRRITATION OF THE RESPIRATORY TRACT AND EYE IRRITATION.

WARNING: ALWAYS KEEP YOUR HANDS AND FEET AWAY FROM THE BLADE. WHEN THE MOTOR OPERATES, THE BLADE TURNS AND CAN CAUSE INJURY.

CAUTION: BEFORE YOU OPERATE THE GROUND TEST UNIT, MAKE SURE THAT THE PRESSURE REGULATOR IS SET TO ZERO. THIS WILL PREVENT DAMAGE TO THE UNIT.

CAUTION: DO NOT USE BLEACH OR CLEANSERS THAT CONTAIN CHLORINE TO CLEAN THE UNIT. THESE CLEANING AGENTS CAN CAUSE CORROSION.

Do an accurate risk analysis to identify the level of risk. Then, make a decision:

- If there is a risk of injury or death to persons, use a "warning."
- If there is a risk of damage to machines, tools, or equipment, use a "caution."
- If there are the two levels of risk together, use a "warning."

Examples:

Non-STE: *CAUTION: EXTREME CLEANLINESS OF OXYGEN TUBES IS IMPERATIVE.*

STE: WARNING: MAKE SURE THAT THE OXYGEN TUBES ARE FULLY CLEAN. OXYGEN AND OIL OR GREASE MAKE AN EXPLOSIVE MIXTURE. AN EXPLOSION CAN CAUSE INJURY OR DEATH TO PERSONNEL.

In the non-STE example, this safety instruction is given as a caution. But if you know about oxygen systems, you also know that oxygen mixed with other materials can cause explosions. Because there is a risk of injury or death here, this safety instruction must be identified as a warning.

Compare the wording in the two safety instructions. The non-STE safety instruction is abstract and only makes a general statement. The warning written in STE gives much more specific information about how to prevent the risk of explosion. The warning uses the words "explosion," "injury," and "death" to make the reader clearly understand how important this safety instruction is.

Rule 7.2 Start a safety instruction with a clear and simple command or condition.

Start a safety instruction with a clear and simple command or condition. Your reader must know what to do to prevent accidents and keep a high level of safety.

Examples in STE:

(The command is underlined.)

WARNING: DO NOT SWALLOW THE SOLVENT. ALWAYS MAKE SURE THAT YOU KNOW THE SAFETY PRECAUTIONS AND FIRST AID INSTRUCTIONS FOR SOLVENTS. SOLVENTS ARE POISONOUS AND CAN CAUSE INJURY OR DEATH TO PERSONNEL.

CAUTION: DO NOT USE BLEACH OR CLEANSERS THAT CONTAIN CHLORINE TO CLEAN THE UNIT. THESE CLEANING AGENTS CAN CAUSE CORROSION.

If your reader must know about a specific condition before the start of a procedure or work step, give this condition first.

Examples in STE:

(The condition is underlined.)

WARNING: WHILE YOU USE THE SPRAY PAINT, POINT THE SPRAY AWAY FROM YOUR FACE. THE SPRAY PAINT CAN CAUSE INJURY TO YOUR EYES.

CAUTION: WHEN YOU ASSEMBLE THE UNIT, DO NOT LET THE PARTS FALL. IF THEY FALL, PERMANENT DAMAGE TO THE PARTS CAN OCCUR.

Rule 7.3	**Give an explanation to show the specific risk or possible result.**

If possible, always tell your reader what can occur if the safety instruction is not obeyed. If the danger is clearly specified, the person who does the task will understand the risk and be more careful. The examples that follow are those given in rule 7.2, in which the underlined explanation clearly shows the specific risk or possible result.

Examples:

(The risk or result is underlined.)

WARNING: DO NOT SWALLOW THE SOLVENT. ALWAYS MAKE SURE THAT YOU KNOW THE SAFETY PRECAUTIONS AND FIRST AID INSTRUCTIONS FOR SOLVENTS. SOLVENTS ARE POISONOUS AND CAN CAUSE INJURY OR DEATH TO PERSONNEL.

CAUTION: DO NOT USE BLEACH OR CLEANSERS THAT CONTAIN CHLORINE TO CLEAN THE UNIT. THESE CLEANING AGENTS CAN CAUSE CORROSION.

WARNING: WHILE YOU USE THE SPRAY PAINT, POINT THE SPRAY AWAY FROM YOUR FACE. THE SPRAY PAINT CAN CAUSE INJURY TO YOUR EYES.

CAUTION: WHEN YOU ASSEMBLE THE UNIT, DO NOT LET THE PARTS FALL. IF THEY FALL, PERMANENT DAMAGE TO THE PARTS CAN OCCUR.

Section 8 - Punctuation and word count

Summary of the rules

Punctuation

Rule 8.1	You can use all standard English punctuation marks except the semicolon (;).
Rule 8.2	Use hyphens (-) to connect closely related words.
Rule 8.3	You can use parentheses: - To make references to illustrations or text - To include letters or numbers that identify items on an illustration or in a text - To identify the steps in a procedure - To include abbreviations - To give the singular and plural forms of a noun at the same time - To explain words or a part of a sentence - To include an alternative.

Word count

Rule 8.4	In a vertical list, a colon (:) has the same effect on word count as a full stop (period) and shows the end of a sentence.
Rule 8.5	When you put text in parentheses, it counts as one word in that sentence.
Rule 8.6	Count each of these as one word: - Numbers - Units of measurement - Abbreviations - Alphanumeric identifiers - Quoted text - Titles, headings, placards, and labels.
Rule 8.7	Hyphenated words count as one word.

Punctuation

Correct punctuation is important because it shows how the different parts of the text are related and prevents ambiguity.

STE does not give you rules about general punctuation. For more information, refer to official reference books. For example:

- The Chicago Manual of Style
- The Gregg Reference Manual
- U.S. Government Printing Office Style Manual
- John Kirkman: Punctuation Matters - Advice on punctuation for scientific and technical writing.

Rule 8.1 You can use all standard English punctuation marks except the semicolon (;).

The semicolon (;) is not approved in STE because it lets you write very long sentences. It is also difficult to use correctly. As an alternative to the semicolon, always write two different sentences.

Examples:

Non-STE: *(1) Examine the removed parts; replace the damaged ones.*
STE: (1) Examine the removed parts for damage.
(2) Replace the damaged part(s).

Non-STE: *The battery is not user-replaceable; it can only be replaced by an approved service provider.*
STE: Do not replace the battery. Only an approved service provider can replace it.

Rule 8.2 Use hyphens (-) to connect closely related words.

A hyphen (-) is a punctuation mark that connects words or parts of words. Use the hyphen to show that two or more words are closely related. This helps the reader to understand words and phrases more easily.

The examples that follow show how to use hyphens to connect closely related words:

1. **Terms that have two or more words and are adjectives before a noun:**
   low-altitude flight, high-pressure chamber, air-conditioned compartment, transmitter-receiver system, quick-release fastener, clamshell-type flap, eighteen-inch monitor, cast-aluminum bracket, three-to-one ratio, trial-and-error method, air-to-air refueling, soap-and-water solution, up-to-date information, run-on torque, break-away torque, cut-in speed, in-flight entertainment system, stiff-bristled brush, fire-resistant material, self-sealing hose
2. **Two-word fractions or numbers:**
   forty-seven, ninety-ninth, one hundred and sixty-two, three-sixteenths, one thirty-second
3. **Terms that contain a capital letter plus a noun, or a number plus a noun, and that usually describe the shape or configuration of something:**
   L-shaped bracket, O-ring, T-shirt, U-beam, Y-coupling, V-band clamp, 3-prong connector, 180-grit abrasive cloth
4. **Verbs that contain a noun or another part of speech as the first element:**
   die-cast, arc-weld, fusion-bond, stop-drill, vacuum-pack, heat-treat, jump-start, air-condition, short-circuit, fast-forward, cold-roll, dry-clean, blow-dry
5. **Terms in which the prefix ends with a vowel and the root word begins with a vowel:**
   de-energize, pre-amplifier, de-ice, anti-icing, pre-engage

A hyphen is different from a dash, which keeps words apart, indicates a range, or shows a pause. A dash is usually longer than a hyphen but is sometimes shown as a hyphen with a space on each side.

Rule 8.3	**You can use parentheses:** **- To make references to illustrations or text** **- To include letters or numbers that identify items on an illustration or in a text** **- To identify the steps in a procedure** **- To include abbreviations** **- To give the singular and plural forms of a noun at the same time** **- To explain words or a part of a sentence** **- To include an alternative.**

In STE, you can use parentheses as follows:

1. To make references to illustrations or text

Examples in STE:

Remove the valve (10, Figure 1).

Install the cover (refer to paragraphs 2 thru 5.)

2. To include letters or numbers that identify items on an illustration or in a text

Examples in STE:

Disconnect the hoses (2) and (12) from the suction ejector (8).

Remove the nuts (74), the washers (76), the bolts (68), the seals (70), and the bonding straps (72).

3. To identify the steps in a procedure

Example in STE:

(1) Install the locking cap (4) on the body (8).
(2) Safety the locking cap (4) with the cotter pin (5).
(3) Install a new retaining ring (6).

4. To include abbreviations

Example in STE:

A Liquid Crystal Display (LCD) is a flat-panel display that uses the light-modulating properties of liquid crystals.

5. To give the singular and plural forms of a noun at the same time

Example in STE:

A. Before you do the test(s), install the component(s).
B. Do the applicable test(s).

6. To explain words or a part of a sentence

Examples in STE:

Increase the pressure slowly (not more than 10 psi each minute.)

Make sure that the BLEED pushbutton switch is released (the ON legend is off.)

7. To include an alternative

Example in STE:

Open the left (right) access panel L42 (R42).

Word count

Rule 8.4	**In a vertical list, a colon (:) has the same effect on word count as a full stop (period) and shows the end of a sentence.**

In a vertical list, a colon (:) divides the first part of the sentence from the subsequent items in the vertical list. This colon has the effect of a full stop (period). Thus:

- In procedural sentences, you can use a maximum of 20 words before the colon.
- In descriptive sentences, you can use a maximum of 25 words before the colon.

Each item in a vertical list that comes after the colon counts as a new sentence. Thus, the limit for each item in a vertical list is:

- 20 words for procedural sentences
- 25 words for descriptive sentences.

Example in STE:

To extinguish a possible fire, portable fire extinguishers are installed in these areas:	(13 words)
- The cockpit	(2 words)
- The cabin	(2 words)
- The cabin sub-compartment	(3 words)
- The crew rest compartment.	(4 words)

Rule 8.5	**When you put text in parentheses, it counts as one word in that sentence.**

When you count words for sentence length, text in parentheses counts as one word of that sentence. But the words that you put between parentheses also make a new sentence. Thus, count them in that different sentence.

Example in STE:

Make sure that the EMER pushbutton switch is released (the EMER legend is off.)
(This sentence has 10 words, because the text in parentheses counts as one word. The sentence in parentheses has 5 words and counts as a different sentence.)

If there is an identifier in parentheses (a number, a letter, or an alphanumeric identifier,) this identifier counts as one word in the sentence. Abbreviations in parentheses also count as one word.

Examples in STE:

Remove the safety pin (10).	(5 words)
Installation of a Business Class (B/C) Seat	(7 words)
Hardware and Software Configuration Check of the In-Flight Entertainment (IFE) System	(11 words)

Rule 8.6	**Count each of these as one word:** **- Numbers** **- Units of measurement** **- Abbreviations** **- Alphanumeric identifiers** **- Quoted text** **- Titles, headings, placards, and labels.**

When you count words for sentence length, count each of these as one word:

1. Numbers

Examples in STE:

Do steps 13 thru 16 a minimum of three times. ("13" and "16" are numbers and each of them counts as one word.)	(10 words)
The spar box has twenty-one ribs. ("Twenty-one" is a number and counts as one word.)	(6 words)

2. Units of measurement

Usually, in technical texts, units of measurement follow the related number (for example, 10 mA). When you write an abbreviation or a symbol for a unit of measurement, count it and the number as one word. When you write units of measurement or their symbols in full, count them as one word.

Examples in STE:

Make sure that the temperature in the room is 10 °C. (10 words)
Make sure that the temperature in the room is 10 degrees Celsius. (12 words)

The unit weighs 20 kg. (4 words)
The unit weighs 20 kilograms. (5 words)

The resistance must be 10 Ω. (5 words)
The resistance must be 10 ohms. (6 words)

3. Abbreviations

In technical texts, you can use abbreviations (also acronyms and initialisms) to make sentences shorter and easier to read. But, these abbreviations only help the reader if the reader knows what they mean. Each abbreviation counts as one word.

Examples in STE:

We do a test of this system each day at 10 a.m. (12 words)
("a.m." is an abbreviation and counts as one word.)

Make sure that the unit has the NATO code. (9 words)
(NATO is an abbreviation (acronym) and counts as one word.)

4. Alphanumeric identifiers

An alphanumeric identifier is a combination of letters and numbers that identifies something. Each alphanumeric identifier counts as one word.

Examples in STE:

Examine the No. 1 bearing installation. (5 words)
("No. 1" is an alphanumeric identifier and counts as one word.)

Tag circuit breaker 36L7. (4 words)
("36L7" is an alphanumeric identifier and counts as one word.)

5. Quoted text

Quoted text is usually given in quotation marks ("....") Words between quotation marks count as one word in a sentence. Sometimes capital letters or differences in font can also show quoted text. Such text also counts as one word.

Examples in STE:

Touch the "Service Overview" arrow to select the function page. (9 words)
("Service Overview" is quoted text and counts as one word.)

Release the SHORT-CIRCUIT TEST switch. (4 words)
(SHORT-CIRCUIT TEST is quoted text and counts as one word.)

C = (A - B) - 0.063 mm. (1 word)
(In STE, a formula is quoted text and counts as one word.)

6. Titles, headings, placards, and labels

In some texts, there are words or groups of words that you cannot change. These are:

- Titles and headings
- Placards (for example, a warning notice in a work area or public place)
- Labels (for example, warning and caution labels that are attached to objects).

Examples in STE:

Refer to Testing and Fault Isolation, page block 1001. (6 words)
("Testing and Fault Isolation" is the title of a section and counts as one word.)

Refer to Requirements after Job Completion for the applicable procedures. (7 words)
("Requirements after Job Completion" is a heading in a data module and counts as one word.)

This procedure is for the inspection of SSI No. 57-21-16, "Outer wing bottom skin lower surface spanwise skin joints at stringer 13 and stringer 20 between Rib 12 and Rib 27 excluding areas covered by flap track fairings 3, 4, and 5." (9 words)
("This procedure is for the inspection of" has 7 words. "SSI No. 57-21-16" is the reference number of the Structurally Significant Item (SSI) and counts as one word. The subsequent text given in quotation marks is the title of the SSI. It is not written in STE and it is not possible to changed it. Thus, this text counts as one word. As a result, the full sentence has a total of 9 words.)

"Interior hazards exist to such a degree that interior operations may be conducted only after full examination, and with extreme caution." (1 word)
(The text given in quotation marks is not written in STE. It comes from a placard and it is not possible to change it. It counts as one word.)

"FRAGILE - Please handle with care." (1 word)
(The text given in quotation marks is not written in STE. It comes from a label on shipping containers, and it is not possible to change it. It counts as one word.)

Rule 8.7 Hyphenated words count as one word.

Groups of words that are not usually adjectives but function together as an adjective before a noun are hyphenated. Such groups of words count as one word.

Examples in STE:

Clean the surface with a soap-and-water solution. (7 words)

Use the trial-and-error method. (4 words)

When you hyphenate long technical names to make them clearer to the reader, a hyphenated group of words also counts as one word in the applicable noun cluster.

Examples in STE:

Cutoff-switch power connection (3 words)

Main-gear-door retraction-winch handle (3 words)

Section 9 - Writing practices

Summary of the rules

Different sentence constructions

Rule 9.1	Use a different construction to write a sentence when a word-for-word replacement is not sufficient.

Correct use of approved words

Rule 9.2	Use each approved word correctly.
Rule 9.3	When you use two words together, do not make phrasal verbs.

Consistent style

Rule 9.4	When you select terminology or wording, always use a consistent style.

General recommendations

GR-1	The conjunction "that"
GR-2	The preposition "with"
GR-3	How to use pronouns
GR-4	The pronoun "this"
GR-5	False friends
GR-6	Latin abbreviations

Different sentence constructions

Rule 9.1 Use a different construction to write a sentence when a word-for-word replacement is not sufficient.

STE is a controlled language with a controlled dictionary.

To help you use the approved words correctly, the dictionary gives approved alternatives for unapproved words. If you find an alternative that has the same part of speech, you can use that word to replace the unapproved word in the sentence (word-for-word replacement).

Example:

Non-STE: *A value of 2 mm is acceptable.*
("Acceptable" is not approved.)
STE: A value of 2 mm is permitted.

When you replace a word, always make sure that the alternative you select does not change the meaning of the sentence. If the meaning changes, or if the alternative does not have the same part of speech, you must use a different construction.

A different construction is necessary because:

1. **You must change the grammatical structure of the sentence to use the alternative that you selected. Usually, the approved examples show you how to do this.**

Examples:

Non-STE: *The oil level on the sight gauge must be visible during the test.*
STE: During the test, make sure that you can see the oil level on the sight gauge.

In this example, the approved verb "see" replaces the unapproved adjective "visible." But to use the verb "see," it is necessary to replace "must be" with "make sure that you can."

Non-STE: *Cycle the unit twice to remove air from the lines.*
STE: Operate the unit for two cycles to remove air from the lines.

In this example, the approved noun "cycle" replaces the unapproved verb "cycle." And the technical name "two" replaces the unapproved adverb "twice."

When you use different constructions, always make sure that you use approved words that do not change the meaning of the sentence.

2. **The word-for-word replacement of the unapproved word gives a meaningless result.**

Example:

Non-STE: *Without this modification, the service life of the unit can be uncertain.*

"Uncertain" is not in the dictionary, but it means "not sure" or "not known." Thus, you can refer to "sure" and "know" in the dictionary. You will find that they are approved, and you can use "not sure" and "not known" ("not" + the approved word.)

But do not write:

Non-STE: *Without this modification, the service life of the unit cannot be sure.*
or
Without this modification, the service life of the unit cannot be known.

These sentences do not make sense and they are not good English.

You must ask yourself what this sentence really means and then use a different construction.

Non-STE: *Without this modification, the service life of the unit can be uncertain.*
STE: Without this modification, it is possible that the service life of this unit will be shorter than usual.

The STE sentence uses approved words in a different way to give the same information clearly. Always make sure that you know what you want to say and that the words that you use give the meaning that you want.

3. The approved alternative that you find changes the meaning of the sentence.

Example:

Non-STE: *Just apply very light pressure to the surface.*
STE: Only apply very light pressure to the surface.
NOT: *Immediately apply very light pressure to the surface.*

"Immediately" is the approved alternative for "just". But if you use the word "immediately" in this context, you will change the meaning of the instruction.

4. The word that you must replace is not in the dictionary.

Example:

Non-STE: *The incidence of water in fuel is dangerous.*
STE: Water in fuel is dangerous.

"Incidence" is not in the dictionary, so you must think of a different word that has the same meaning, for example, "presence". But "presence" is not approved, and its suggested alternative is "be". You can use a different construction to keep the same meaning.

How to use a different construction

When you cannot do a word-for-word replacement, use different words to get the same result.

Example:

Non-STE: *If the air intake flap has failed, deactivate air inlet valve.*
STE: If the air intake flap is not serviceable, lock the air inlet valve in the fully open position.
NOT: *If the air intake flap is not serviceable, disconnect (or stop, or isolate) the air inlet valve.*

In this example, "deactivate" is not approved. The alternatives in the dictionary are "disconnect," "stop," and "isolate." In some contexts, these alternatives are satisfactory.

But in this example:

- "Disconnect" is technically incorrect.
- "Stop" and "isolate" are meaningless.

Answer these questions before you write a sentence:

- "What is the meaning of the word "deactivate" in this context?"
  or
- "What is the action that the reader must do?"

You will then understand that the "deactivation" is the result of an action: lock the air inlet valve in the fully open position. Thus, if you tell the reader to do that, your instruction will be correct, and the reader will do the task correctly.

In the examples, you saw:

- How you can change an unapproved word in a sentence.
- How you can write a new sentence that has the same meaning.

Often you must select different words, use different verb forms, or write completely new sentence constructions. Sometimes you must break long sentences into shorter ones, delete information which is not necessary, or ask an engineer for further explanation. And frequently you must fully reorganize your text. Then, when you change one part of a text, you must make sure that it has no negative effect on other parts of the text.

Example:

Non-STE: *If cracks are detected during this procedure, the operator must perform the repair within a certain number of flight hours depending on crack length. Refer to following table:*

Crack length detected	*Time before repair*
L> ...	*1000 flight hours*
L = ...	*2000 flight hours*
L< ...	*3000 flight hours*

STE: If you find cracks, refer to the table that follows:

If the crack is of this length	Do the repair before
L> ...	1000 flight hours
L = ...	2000 flight hours
L< ...	3000 flight hours

In the non-STE example:

a) The underlined words are not approved in STE.
b) The first verb form "are detected" is in the passive voice.
c) The first sentence is 24 words long.
d) An article is missing in front of "crack length" and "following table".
e) The instruction is not in the imperative form.

As shown in the dictionary, you can replace:

- "perform" with "do"
- "within" with "in"
- "certain" with "some"
- "detect" with "find."

You can also include the missing articles. Thus, the sentence becomes:

If cracks are found during this procedure, the operator must do the repair in some flight hours depending on the crack length. Refer to the following table:

This keeps the same construction, but it is not good English.

To replace "are found" (passive voice), you must change the construction to: "If you find cracks ...". But this does not change the construction of the remaining part of the sentence.

To replace "following," you must make a small change to the construction: "Refer to the table that follows:"

To replace "the operator must do the repair", you must change the construction to: "do the repair."

To replace "depending" you must make a major change to the sentence. This is because you cannot do a word-for-word replacement with the alternative given in the dictionary: "if". ("If" what?).

You can then write the first sentence as follows:

> *If you find cracks during this procedure, do the repair before the number of flight hours applicable to the crack length.*

But there are still 21 words and the technical meaning is not easy to understand. Also, part of the information is in the table. Thus, it is not necessary to give it two times. A good solution is to put the instruction in the table heading, as shown in the STE version.

Always remember that the primary objective of STE is that readers immediately understand each sentence that they read.

Correct use of approved words

Rule 9.2	**Use each approved word correctly.**

Some STE approved words have "restricted" meanings.

Before you use a word, read its definition in the approved meaning column of the dictionary. Words frequently have many different meanings in English. In STE, approved words usually only have one approved meaning. Other meanings that the word can have in standard English are not approved.

Make sure that the word that you select also has the meaning which is correct for that specific context.

Examples:

Non-STE: *Wear protective clothing.*
STE: Use (or put on) protective clothing

The word "wear" is approved as a verb, but with the meaning "to become damaged by friction."

Non-STE: *This regulation extends to all units.*
STE: This regulation is applicable to all units.

The verb "extend" has the approved meaning "to increase, or cause to increase, in dimension or range." In the non-STE example, it is not used with the approved meaning.

Non-STE: *When the pressure goes down, lift the cover.*
STE: When the pressure decreases, lift the cover.

"Goes down" is a phrase that describes a physical movement, as when an indicator (for example, a gauge needle or a flag) goes down.

"Decrease" is better because it describes the pressure, not the indicator that monitors the pressure.

Non-STE: *Make sure that the probe went through more than 500 degrees C.*
STE: Make sure that the temperature of the probe increased to more than 500 degrees C.

You can use "go through" only for something that physically moves through something. You cannot use it to show a requirement or a condition.

Non-STE: *Move the tube to see if the inner connection is tight.*
STE: Move the tube to make sure that the inner connection is tight.

You can use "see" only for something that you can see with your eyes and not as a word for "come to know."

Non-STE: *The indicator turns green.*
STE: The color of the indicator changes to green.

You can use "turn" only with the meaning "to move or cause something to move around an axis or a point."

Non-STE: *Do not let the pressure go below (or above) 20 psi.*
STE: Do not let the pressure become less than (or more than) 20 psi.

"Above" and "below" are for physical positions only and not for limits.

Also, make sure that you use approved words as their approved part of speech. In English, words usually do not have different forms that immediately show how they function in a sentence. Thus, readers can frequently understand the same word in different ways. To make sentences clearer, an approved word can usually only have one function (part of speech). Use each approved word as the approved part of speech.

Examples:

Non-STE: *When you work with cleaning agents, use breathing equipment.*
STE: When you do work with cleaning agents, use breathing equipment.

The word "work" is approved as a noun, but not as a verb.

Non-STE: *Install the cover with the help of a second person.*
STE: Install the cover with the aid of a second person.

The word "help" is approved as a verb, but not as a noun.

Non-STE: *Be careful not to damage the sleeve.*
STE: Be careful not to cause damage to the sleeve.

The word "damage" is approved as a noun, but not as a verb.

There are a few words which are approved as more than one part of speech and have more than one meaning. This is because these words are important and commonly occur in technical English.

The word "flush" for example, frequently occurs in technical English as a verb (to let water flow quickly, through a pipe) or as an adjective (when one surface completely touches another surface).

Examples:

STE: Flush the pipes with a disinfectant solution.
("Flush" is a verb here.)

STE: Make sure that the door is flush with the adjacent surfaces.
("Flush" is an adjective here.)

But because the verb and the adjective have different positions in a sentence and they occur in different contexts, it is very difficult to confuse them.

Rule 9.3 When you use two words together, do not make phrasal verbs.

In English, a verb and one or more prepositions sometimes go together to make a "phrasal verb." This phrasal verb then has a meaning which is different from the meanings of its parts. Phrasal verbs usually have two meanings, the original, more concrete meaning, and a more abstract one, too.

General examples:

Put out the cat.
(Concrete meaning: "Put the cat outside of the house.")
Put out the fire.
(Abstract meaning: "Extinguish the fire.")

To prevent such ambiguity, it is not permitted in STE to use approved words together to make a new phrase that has a different, more abstract meaning.

Examples:

Non-STE:	*After you put out the fire, close the valve on the fire extinguisher.*
STE:	After you extinguish the fire, close the valve on the fire extinguisher.

"Put" and "out" are approved words in STE. In the example, "put out" has a different meaning from the approved meanings for "put" and "out" given in the dictionary. The word "extinguish" has the approved meaning "To stop burning" and it is the word most usual in technical documentation.

Non-STE:	*This compound can give off poisonous fumes.*
STE:	This compound can release poisonous fumes.

"Give" and "off" are approved words in STE. In the example, "give off" has a different meaning from the approved meaning for "give" and "off" given in the dictionary. Here you must think of another word that has the same meaning. "Release" has the approved meaning "to make free," or "to let go." And it is also possible to use this word in this context as a replacement for "give off."

You will not usually find phrasal verbs listed as "not approved" in the dictionary. Remember, when you re-write a sentence in STE, always make sure that the new sentence is still good standard English.

Consistent style

Rule 9.4 When you select terminology or wording, always use a consistent style.

In procedures, you will frequently give the same information again and again. For example, most procedures describe how to remove or install components or parts. When you decide on a specific wording for a work step, use the same wording each time that type of step occurs. The reader will recognize the wording and will quickly know what to do. Different words and wordings can cause confusion and be a waste of time.

Example:

These two sentences each give the same instruction and each correctly obeys the rules of STE.

STE: Lubricate the two bolts with a small quantity of oil.
STE: Apply a small quantity of oil to the threads of the two bolts.

If you use these two sentences in different procedural steps to give the same instruction, you do not help the reader. Select one sentence or sentence style and then use it again and again when the context (and meaning) is the same.

Non-STE:

1. *Put the housing (8) in its installation position on the main body (9).*
2. *Lubricate the two bolts (10) with oil.*
3. *Install the two bolts (10) in the body (9).*
4. *Torque-tighten the two bolts to 4 Nm.*
5. *Put the transducer (11) on the body assembly (9).*
6. *Apply a small quantity of oil to the threads of the four bolts (12).*
7. *Attach the transducer (11) to the body assembly (9) with the four bolts (12).*
8. *Torque each of the four bolts (12) to 0.8 Nm.*

STE:

1. Put the housing (8) in its installation position on the body assembly (9).
2. Apply a small quantity of oil to the threads of the two bolts (10).
3. Install the two bolts (10) in the body assembly (9).
4. Torque each of the two bolts to 4 Nm.
5. Put the transducer (11) on the body assembly (9).
6. Apply a small quantity of oil to the threads of the four bolts (12).
7. Attach the transducer (11) to the body assembly (9) with the four bolts (12).
8. Torque each of the four bolts (12) to 0.8 Nm.

In the non-STE example, you can see that there are different wordings:

- Different terms are given for item (9) ("main body", "body", and "body assembly")
- Different wordings are used for tightening the bolts ("torque" and "torque-tighten")
- Different sentences are used for the lubrication steps ("lubricate the XX bolts" and "apply a small quantity of oil to the threads of the XX bolts.")

In the STE version, each time the same component is mentioned, it has the same name, and the same context always has the same wording. This makes the text much clearer to read. The reader reads the work steps one at a time and does the applicable task. To recognize the content quickly and do each work step correctly, it is important that the wording is not always new and different.

In descriptive writing, the reader must understand the text as a whole. Thus, it is important to use words and sentences correctly and give a logical structure to the text. The text is then easier to read and to remember.

General recommendations

The general recommendations (GR) given in this section are not STE rules. They can help you avoid some typical mistakes that writers frequently make.

GR-1 The conjunction "that"

In English, you can use the conjunction "that" to connect additional information (in a subordinate clause) to a main clause. You can use it after verbs. For example, "make sure," "show," and "recommend." Native English speakers frequently omit it when they speak quickly, and this style is also found in writing. But this word helps the reader to understand where the main clause ends, and the subordinate clause starts. Because the conjunction "that" can prevent ambiguity, try not to omit it.

This conjunction also helps with translation because, in many other languages, it is not possible to omit the equivalent word.

Examples:

Do not write: *Make sure the valve is open.*
WRITE: Make sure that the valve is open.

Do not write: *The manufacturer recommends you prepare the mixture in an area with good airflow.*
WRITE: The manufacturer recommends that you prepare the mixture in an area with good airflow.

Do not write: *The gauge shows the reservoir is full.*
WRITE: The gauge shows that the reservoir is full.

GR-2 The preposition “with”

In STE, the preposition “with” has three approved meanings. It is a function word that shows “association or relationship,” “help or sharing,” or “a means or instrument.” This word can sometimes cause ambiguity.

For example, the sentence: “Install the panel with the green fasteners.” can have different meanings:

- Install the panel that has green fasteners.
- Install the panel together with the green fasteners.
- Use the green fasteners to install the panel.

Usually, the context of the sentence will give the reader sufficient information to know which meaning is correct.

But when you use the word “with,” it is a good idea to read your sentence again to make sure that it is not ambiguous, there are no errors, and that it says what you wanted to say.

Examples:

Do not write: *Lift the aircraft at the maximum takeoff weight with passengers.*
WRITE: Lift the aircraft at the maximum takeoff weight (passenger weight included.)

The context (or experience) tells you that you will not ask the passengers to help you lift the aircraft. But because you can understand this sentence in two ways, it becomes a joke. And this is not the function of technical documentation.

Do not write: *Make sure that there is no contact with stop (1) with hydraulic pressure applied.*
WRITE: When you apply hydraulic pressure, make sure that the lever does not touch the stop (1).
(Write the condition first.)

Do not write: *Seal the opening with tool TS9867.*
WRITE: Use tool TS9867 to seal the opening.
(Write this type of sentence only if you think that the sentence is ambiguous. In STE, you must give the action verb which is not “use”, but “seal”.) You can write:
Seal the opening. To do that, use tool TS9867.

GR-3 How to use pronouns

Pronouns refer back to a person, place, or thing, that you already mentioned in a text.

Examples are "it," "they," "that," "these," and "those." If you use them correctly, your text will be easy to read.

In STE, the approved pronouns are in the dictionary. Do not use a pronoun if it is not the dictionary (for example, "she" or "he").

If a pronoun can refer to one or more nouns in a text, it can make a sentence ambiguous. If a sentence is ambiguous, replace the pronoun with the word that it refers to. This will make the sentence clear and easier to read.

Examples:

Do not write: *If you engage the pins incorrectly with the seats, they can become damaged.*

WRITE: If you engage the pins incorrectly with the seats, the pins can become damaged.

or

If you engage the pins incorrectly with the seats, the seats can become damaged.

or

If you engage the pins incorrectly with the seats, the pins and seats can become damaged.

GR-4 The pronoun "this"

When you use the pronoun "this" in a sentence, make sure that the reader knows what the word refers to. If "this" can refer to more than one thing, describe the applicable context again.

Examples:

Do not write: *Make sure that the cover is not locked (this can cause damage to the probe.)* (What can cause damage to the probe? The cover in the locked condition? Or the cover in the unlocked condition?)

WRITE: Make sure that the cover is not locked. If the cover is locked, this can cause damage to the probe.

or

If the cover is locked, damage to the probe can occur.

Do not write: *Do not use crocus cloth on aluminum parts. If you do this, you can cause corrosion on aluminum parts. Crocus cloth contains ferrous oxide.*

WRITE: Do not use crocus cloth on aluminum parts. Crocus cloth contains ferrous oxide and can cause corrosion on aluminum parts.

GR-5 False friends

A false friend is a word or expression that has a similar form to one in a person's native language but a different meaning. For example, the English word "disposition" and the Italian word "disposizione", or the Spanish word "disposición."

If you are a non-native English speaker, when you use a word in a sentence, make sure that this word has the correct meaning in English (not the meaning of the similar word in your native language.)

Example:

Do not write: *Obey the dispositions of the manufacturer when you use this adhesive.*
("disposition" in English does not mean an "instruction.")
WRITE: When you use this adhesive, obey the manufacturer's instructions.

GR-6 Latin abbreviations

In English texts, it is possible to find some Latin abbreviations (for example, "e.g.," "i.e.," and "etc."). We recommend that you do not use them in STE because they can confuse your readers. Always use English words.

Examples:

Do not write: *Discard the standard parts (e.g. washers, screws, bolts, and nuts) each time you remove them.*
WRITE: Discard the standard parts (for example, washers, bolts, and nuts) each time you remove them.

Do not write: *These wires can have insulation of different colors (blue, green, red, etc.)*
WRITE: These wires can have insulation of different colors. They can be blue, green, read, or other colors.

Frequently, these abbreviations are not necessary. You can omit them.

Do not write: *These wires can have insulation of different colors (blue, green, red, etc.)*
WRITE: These wires can have insulation of different colors.

Blank Page

PART 2 - SYSTEM GENERAL-TECHNICAL NAMES(TN)

Blank Page

GENERAL TERMS

Line technician

The Line Technician performs the day-to-day maintenance activities which ensure the airplane is ready to fly. These activities may be performed during turnaround service or in the hangar. They include servicing the mechanical and electronic systems, conducting periodic checks, trouble shooting, testing and inspecting using Built-In-Equipment (BITE) or external test equipment. The Line Technician's activities include removal and replacement of faulty line replaceable units (LRU) and certain technicians may specialize in mechanical, electrical or avionic sub specialties.

Shop technician

The Shop Technician repairs faulty LRU's removed by the Line Technician. He uses BITE and automatic test equipment (ATE), bench testing and trouble shooting techniques to repair the LRU's. Repairs may involve replacing faulty components with in the LRU or returning them to the manufacturer or vendor for additional rework or overhaul.

Engineer

The Engineer analyses data from maintenance records and determines that the airplane systems are performing to the design specifications of maintainability and reliability. When required, the engineer assists the airline's technicians, conducting in-depth analyses of maintenance problems. Engineers assist the airline to comply with engineering changes demanded by government agencies or those actions, called service bulletins, recommended by the airplane manufacturer. Engineers ensure that any changes made to the airplane's systems or structure are properly documented and recorded. These actions are part of a process called configuration control.

Airworthiness

The status of an aircraft, engine, propeller or part when it conforms to its approved design and is in a condition for safe operation.

Continuing airworthiness

The set of processes by which an aircraft, engine, propeller or part complies with the applicable airworthiness requirements and remains in a condition for safe operation throughout its operating life.

Ground handling

Services necessary for an aircraft's arrival at, and departure from, an airport, other than air traffic services.

Operational personnel

Personnel involved in aviation activities who are in a position to report safety information.
Note: Such personnel include, but are not limited to: flight crews; air traffic controllers; aeronautical station operators; maintenance technicians; personnel of aircraft design and manufacturing organizations; cabin crews; flight dispatchers, apron personnel and ground.

Flight time-aeroplane

The total time from the moment an aeroplane first moves for the purpose of taking off until the moment it finally comes to rest at the end of the flight.
Note: Flight time as here defined is synonymous with the term "block to block" time or "chock to chock" time in general usage which is measured from the time an aeroplane first moves for the purpose of taking off until it finally stops at the end of the flight.
Flight time means: Pilot time that commences when an aircraft moves under its own power for the purpose of flight and ends when the aircraft comes to rest after landing.

Flight time-helicopter

The total time from the moment a helicopter's rotor blades start turning until the moment the helicopter finally comes to rest at the end of the flight, and the rotor blades are stopped.

Time in service

With respect to maintenance time records, time-in-service means the time from the moment an aircraft leaves the surface of the earth until it touches it at the next point of landing.

Maintenance

The performance of tasks on an aircraft, engine, propeller or associated part required to ensure the continuing airworthiness of an aircraft engine, propeller or associated part including any one or combination of overhaul, inspection, replacement, defect rectification, and the embodiment of a modification or repair.

Maintenance records

Records that set out the details of the maintenance carried out on an aircraft, engine, propeller or associated part.

Maintenance release

A document which contains a certification confirming that the maintenance work to which it relates has been completed in a satisfactory manner in accordance with appropriate airworthiness requirements.

Powerplant

The system consisting of all the engines, drive system components (if applicable), and propellers (if installed), their accessories, ancillary parts, and fuel and oil systems installed on an aircraft but excluding the rotors for a helicopter.

Engine

A unit used or intended to be used for aircraft propulsion. It consists of at least those components and equipment necessary for functioning and control, but excludes the propeller/rotors (if applicable).

ATA SPECIFICATION 100

The air transport association of America (ATA) Specification 100 establishes a standard presentation and policy so that the manufacturer's data fulfills the needs of airline maintenance personnel. The specification tells how to number and organize the manual chapters.

ATA Number - Chapter name	ATA Number - Chapter name
ATA 05~12 Aircraft General	**ATA 51~57 Structure**
ATA 05 TIME LIMITS-MAINTENANCE CHECKS	ATA 51 STANDARD PRACTICES-STRUCTURES
ATA 06 DIMENSIONS AND AREAS	ATA 52 DOORS
ATA 07 LIFTING AND SHORING	ATA 53 FUSELAGE
ATA 08 LEVELING AND WEIGHING	ATA 54 NACELLES/PYLONS
ATA 09 TOWING AND TAXI	ATA 55 STABILIZERS
ATA 10 PARKING, MOORING, STORAGE	ATA 56 WINDOWS
ATA 11 PLACARDS AND MARKINGS	ATA 57 WINGS
ATA 12 SERVICING - ROUTINE MAINTENANCE	**ATA 60~67 Propeller/Rotor**
ATA 20~50 Airframe systems	ATA 60 STANDARD PRACTICES-PROP/ROTOR
ATA 20 STANDARD PRACTICES - AIRFRAME	ATA 61 PROPELLERS/PROPULSORS
ATA 21 AIR CONDITIONING	ATA 62 MAIN ROTORS
ATA 22 AUTO FLIGHT	ATA 63 MAIN ROTOR DRIVES
ATA 23 COMMUNICATIONS	ATA 64 TAIL ROTORS
ATA 24 ELECTRICAL POWER	ATA 65 TAIL ROTOR DRIVE
ATA 25 EQUIPMENT/FURNISHINGS	ATA 66 FOLDING BLADES/PYLON
ATA 26 FIRE PROTECTION	ATA 67 ROTOR FLIGHT CONTROL
ATA 27 FLIGHT CONTROLS	
ATA 28 FUEL	**ATA 70~80 Powerplant systems**
ATA 29 HYDRAULIC POWER	ATA 70 STANDARD PRACTICES- ENGINE
ATA 30 ICE AND RAIN PROTECTION	ATA 71 POWER PLANT
ATA 31 INDICATING / RECORDING SYSTEM	ATA 72 ENGINE
ATA 32 LANDING GEAR	ATA 73 ENGINE - FUEL AND CONTROL
ATA 33 LIGHTS	ATA 74 IGNITION
ATA 34 NAVIGATION	ATA 75 BLEED AIR
ATA 35 OXYGEN	ATA 76 ENGINE CONTROLS
ATA 36 PNEUMATIC	ATA 77 ENGINE INDICATING
ATA 37 VACUUM	ATA 78 EXHAUST
ATA 38 WATER/WASTE	ATA 79 OIL
ATA 49 AIRBORNE AUXILIARY POWER	ATA 80 STARTING
ATA 50 CARGO AND ACCESSORY COMPARTMENT	ATA 82 WATER INJECTION

TYPES OF TECHNICAL MANUAL

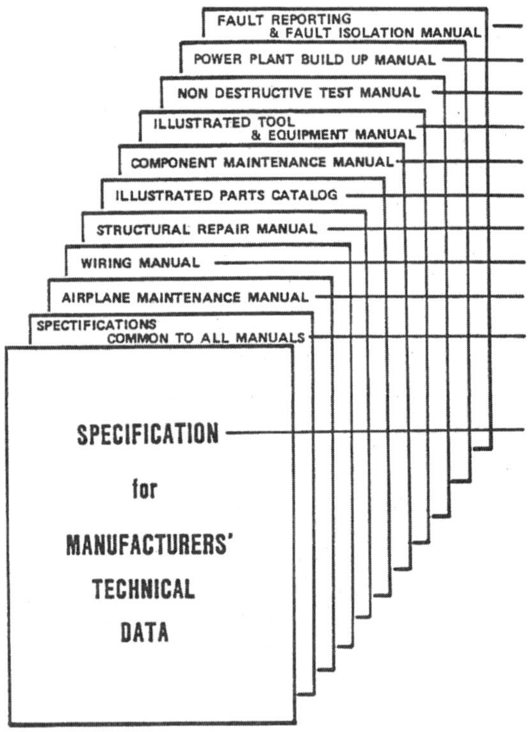

MAINTENANCE MANUAL

The maintenance manual contains the information necessary to allow the mechanic, or line technician, to SERVICE, TROUBLESHOOT, FUNCTIONALLY CHECK, and REPAIR the airplane's systems and equipment.
The manual includes a system description, instructions in how to remove and replace line replaceable units (LRU), and test procedures.

FAULT REPORTING MANUAL AND FAULT ISOLATION MANUAL

The Fault Reporting Manual (FRM) and Fault Isolation Manual (FIM) are used by the flight crews and maintenance personnel to analyze and isolate airplane faults.
The Fault Reporting Manual is used by the flight crews. It contains information to analyze and report faults or malfunctions while in flight.

RAMP MAINTENANCE MANUAL

This customized manual is a condensed Maintenance and Fault Isolation Manual. It contains the information most frequently needed by the line technician during TURNAROUND maintenance.

BUILT IN TEST EQUIPMENT (BITE) MANUAL

The built in test equipment (BITE) manual is customized.
Some line replaceable units (LRUs) have built-in test equipment (BITE). BITE permits testing of the LRU and related components without using additional test equipment. The BITE Manual provides fault isolation procedures and operating instructions for LRUs with BITE.

WIRING DIAGRAM MANUAL

This is a customized manual that describes the wiring and electrical equipment installed in the airplane. It is used for trouble shooting, repairing wiring, and replacing electrical equipment.

SYSTEM SCHEMATICS MANUAL

This is another customized manual that contains schematic diagrams for all of the airplane systems. The schematic diagrams in this manual show system layout, system function, circuit operation, component identification, and location.

COMPONENT MAINTENANCE MANUAL

This manual is not customized. It contains instructions for restoring manufacture designed components to a serviceable condition. It is used by the overhaul shop.

STRUCTURAL REPAIR MANUAL

This is a non-customized manual, contains instructions on how to repair airplane structure in the field.

ATA 21 - AIR CONDITIONING

The purpose of pressurization system is to create a comfortable environment for passengers and crew by pressurizing the cabins and controlling the rate of air flow to ambient.

PRESSURIZATION

Purpose

The purpose of airplane pressurization is to provide a comfortable environment by increasing cabin pressure to maintain a cabin altitude as close as possible to Sea Level.

Physical Description

A one inch column of air extending from the upper atmosphere down to sea level will exert a pressure of 14.7 psi on the Earth. This pressure is measured at 29.92 inches of mercury and referred to as Sea Level pressure or Standard Day, barometric pressure. As altitude increases, barometric pressure decreases.
At 10,000 feet above Sea Level barometric pressure has decreased to 10.11 psi, and at 30,000 feet the pressure is only 4.36 psi. From Sea Level to 30,000 feet, barometric pressure decreases approximately 70%.

Operation

In order to provide a comfortable environment for passengers and crew, the airplane cabin must be as close to Sea Level as possible. To provide a low cabin altitude, pressure inside the airplane must be greater than outside pressure. this is referred to as **differential pressure**.

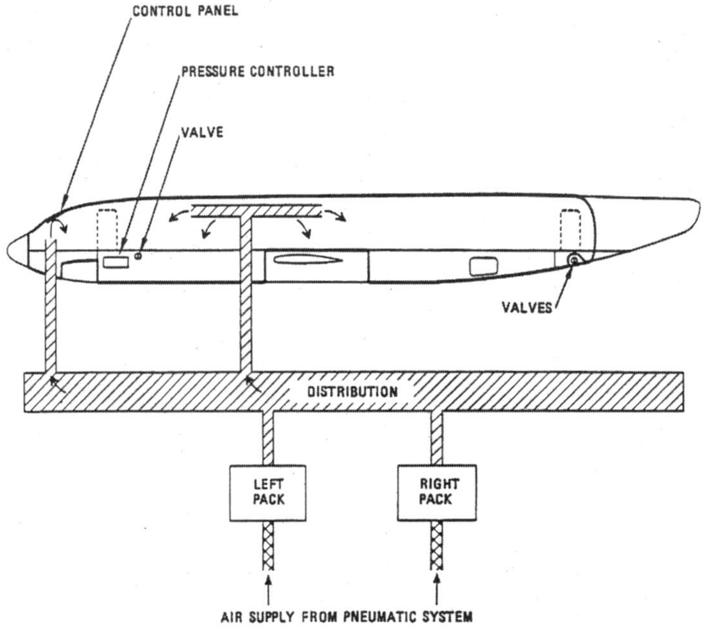

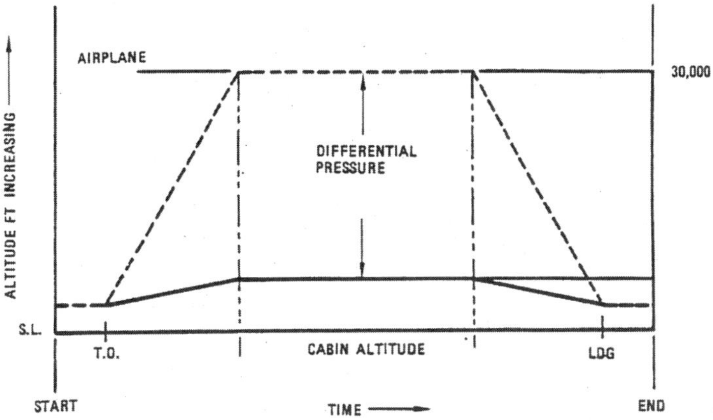
AIRPLANE
30,000
DIFFERENTIAL
PRESSURE
ALTITUDE FT INCREASING
S.L.
T.O.
CABIN ALTITUDE
LDG
START
TIME
END

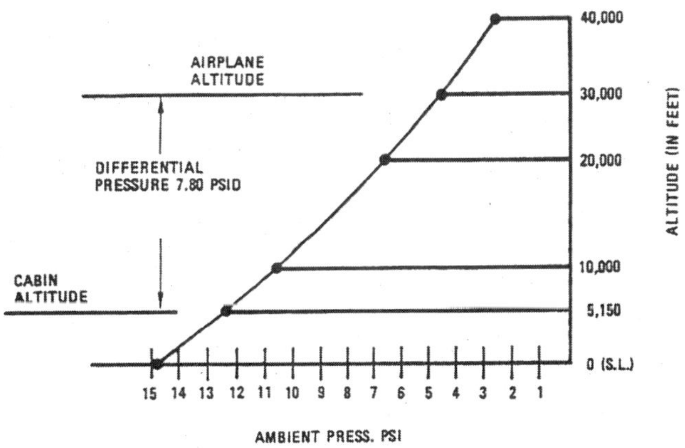
40,000
AIRPLANE
ALTITUDE
30,000
20,000
DIFFERENTIAL
PRESSURE 7.80 PSID
ALTITUDE (IN FEET)
CABIN
ALTITUDE
10,000
5,150
0 (S.L.)
15 14 13 12 11 10 9 8 7 6 5 4 3 2 1
AMBIENT PRESS. PSI

ATA 22 - AUTO FLIGHT

An aircraft autopilot(automatic pilot) system controls the aircraft without the pilot directly operating the controls. Such system is developed to reduce the work load of human pilots in order to lessen their fatigue and reduce operation errors during long flights.

AFCS(Automatic Flight Control System)

General description
An aircraft autopilot with many features and various autopilot related systems integrated into a single system is called an automatic flight control system (AFCS). Because of advances in digital technology for aircraft, modern aircraft of any size may have AFCS.

Operation
AFCS capabilities vary from system to system. Some of the advances beyond ordinary autopilot systems are the extent of programmability, the level of integration of navigational aids, the integration of flight director and auto throttle systems, and combining of the command elements of these various systems into a single integrated flight control human interface.

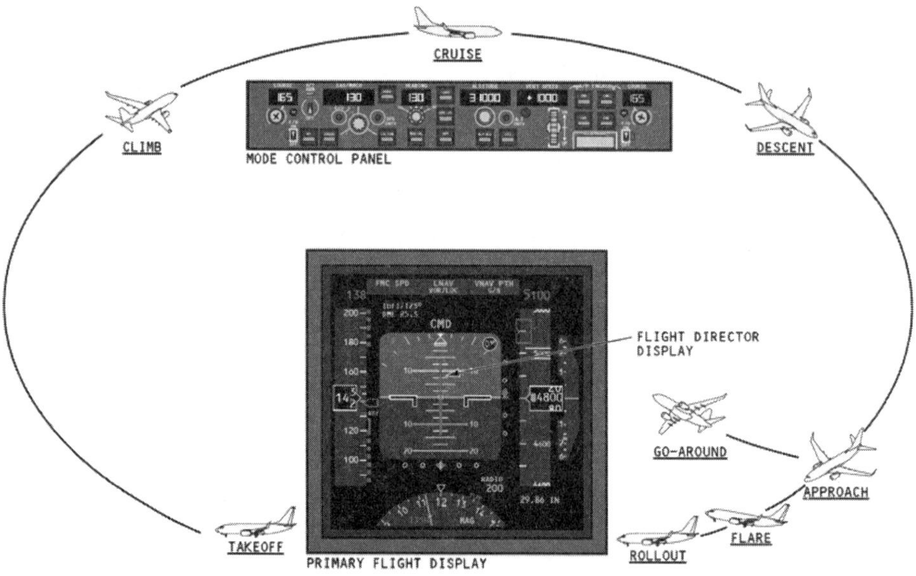

DUTCH ROLL

Lateral instability. Alternating (left-right) yawing movement which induces a roll. An undesirable, low-amplitude coupled oscillation about both the yaw and roll axes that affects many swept wing airplanes. Dutch roll is minimized by the use of a yaw damper.

YAW DAMPER

A yaw damper (sometimes referred to as a stability augmentation system. SAS) reduces the undesirable tendencies of an aircraft to oscillate in a repetitive rolling and yawing motion, a phenomenon known as the Dutch roll. This yaw damper system acts on the rudder automatically to stabilize the aircraft.

To engage / to disengage

In AP system, when the switch is engaged instead of ON, normally it means to clutch and gear or to fit and lock together(contrary: disengage). Instinctive disconnect push-buttons on the throttle levers let the pilot to override the automatic system at any time.

STANDBY

For a system which is not operating, but is ready to operate if the primary fails.

FLARE

Final nose-up movement immediately prior to touchdown. The Landing flare, in a fixed wing aircraft, is the transition phase between the final approach and the touchdown on the landing surface.

GO-AROUND(GA)

Decision not to land: increase of thrust and altitude. In aviation, ago-around is an aborted landing of an aircraft that is on final approach or has already touched down. A go-around can either be initiated by the pilot flying or requested by air traffic control for various reasons, such as an unstabilized approach or an obstruction on the runway.

FLY-BY-WIRE

A method of control used by some modern aircraft in which control movement or pressures exerted by the pilot are directed into a digital computer where they are input into a program tailored to the flight characteristics of the aircraft. The computer output signal is sent to actuators at the control surfaces to move them the optimum amount for the desired maneuver.

ATA 23 - COMMUNICATIONS

Traditional aircraft communications are based on analog voice on either a Very High Frequency (VHF) or High Frequency (HF) radio waves. In the mid1980s, the use of data-based communications became a reality. Airspace management is transcending into the computer age as new requirements evolve and the choice of communications technologies expands. Accordingly, the regulating the world's air traffic flow can safely become more automated. Aircraft are currently equipped with communications technologies that transport data through satellite plus while they are on the ground. The mobile communication and broadband networks can receive or transmit strategic information regarding aircraft situation and even maintenance trends.

Aircraft began to be equipped with computers in the 1970s and this led to the development in 1978 of a data communications system called the "Aircraft Communications Addressing and Reporting System" (ACARS).

AIRCRAFT COMMUNICATIONS ADDRESSING AND REPORTING SYSTEM (ACARS)

Purpose
The purpose of ACARS is a data link communication system. It lets you transmit messages and reports between an airplane and an airline ground base. A message or report from the airplane to the airline ground base is called a downlink. A message or report from the airline ground base to the airplane is called an uplink. ACARS automatically sends reports at scheduled times in the flight to reduce crew workload.

Operation
Aircraft with ACARS can exchange data messages through a network of automated ground stations incorporating internal computers. Airlines first used the data link system to send movement reports to the ACARS service processors using the telex formats that operators previously used to send those reports. ACARS are widely used today on the recent aircraft.

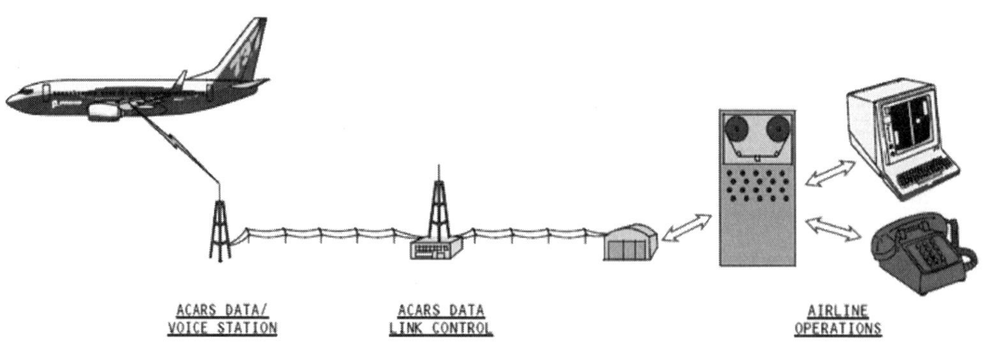

VERY HIGH FREQUENCY (VHF) COMMUNICATION

VHF communication systems are the most widely used for maintaining contact between ground and aircraft. This employs “Line Of Sight” transmission, which translates to a range of about thirty miles for an aircraft operating at 1,000 feet above the ground, or about 135 miles with an aircraft operating at 10,000 feet. The advantage of VHF over HF is that the link quality is much better and there is greater reuse of the frequency channel.

HIGH FREQUENCY (HF) COMMUNICATION

HF signals are reflected to the earth by ionized layers 60 to 200 miles above the earth’s surface. The HF system on an aircraft provides two-way voice communication with or digitally coded signals for ground stations or other aircraft. The HF radio control panel is located where it is easily accessible to the pilot or copilot.
The difference between HF and VHF is that the HF communications are best suited for long distance communication between ground operators and base stations. VHF communications are best used for transmission of data between radio operators within line of sight over flat ground.

To tune in

To adjust the controls of a radio or the receiver to detect the desired radio signal carrier frequency that a particular radio station uses.

INTERPHONE

There are 3 interphone systems used in airplane: flight interphone, cabin interphone, service interphone and it is a system providing communications between the different flight crew members, cabin crew and ground crew.

SERVICE INTERPHONE JACKS

Electrical receptacles for connection of headset to enable communication between maintenance personnel and flight crew.

COCKPIT VOICE RECORDER (CVR)

The cockpit voice recorder(CVR), one of black boxes makes a record of flight crew communication and flight compartment sounds. Data is erased automatically when new data is recorded by the voice recorder. Only the newest data is kept in the memory. The voice recorder unit keeps the last 120 minutes of communication data in memory.

FLIGHT DATA RECORDER(FDR)

A flight recorder is an electronic recording device placed in an aircraft for the purpose of facilitating the investigation of aviation accidents and incidents. Flight recorders are also known by the misnomer black box, in fact, painted bright orange in color to aid in their recovery after accidents.

ATA 24 - ELECTRICAL POWER

The electrical power is used for the control, operation, and indication of the various airplane systems on the ground and in flight.

ELECTRICAL POWER DISTRIBUTION

Purpose

The electrical power distribution provides power from the supply to the user systems.

General Description

The 115 volt AC electrical power is obtained from airplane generators or external source. Voltage regulator is used for control of the generator. Transformer is used to change AC voltage levels. Power is connected to the bus by means of breakers, 28 volts AC is obtained from 115 AC by use of a transformer.

The battery supplies 28 volts DC to the busses through relays. A battery charger uses 115 volts AC and recharges the battery.

Transformer rectifier converts 115 volts AC to 28 volts DC. The standby power is obtained normally from 115 volts AC and alternatively from the battery through a static inverter.

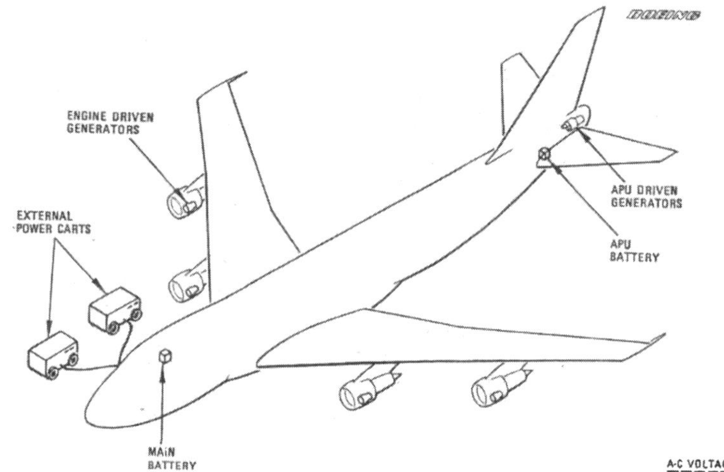

VOLTAGE REGULATOR

Purpose

The voltage regulator is used for controlling the output current from a generator.

Physical Description

The voltage regulator consists of a transformer

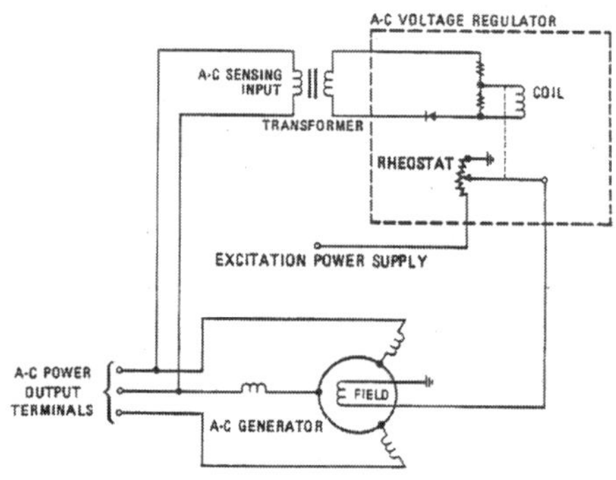

VOLTAGE REGULATOR

power supply, coil and rheostat.

Operation
The output from the generator depends on the power supply to the generator field (excitation power). A transformer senses the output of the generator and supplies power to the coil.
If output changes, the coil power supply changes. The coil moves a wiper on the rheostat, thus supplying variable excitation power which is inversely proportional to the generator output.

AC EXTERNAL POWER RECEPTACLE

Purpose
The receptacle is used for connections of external AC power to the airplane.

Physical Description
The standard AC receptacle includes four AC pins, for phases A, B, C, and N (neutral or AC ground) and two shorter interlock pins E and F. The DC for the interlock circuit is supplied from AC external power by the TR unit inside the airplane. The relays connecting the AC external power to the bus system depend upon an external interlock circuit joining the two short pins E and F. If the external power cord is inadvertently removed while AC power is still flowing, the E and F pin interlock circuit will be broken before the longer AC pins separate from the cord. Wien the interlock circuit is broken, all relay holding external AC power on the electrical system will relax. Therefore, there will be no current flowing, and therefore no arcing or flash when the cord pulls away from the large pins.

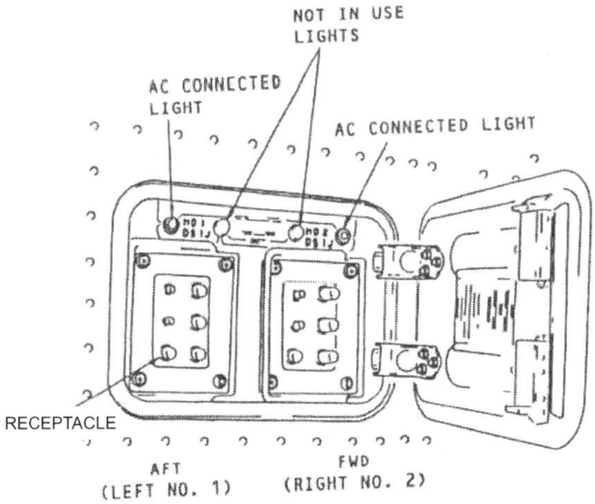

AC EXT PWR Receptacle

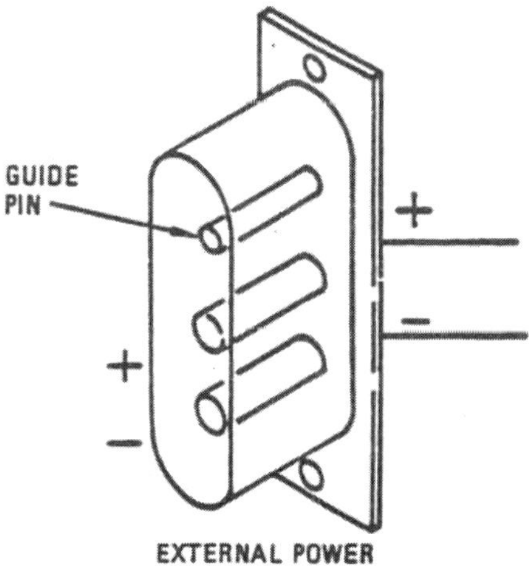

DC EXT PWR Receptacle

DC EXTERNAL POWER RECEPTACLE

Purpose
The receptacle is used for connection of external DC power to the airplane.

Physical Description
The receptacle consists of three pins. These are identified as positive, negative and a guide pin. The guide pin ensures correct connection of DC power to the receptacle.

TRANSFORMER RECTIFIER

Purpose
The transformer-rectifier converts 115 volts AC at 400 Hz, 3 phases to 28 volts DC. Maximum output rating is 75 amperes.

Physical Description
The transformer-rectifier consists of an input power transformer to convert high-voltage 3-phase power to low-voltage 6-phase power, twelve silicon power rectifiers to convert the 6-phase power to direct current.

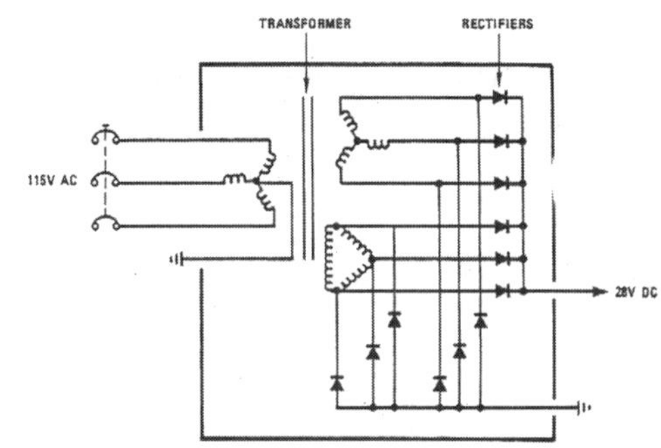

Operation
The 115 volts AC, 400 Hz, 3 phase power is supplied to the transformer section. Output of the transformer is 28 volts DC. The diodes rectify (convert) the 28 volts AC to 28 volts DC power.

PRINTED CIRCUIT CARD

Purpose
The printed circuit card is used in electronic modules to provide a compact method of installing electronic components.

Physical Description
The card consists of a fiberglass board with the various electronic components mounted on the card. The electrical connections between the components are embedded in the card. A connector at the side of the card connects the card to the circuits in the module.

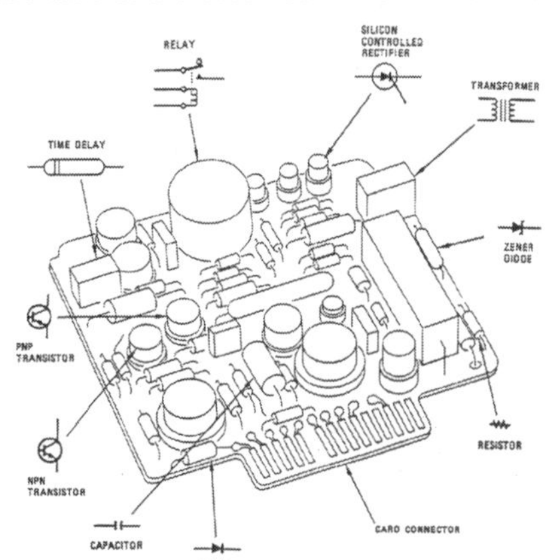

PRINTED CIRCUIT CARD

ATA 26 - FIRE PROTECTION

The fire extinguishing system is used for suppressing a fire on the engines and the auxiliary power unit. The fire extinguisher agent is stored in spherical containers on the airplane.

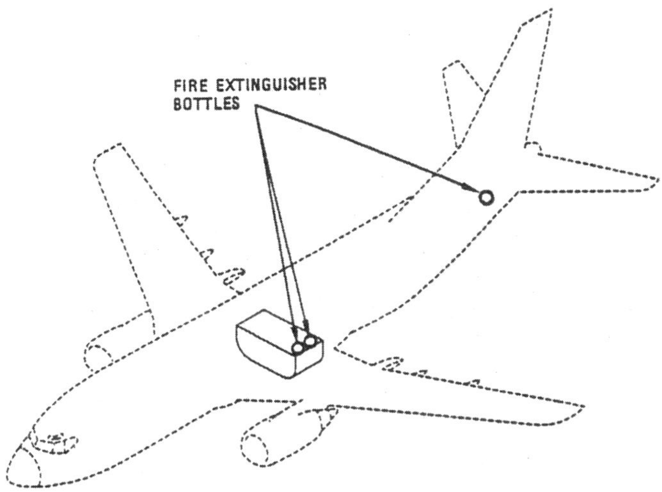

FIRE EXTINGUISHING

EXTINGUISHER BOTTLE

Purpose

The extinguisher bottles are used for storage of the extinguishing agent for fire suppression.

Physical Description

The extinguisher bottle consists of a spherical vessel, discharge systems consisting of a squib (detonator), heating element and rupture disc.

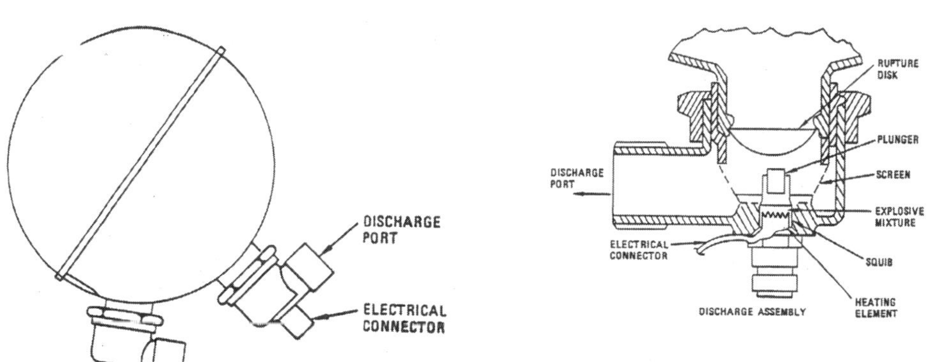

EXTINGUISHER BOTTLE

Operation

When the bottle is required to be discharged, electrical power is supplied to the heating element, heat developed by the element, detonate. The forces the disc which the from the bottle allows the squib to force of detonation plunger against the ruptures, releasing extinguishing agent

PORTABLE FIRE EXTINGUISHERS

Purpose

There are three types of portable extinguishers stowed on-board the airplane: carbon dioxide, dry chemical and water.

Carbon dioxide (CO2) extinguishers are best for controlling electrical fires. The cylinder has a horn to direct the gas at the fire.

Dry chemical extinguishers work best on fuel or oil fires, but will also control electrical fires.

Water extinguishers are used to extinguish fires on clothing, seats or paper. Water extinguishers are never to be used for electrical fires.

Carbon Dioxide (CO2) Type

The carbon dioxide extinguisher is identified by the horn nozzle, and is intended primarily for use in extinguishing electrical fires. Operation is controlled by a trigger mechanism in the handle. Until operated, the trigger is lockwired and sealed. The trigger has an overtemperature warning indicator legend (REPLACE DISC) which is covered by the handle when the bottle is in the ready condition and exposed when the bottle has been discharged due to an overtemperature condition.

Water Type

The water type extinguishers are intended for extinguishing nonelectrical fires, and contain an anti freeze agent to retain serviceability at low temperatures. Discharge is effected by pressure from a carbon dioxide cartridge in the extinguisher handle. The pressure is released by twisting the handle of the extinguisher to puncture the cartridge. Operation is controlled by actuation of a thumb-operated lever. Until operated, the handle is lockwired and sealed.

Dry Chemical Type

The dry chemical type extinguishers are intended for use in extinguishing all fires in the passenger cabin areas. A pressure indicator on the bottom of the cylinder indicates condition of the cylinder. The dry chemical-type extinguisher is discharged by actuating the operating lever. After use of the extinguisher, all residue should be cleaned up with a vacuum cleaner or damp cloth. Normally it has no harmful effects except for slight stripping and staining of zinc chromate primer; however, due to high temperatures, surfaces may be pitted.

ATA 27 - FLIGHT CONTROLS

The flight controls provide maneuvering control about the lateral, longitudinal, and vertical axes. They also provide increased lift for takeoff and landing as well as increased aerodynamic drag both in flight and on the ground.

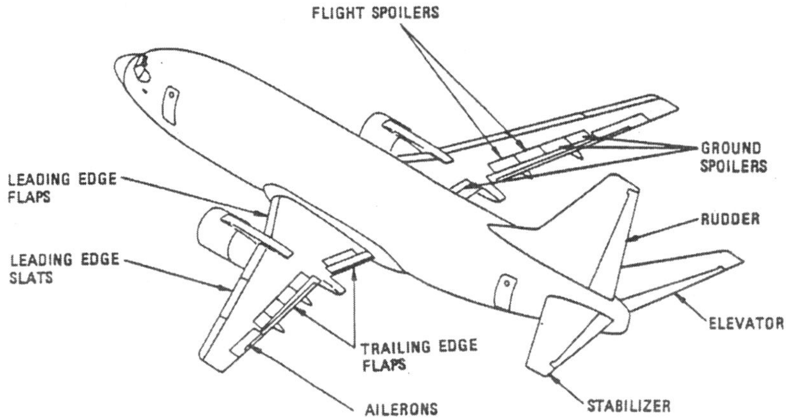

FLIGHT CONTROLS

BASIC CABLE SYSTEM

Purpose

Cable systems are used to transfer control motions from one point to another. For example, from the flight compartment to the flight controls.

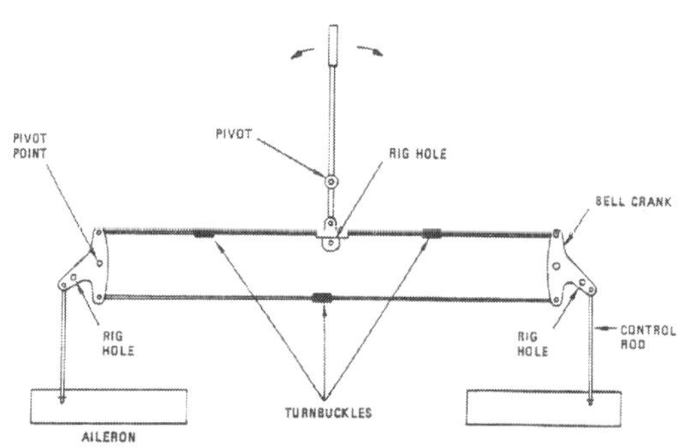

Physical Description

Most cable systems are a closed loop. The essential parts of a cable system are the input, the output, the cables, and a means of adjusting the system. The input to the system is the control lever and the output is the movement of the ailerons. The system is adjusted by adjusting the cable turnbuckles and the control rods.

Operation

Cables are adjusted to achieve the proper cable tension and the proper alignment of inputs and outputs. The tension and position of the cranks is adjusted by turnbuckles. The proper alignment is determined by checking for a free fit of rig pins in the rig holes. After turnbuckles are adjusted, they are locked in position by means of locking clips.

CONTROL CABLES

Purpose

Control cables are used to transfer a control motion from one point in the airplane to another.

Physical Description

The material normally used in fabricating control cables is tin plated carbon steel. Corrosion resistant steel is infrequently used. The basic element of a cable is an individual wire. The wires are spirally wound to form a strand. One strand becomes the center strand, and others are wound around it to form the cable. Cables are identified by diameter and the number of wires and strands used in fabrication.

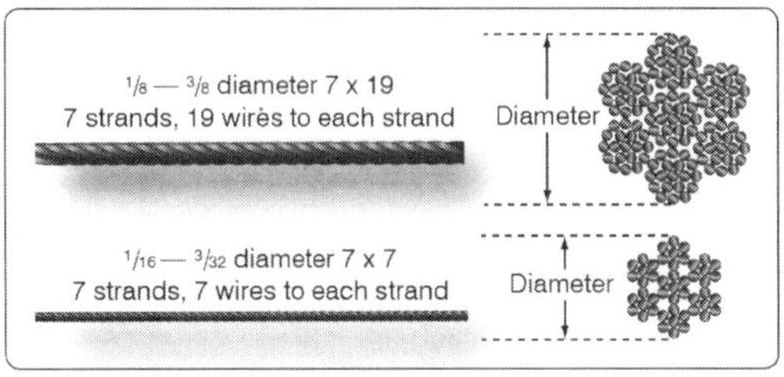

CONTROL CABLE

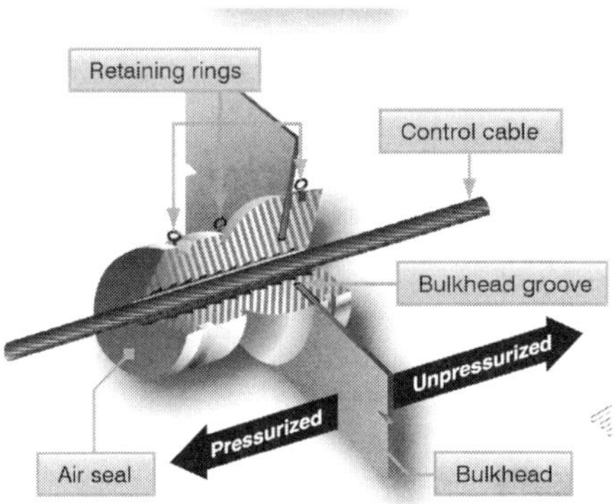

CABLE PRESSURE SEALS

CABLE PRESSURE SEALS

Purpose

Cable pressure seals are used where control cables pass from pressurized to unpressurized areas.

Physical Description

The seal assembly consists of a split ball made from a non-metallic material supported by a seal plate and seal cover. The diameter of the hole through the split ball is slightly larger than the cable. The seal plate is fitted with a sealing gasket.

Operation

The seals prevent pressure loss while allowing free movement of the cable through the bulkhead. The mounting holes in the seal cover and seal plate are oversized to allow the assembly to be positioned so that the cable is not deflected. Final position of the seal is established after the cable is at final tension. The prescribed seal size must be used, as a larger seal will allow excessive air leakage and a small seal will cause early cable wear.

CABLE FAIRLEADS AND GROMMETS

Purpose

Fairleads and grommets prevent cables from contacting surrounding structure.

Physical Description

Fairleads and grommets are made from non-metallic materials such as nylon or phenolic. They are placed around cables in areas where cables pass close to metal structure.

Operation

Cables should not normally contact grommets or fairleads. They should be placed so that a slack or deflected cable will contact the fairlead or grommet before the cable contacts metal structure.

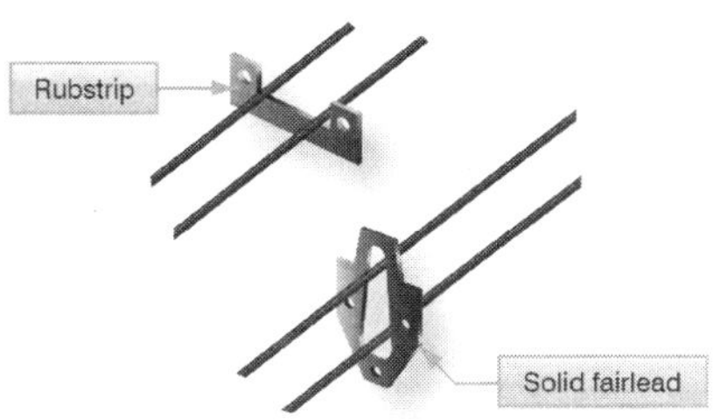

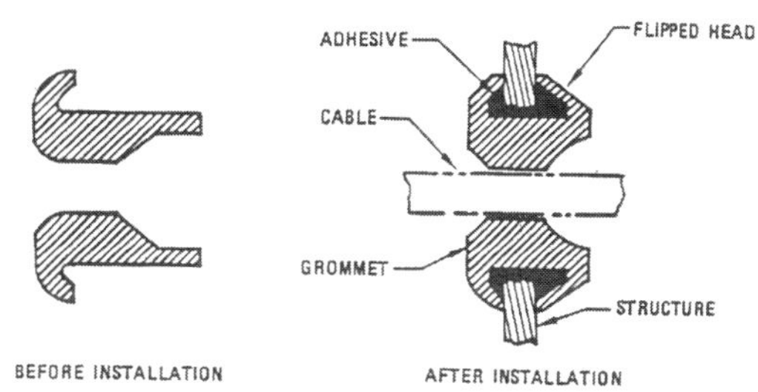

LINEAR VARIABLE DISPLACEMENT TRANSFORMER

Purpose

The linear variable displacement transducer (LYDT) is used to provide electrical signals which vary in proportion to changes in mechanical displacement.

Physical Description

The transducer consists of a housing with fixed transformer coils and a movable armature which is the mechanical input. The device shown is a dual LVDT.

Operation

The primary coil of the transformer is excited by an alternating current. The secondary coil is split into two windings. When the armature is positioned so that it is equally spaced over the dual secondary coils, the magnetic coupling is equal resulting in the same voltage being induced into each secondary coil. Since the voltages are subtractive, the resulting output is near zero. As the armature is moved, the induced voltages increase in one coil and decrease in the other. The resulting output voltage is representative of armature position.

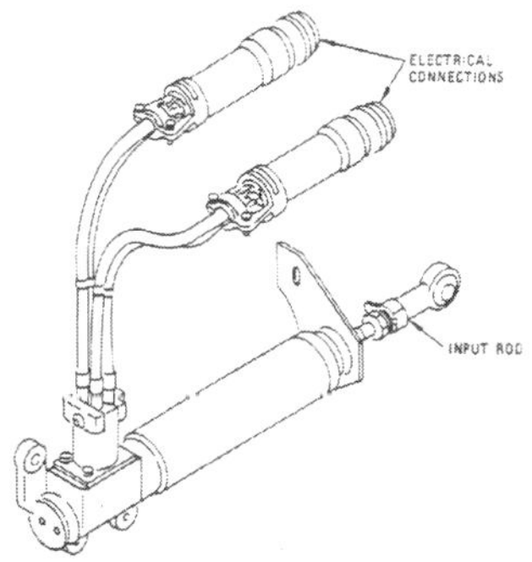

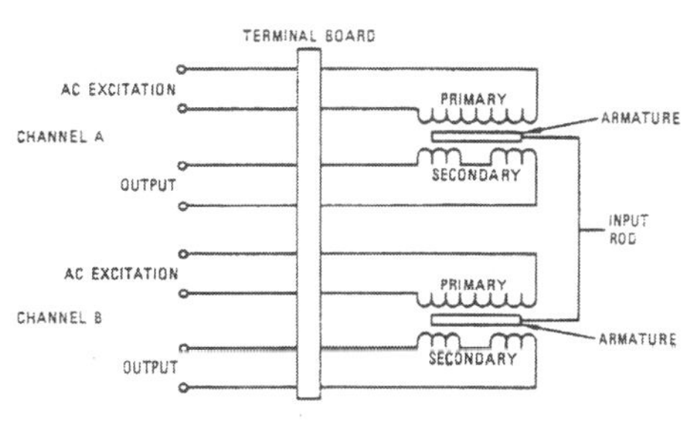

LINEAR VARIABLE DISPLACEMENT TRANSFORMER (LVDT)

ATA 28 - FUEL

The purpose of the fuel system is to store and deliver fuel to the engines and APU. The fuel system consists of 3 tanks, lines, fittings and operating components which are compatible with all fuels meeting the engine and APU specifications.

Tank	Gallons	Pounds*
Left main tank	9,560	64,000
Right main tank	9,560	64,000
Center tank	26,100	174,900
Total	45,200	302,900

* Usable fuel at level attitude
Fuel density = 6.7 pounds per U.S. gallon.

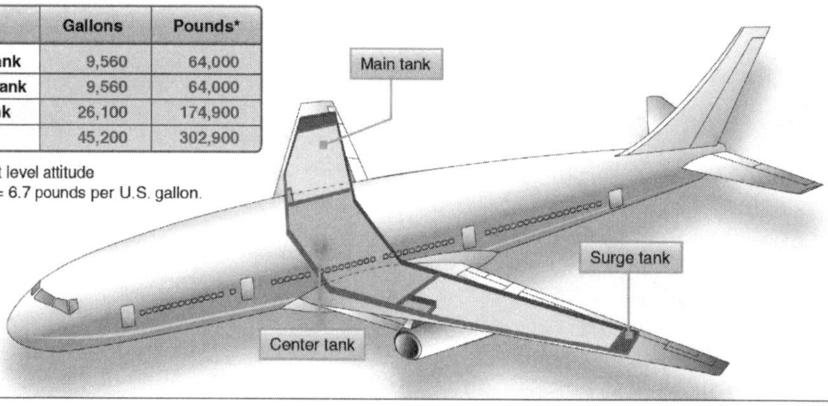

Boeing 777 fuel tank locations and capacities.

General Description
All fuel is stored within vented areas of the wings and wing center section. The fuel storage area is divided into integral fuel tanks. All tanks utilize a sealed-wing structure to retain the fuel.
Fuel is loaded on airplane from a ground source through a fueling receptacle in the pressure fueling station. The tanks can be filled simultaneously or separately.
Electric motor-driven boost pumps and fuel lines deliver fuel from a tank to one or both engines. Fuel from boost pumps can also be delivered through a valve into manifold for removal from the airplane.
Fuel scavenge pump reduces the quantity of unusable fuel in the tank. Fuel quantity indicators indicate the quantity of fuel on the airplane.

VENT SCOOP

Purpose
The vent scoop provides an opening between the fuel tank and ambient for passage of air and fuel vapors.

Physical Description
The vent scoop consi sts of a circular plate with a flush type inlet/outlet and a stand pipe located in the fuel tank.

Operation
When a pressure difference exists between the interior of the tank and ambient conditions, air will flow into the tank or fuel vapors will flow to the outside from the tank.

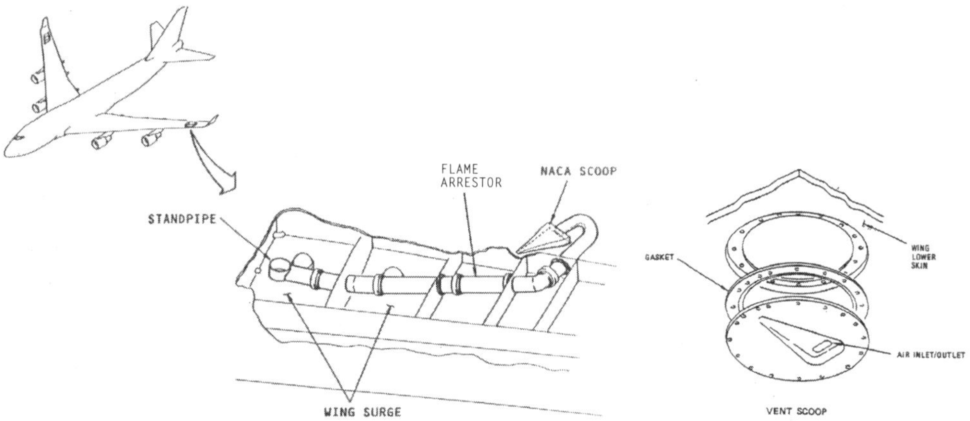

FLAME ARRESTOR

Purpose
The flame arrestor prevents propagation of flames in fuel vent system.

Physical Description
The flame arrestor consists of a housing enclosing a honeycomb core.

Operation
During normal operation fuel vapors are vented to outside through the vent system. If a fire condition should occur, the flame is stopped by the arrestor due to small passages of the honeycomb core.

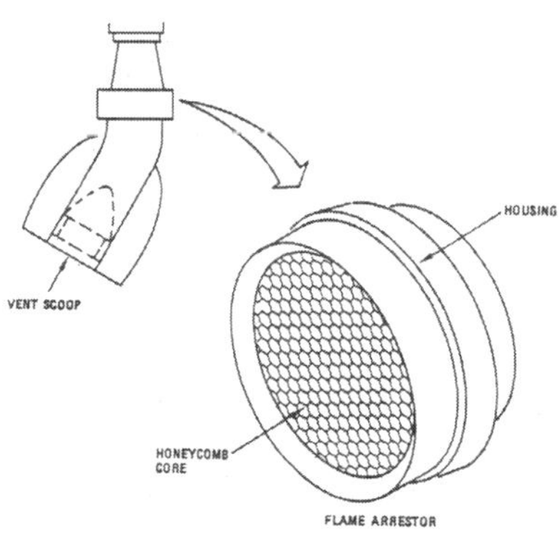

FLAME ARRESTOR

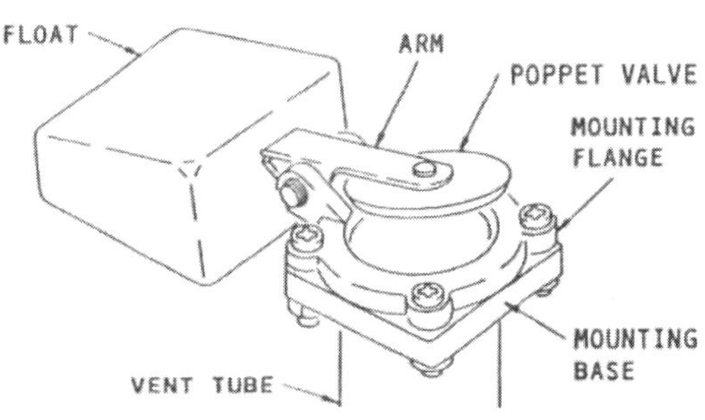

FLOAT VALVE

FLOAT VALVE

Purpose
The float valve is for preventing fuel entering the vent system.

Physical Description
The valve consists of a float mounted on an arm hinged to a mounting base. A valve stopper attached to the float arm incorporates a seal which contacts and seals the vent port when valve is closed.

Operation
The valve is held in the normally open position by the weight of the float and float arm against a stop attached to the valve mounting base. The stop limits the downward travel of the float. As fuel rises in the tank, the float rises closing the valve aperture. When the fuel level recedes, the valve reopens.

BOOST PUMP

Purpose
The boost pump delivers fuel under pressure from tank to the engines. The pump is also used for defueling the tank and for fuel transfer between the tanks.

Physical Description
The boost pump consists of a three-phase induction motor-driven centrifugal impeller, inlet port with pump removal valve and handle, drain cock, vent port, and a discharge port.

Operation
The electric motor rotates the shaft with vanes and impeller. Fuel enters the pump through the inlet and pump removal valve. The rotating impeller pressurizes the fuel and sends it to the manifold. A small portion of the fuel flows through the cooling flow passages to cool and lubricate the motor and the bearings. The rotating vanes help to discharge the cooling and lubricating flow.

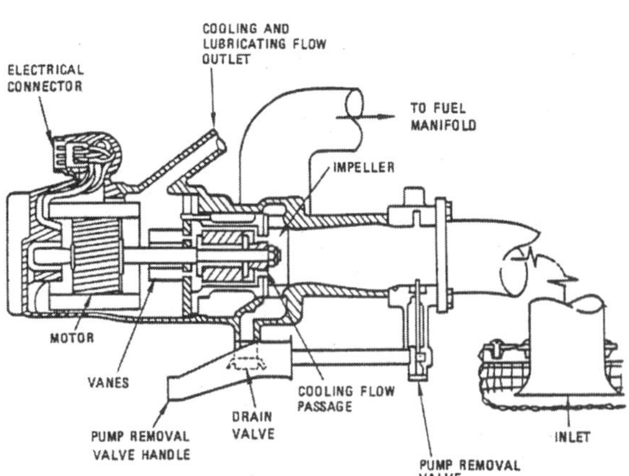

BOOST PUMP

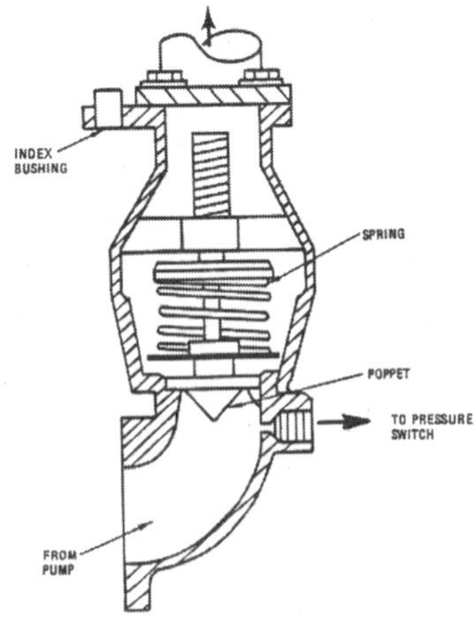

CHECK VALVE

CHECK VALVE

Purpose
The fuel boost pump check valve closes the boost pump outlet line when the pressure drops below a predetermined amount, thus preventing reverse fuel flow to the boost pump.

Physical Description
The valve consists of a spring-loaded-closed valve mounted in a housing. The housing has three ports: fuel in, fuel out and a pressure sensing port. The pressure sensing port is located upstream of the valve.

Operation
When the output pressure from the pump is higher than the spring force□the valve opens and fuel passes to the manifold. When pressure decreases, the spring closes the valve preventing reverse flow from the manifold to the pump.

JET PUMP

Purpose
The jet pump is for siphoning fuel from one tank and delivering it to another tank.

Physical Description
The jet pump consists of a housing enclosing a nozzle and a removal valve.
The housing has two inlet and one outlet ports.

Operation
High pressure fuel enters the housing and passes through the nozzle and outlet. This flow creates a low pressure condition in the inlet surrounding the nozzle and fuel is siphoned from the tank.

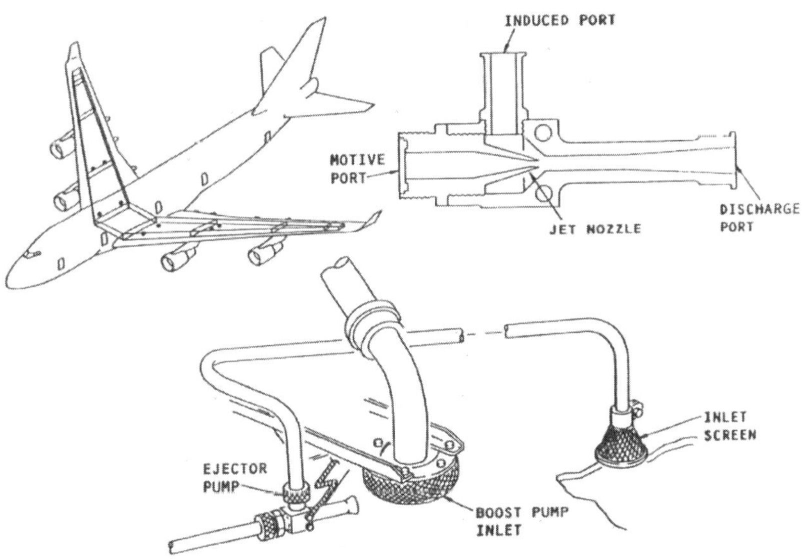

BYPASS VALVE

Purpose
The valve is used for fuel flow bypassing the fuel pump.

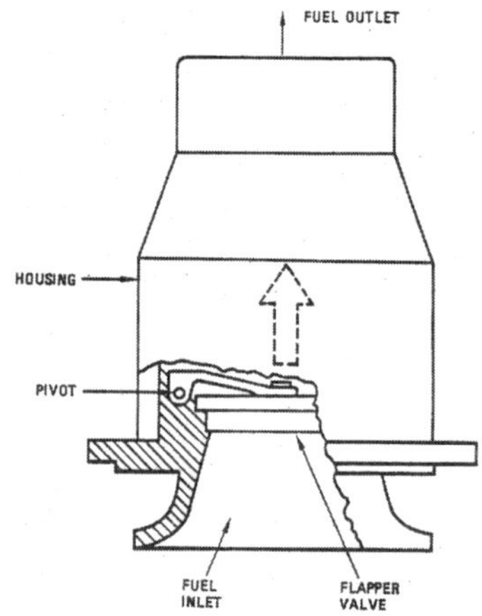

Physical Description
The valve consists of a flapper valve in a housing.

Operation
With the fuel pump operating, the fuel pressure acts on top of the flapper, keeping it closed. When the pump pressure is not available, depth of fuel in the tank forces the flapper to open, fuel is then bypassing the pump.

MOTOR OPERATED VALVE

Purpose
The valve allows fuel flow through the fuel manifold.

Physical Description
The valve is an electric motor-driven slide shutoff valve with a manual override handle. The manual override handle provides a visual check of the valve position and a manual method of positioning the valve when the electric motor is not energized.
The gate assembly consists of a housing, a splined input shaft, a valve actuating arm, a sliding gate, and gate seals. One end of the valve actuating arm is attached to the splined input shaft. The other end of the arm carries a roller which engages in a slot in the end of the sliding gate. Spring-loaded seals, held in carriers attached to the housing, provide fuel line seals on both sides of the sliding gate.

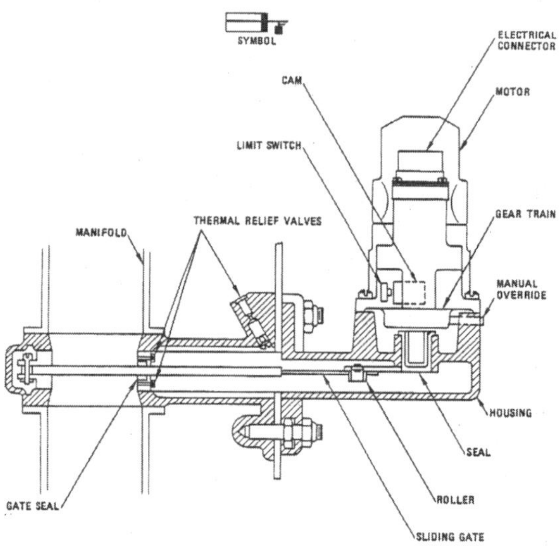

The port adapter assembly consists of a housing and a thermal relief valve. The assembly, which is mounted inside the fuel tank, provides the mounting surfaces for the fuel lines and a fuel tight enclosure for the gate assembly. A thermal relief valve vents excessive pressure inside the valve body back into the fuel tank.

Operation
When the motor is energized, it drives the gear train which in turn moves the actuating arm. The arm moves the sliding gate to open or close the valve.

TANK UNIT

Purpose

The tank units are mounted throughout each fuel tank to sense fuel level variations and, as a result of capacitance change, transmit electrical signals to the fuel quantity indicators.

Physical Description

The tank unit consists of two coaxial cylinders which are rigidly spaced at intervals through their length by insulating centering spacers. The inner electrode is a variable diameter tube. The outer electrode is a tube of constant diameter and is connected to the 400-Hz excitation voltage lead. The diameter of the inner electrode is profiled to provide an output signal which is linear with respect to the volume of fuel in the tank.

Openings at the ends of tank units allow fuel to assume the same level between the two cylinders as that in the fuel tank.

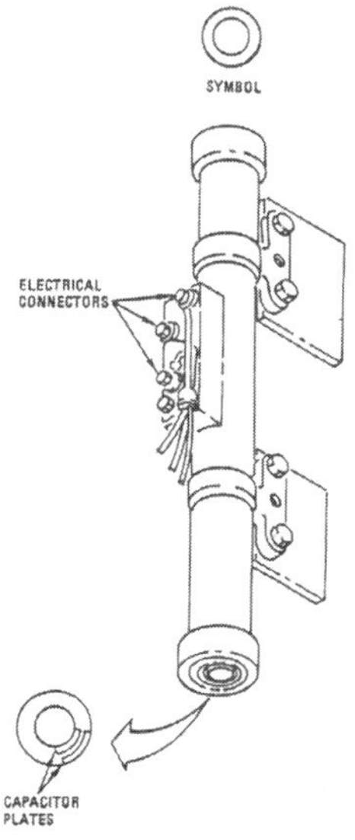

TANK UNIT

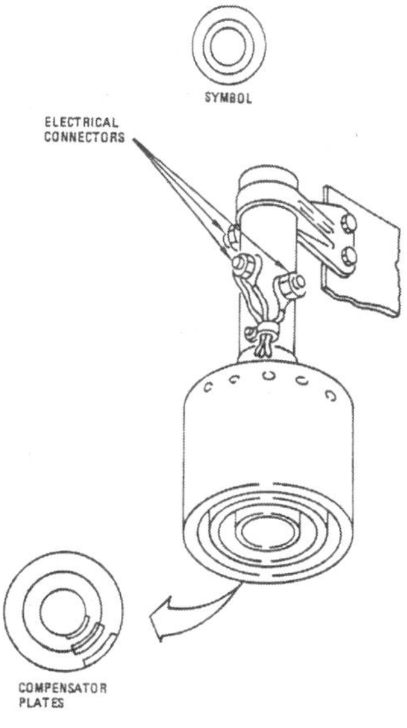

COMPENSATOR

COMPENSATOR

Purpose

The compensator unit in each tank acts as a sensing element to maintain a linear relationship between fuel dielectric properties and fuel density.

Physical Description

The compensator consists of a low impedance plate and two high impedance plates, both protected by a mechanical shield. The compensators are located so that they are completely submerged in fuel at all times. Leads from the compensator units are contained in the same harness used for the tank units.

ATA 29 - HYDRAULIC POWER

The hydraulic system provides the power for operation of the flight controls, landing gear, and thrust reversers.
Hydraulic power is developed by pumps driven mechanically, electrically or pneumatically. The pressurized fluid is delivered to the using systems by metal tubing and flexible hoses.

General Description
A hydraulic system consists of reservoirs, pumps, valves and filters. Fluid is supplied to the pumps from a reservoir which is normally pressurized to ensure a positive flow to the pumps. The pumps are driven mechanically, electrically or pneumatically. Filters are used to remove contaminants from the hydraulic fluid. Fluid flow is controlled by directional valves, shut off valves, check valves, and regulators. Pumps are cooled by heat exchangers immersed in the fuel tanks. Pressure switches, pressure transmitters, temperature switches and sensors provide control and monitoring information.

VARIABLE DISPLACEMENT HYDRAULIC PUMP

Purpose
The pump provides pressure for the hydraulic system.

Physical Description
The basic components of the pump are the case, a cylinder barrel, a yoke assembly, and a means of controlling the angular position of the yoke.
The pump design may be such that the cylinder barrel or the yoke rotates in response to the input shaft.

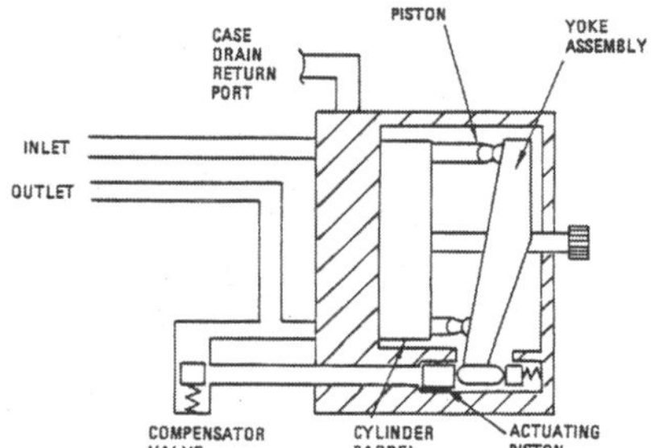

VARIABLE DISPLACEMENT HYDRAULIC PUMP

FILTER

Purpose
Filters are used to remove solid contaminants from fluids.

Physical Description
The filter consists of a bowl and filter element. The filter incorporates a check valve and a magnetic plug.

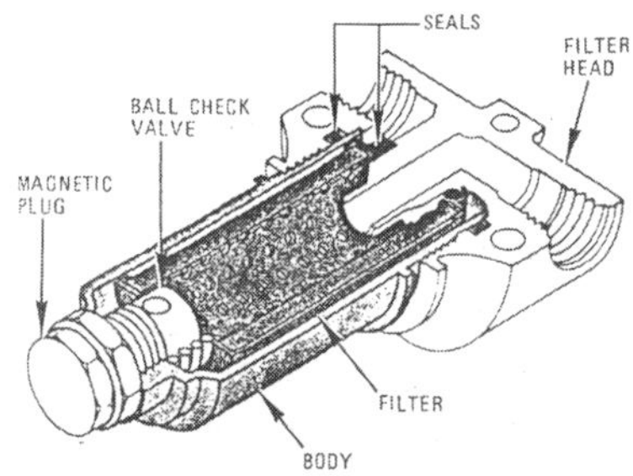

Operation
In the basic filter, all fluid entering the filter body is passed through the filter element. Any solid contaminants larger than the mesh of the element will be trapped by the filter element. The check valve serves to prevent reverse flow through the filter which would dislodge the trapped particles back into the system. The magnetic plug traps any ferrous materials passing through the filter. Depending on the location of the filter in the system, trapped metal particles will provide an indication of system problems.

FLARELESS FITTINGS

Purpose
Flareless fittings are used as tube connectors. When properly installed, they form a strong, fluid-tight joint.

Physical Description
The permaswage fitting is a permanent type fitting. The union shown will join two tubes which cannot be disconnected and then reconnected. It is often used for repairs. The other two fittings shown can be disconnected and reconnected by using a separate union. The flairless fitting remains attached to the tube.

Operation
The fitting can be swaged to the tube using a proper size steel union and B nut. When properly installed, the end of the sleeve will bite into the tube forming the seal. The other two fittings shown require specific swaging tools to install the fittings.

1. union and B nut usage

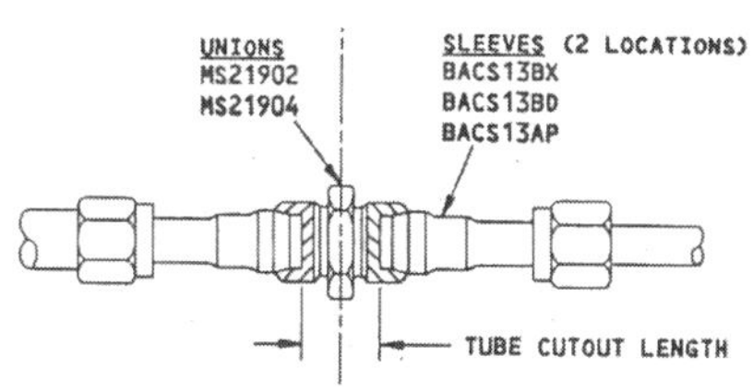

2. flareless fitting usage

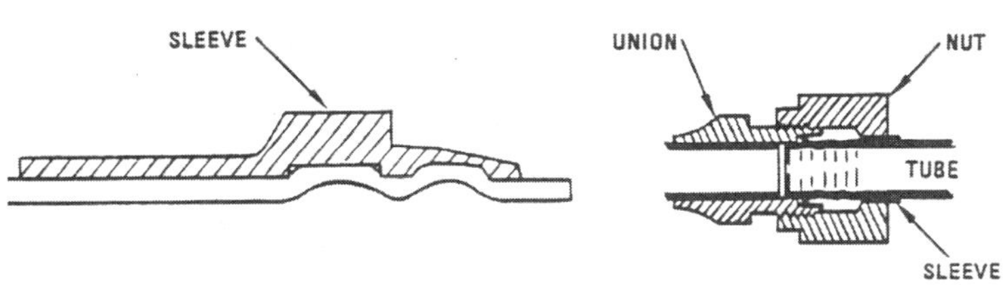

3. permaswage fitting usage

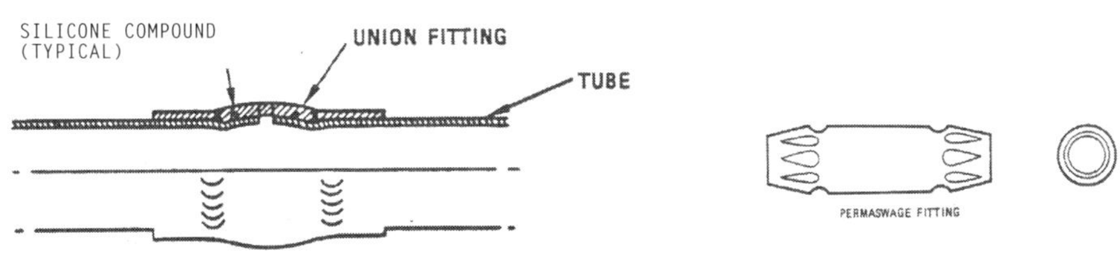

LINE DISCONNECT VALVE(quick disconnect valve)

Purpose
The line disconnect valve is used to allow hydraulic lines to be disconnected without fluid loss.

Physical Description
Each half of the disconnect valve contains a spring operated poppet valve.

Operation
When the two halves of the valves are separated, each poppet valve is forced to its seat by the spring which prevents fluid loss. When the valve halves are joined, the pistons of the poppet valves contact and the poppets are forced away from the valve seats. Therefore that allows uninterrupted fluid flow.

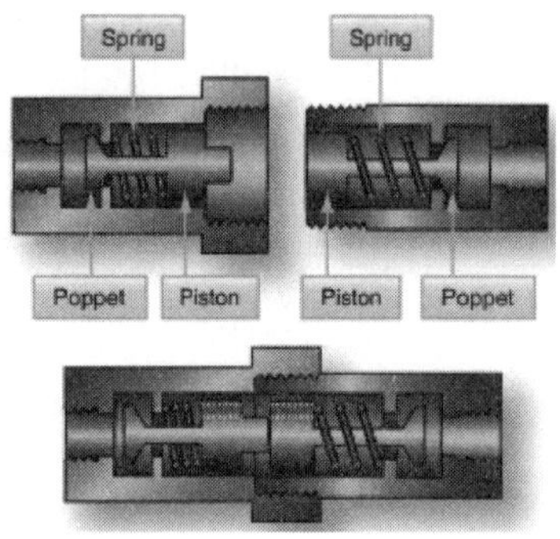

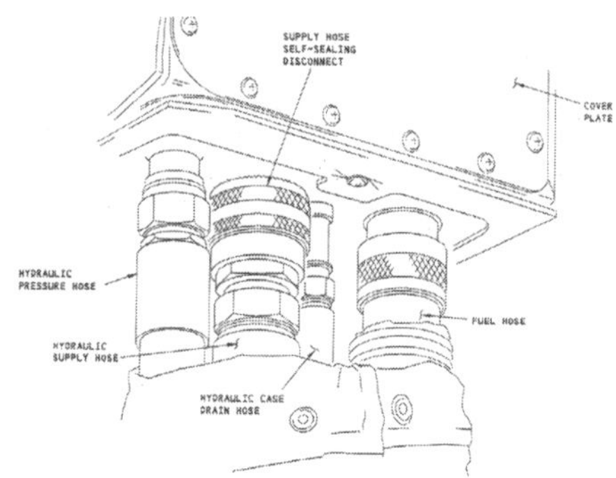

CHECK VALVE

Purpose
The check valve is used to allow flow in one direction only. It is frequently used to isolate one system from another.

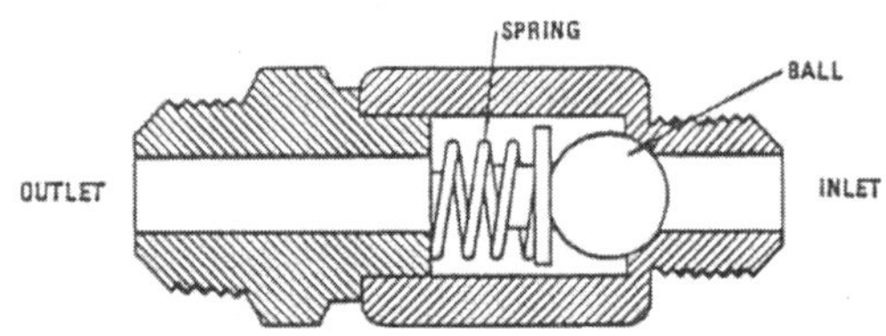

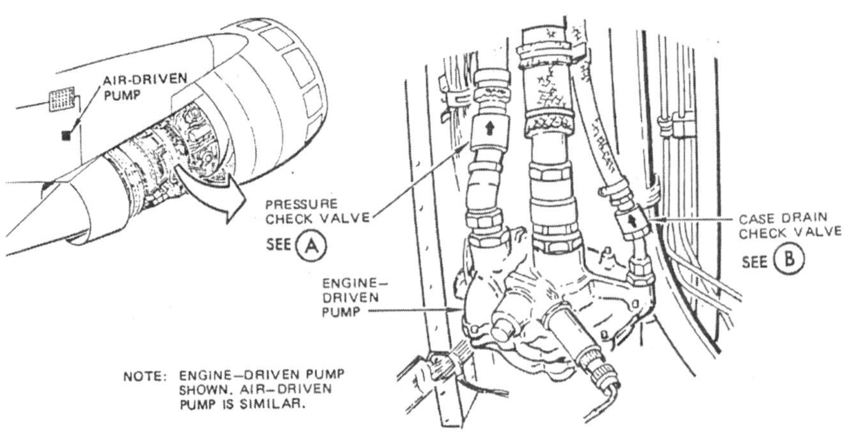

Physical Description
The valve consists of a body, a check ball, and a spring.

Operation
The check ball is held in position against the inlet port by a light spring. Any pressure appearing at the outlet port merely serves to further seat the ball, closing the valve. Since the spring is of low tension, normal system pressures easily unseat the ball al1owing normal fluid flow in the intended direction.

HYDRAULIC FUSE

Purpose
The fuse is designed to prevent hydraulic fluid loss in the event of a down stream line rupture or other failure.

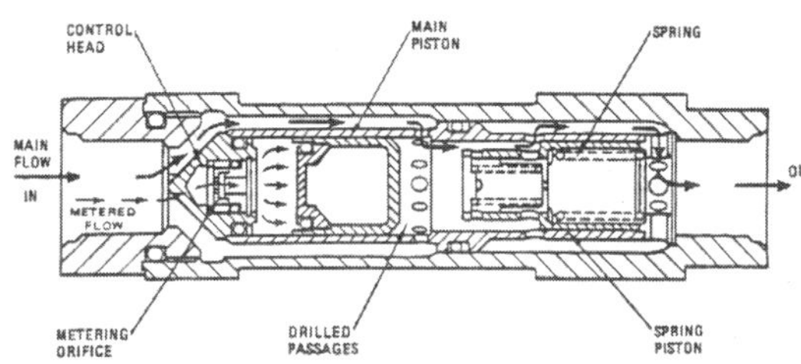

Physical Description
The fuse consists of a case, main piston, spring piston, springs and control orifices.

Operation
Fluid entering the fuse is divided into two flow paths by the control head. The main flow forces the spring piston to the right which uncovers holes in the sleeve surrounding the spring piston. The main flow is now through the sleeve holes on the left and out the sleeve holes to the right. Since the inlet holes are smaller than the outlet holes, there is a pressure drop in the chamber behind the spring piston. The pressure drop is proportional to the flow rate.

The flow through the metering orifice causes the main piston to move to the right. The rate of movement of the main piston is dependent on the pressure drop in the chamber between the two pistons. The pressure drop is proportional to the flow rate through the fuse.
The fuse design is such that the work to be accomplished is completed before the main piston travels fully to the right (shut off) position. If there is a downstream leak, the main piston will continue to travel to the right and shut off further flow.
The fuse remains in this condition until the upstream pressure is reduced, at which time the spring returns both pistons to the left. The fuse is then prepared for another cycle.

PRESSURE RELIEF VALVE

Purpose
The pressure relief valve is used to prevent pressures in a system from exceeding a prescribed limit.

Physical Description
The valve consists of a body, check valve, and spring.

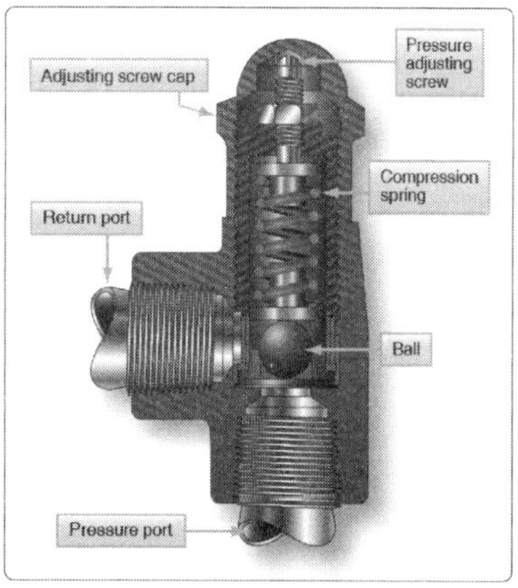

Pressure relief valve.

Operation
The pressure of the spring is selected or adjusted so that as long as a preset hydraulic pressure is not exceeded, the ball remains seated closing the pressure port. When the pressure becomes excessive, the ball is unseated which allows the pressure to enter the return lines.
If a pressure adjusting screw is installed, then adjustments should be made only on a hydraulic test bench using the manufacturer's instructions.

SHUTTLE VALVE

Purpose
A shuttle valve is used to select a single pressure source from two available sources. Selection is based on the presence of one source and the absence of the other, or on one source being greater than the other.

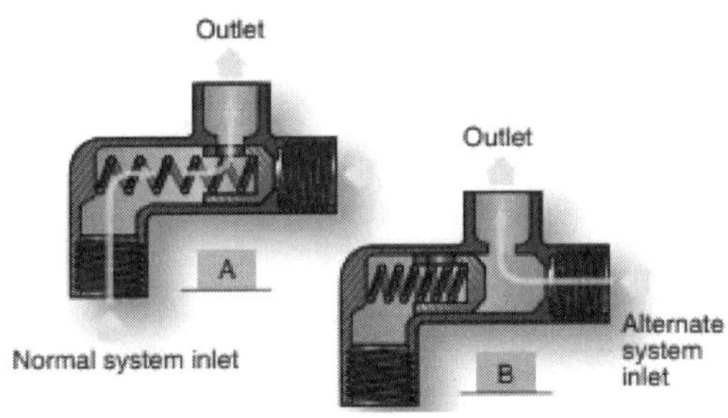

Physical Description
The shuttle valve consists of a valve body, two inlet ports, one outlet port, and a floating piston and spool. There are no control inputs other than the operating pressures.

Operation
In the example shown, the alternate system pressure source is greater than normal system. The greater force operating on the right piston moves the piston and the spool to the left, allowing pressure source alternate system to flow through the valve. Pressure normal system is blocked. Operation of the shuttle valve may be biased to favor one source by making the left and right piston faces different sizes or by adding an internal spring.

PRESSURE OPERATED TRANSFER VALVE

Purpose
The pressure operated transfer valve is used to change fluid flows by means of a small control signal.

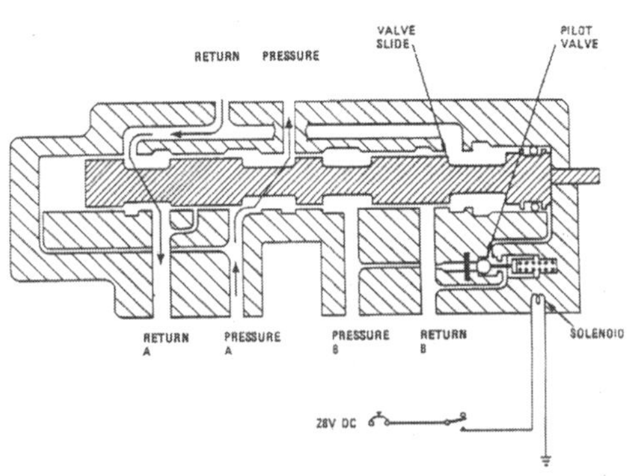

Physical Description
The valve consists of a body, a pressure operated spool, and a solenoid operated pilot valve.

Operation
As shown, the valve is set to connect pressure source A to the output pressure line and the return line to return port A. When the switch is closed, the solenoid operated pilot valve is opened to allow pressure B to flow to the head of the spool. The B pressure forces the spool to the left which now connects the B pressure to the pressure output and the return line to the B return.

ROTARY SELECTOR VALVE

Purpose

The rotary selector valve is used to port hydraulic fluid to different systems or locations.

Physical Description

A rotary selector valve consists of a valve body and a rotatable ball with a drilled passage. The valve body is drilled for one or more inputs and one or more outputs. The valve is configured to port fluid to one of two ports, or to be closed.

Operation

The ball can be rotated to align the drilled passage with the input port and outlet port number 1. By reversing the direction of rotation, the passage can be aligned with the inlet port and the second outlet port.

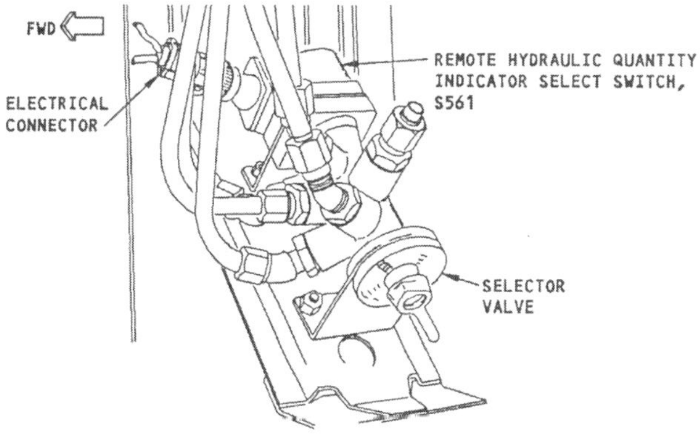

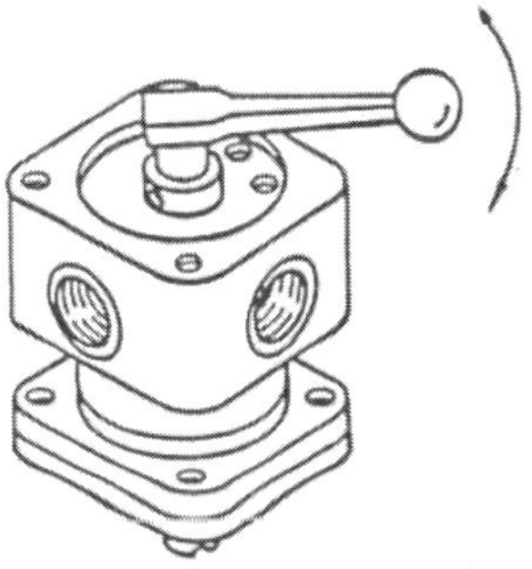

HIGH PRESSURE AIR VALVE

Purpose

The valve is used to service devices requiring high pressure air or nitrogen.

Physical Description

The valve consists of a body, a stem with valve seat, a swivel nut to open and close the valve, and a protective cap. It may also include a spring loaded core.

Operation

The valve is screwed into the device requiring the high pressure air by means of the body nut. The swivel nut is used to seat or unseat the valve. If installed, the spring loaded core traps any air which may escape through the main valve. To open the valve, any trapped air is released by removing the spring loaded core (if installed) after first ensuring that the swivel nut is tight clockwise. To release the air in the device, the swivel nut is turned slowly counterclockwise.

CAUTION: DO NOT TURN THE BODY NUT. THE AIR PRESSURE COULD CAUSE THE VALVE TO LEAVE THE DEVICE WITH GREAT FORCE.

After all of the air has been released by turning the swivel nut, the valve may be removed for any required servicing such as fluid addition or removal.
After fluid servicing, the valve body is reassembled to the device with the swivel nut still open. The high pressure air source is connected to the valve and the air source opened to fill the device. When air servicing is complete, the swivel nut is closed by turning it clockwise and the air source is disconnected.

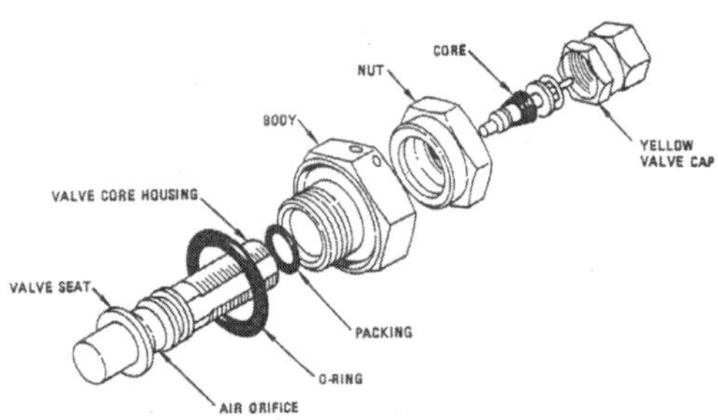

HIGH PRESSURE AIR VALVE

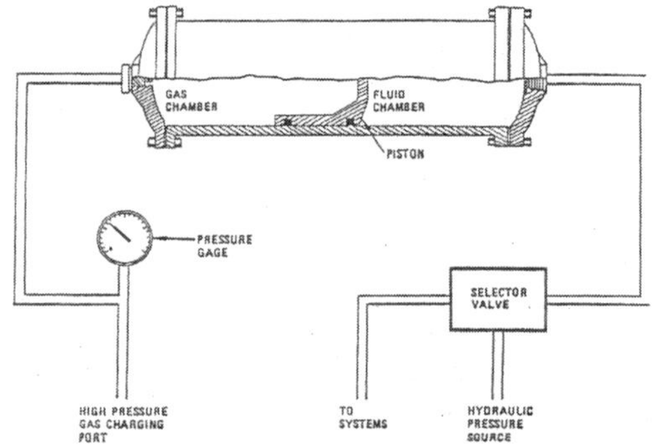

HYDRAULIC ACCUMULATOR

HYDRAULIC ACCUMULATOR

Purpose
A hydraulic accumulator is used to store hydraulic fluid under pressure.

Physical Description
The accumulator consists of a sealed container which is separated into two chambers by a floating piston. One chamber contains the hydraulic fluid and the other chamber contains a compressed gas. The gas is clean, dry air or nitrogen.

Operation
In use, compressed gas is supplied to the gas chamber through the gas charging valve. Hydraulic fluid is supplied to the second chamber under pressure. As the hydraulic pressure increases, the gas is further compressed by action of the floating piston. When fully charged, there exists a quantity of hydraulic fluid under pressure ready to perform some function in the event that normal hydraulic pressure is not available.

ATA 31 - INDICATING / RECORDING SYSTEM

Since the beginning of manned flight, it recognized that supplying the pilot with information about the aircraft and its operation is useful and leads to safer flight. The Wright Brothers had very few instruments on their Wright Flyer, but they did have an engine tachometer, an anemometer, and a stop watch. They obviously concerned about the aircraft's engine and the progress of their flight. From that simple beginning, various instruments developed to inform flight crews of different parameters. Instrument systems now exist to provide information on the condition of the aircraft, engine, components, the aircraft's attitude in the sky, weather, cabin environment, navigation, and communication.

ELECTRONIC INSTRUMENTS

General description

With the development of electronic technology, airplane instrument systems developed from analog instrument to digital instrument. Various types of indicators are displayed on 6 LCD of instrument panel. The 6-LCD integrated instrument system is called differently. European airplane manufacturers call it EIS(Electronic Instrument System) while American aircraft manufacturer call it IDS(Integrated Display System in B747, PDS in B777 and CDS in B737). EIS or IDS has EFIS(Electronic Flight Instrument System) and EICAS(Engine Indicating and Crew Alerting System).

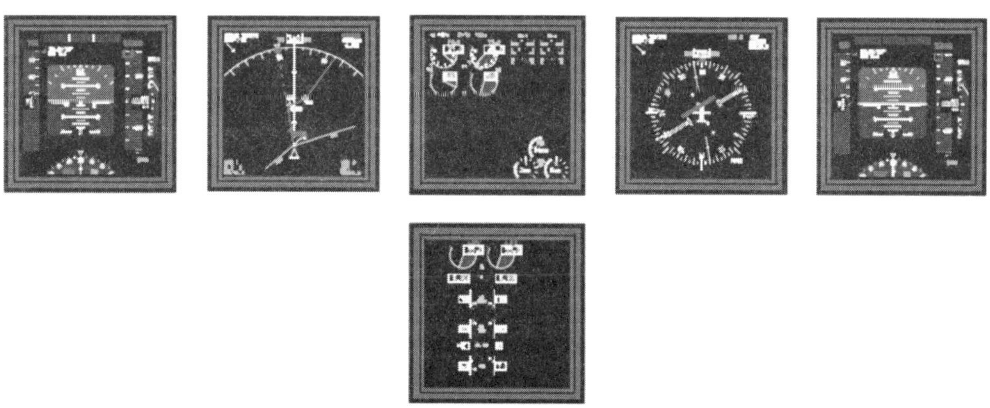

— The above figure is B737 CDS and consists of PFD, ND, main EICAS and auxiliary EICAS.

PITOT-STATIC SYSTEM

General description

A pitot-static system is a system of pressure-sensitive instruments that is most often used in aviation to determine an aircraft's airspeed, Mach number, altitude, and altitude trend. A pitot-static system generally consists of a pitot tube, a static port, and the pitot-static instruments. Other instruments that might be connected are air data computers, flight data recorders, altitude encoders, etc.

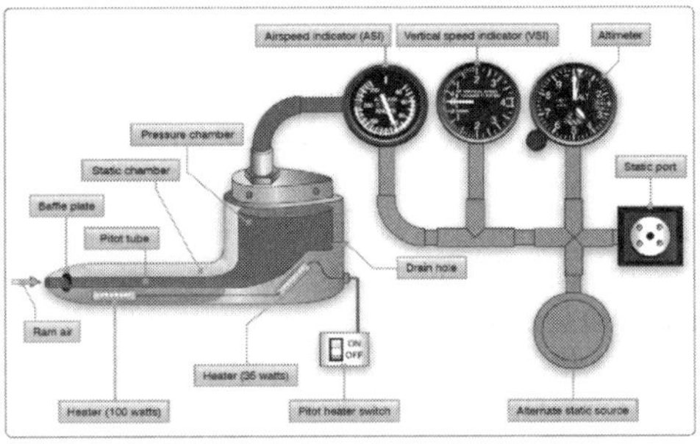

PITOT PRESSURE

The pitot pressure is obtained from the pitot tube. The pitot pressure is a measure of ram air pressure(the air pressure created by vehicle motion or the air ramming into the tube), which, under ideal conditions, is equal to stagnation pressure, also called total pressure. The pitot tube is most often located on the wing or front section of an aircraft, facing forward, where its opening is exposed to the relative wind. When airspeed increases, the ram air pressure is increased, which can be translated by the airspeed indicator.

STATIC PRESSURE

The static pressure is obtained through a static port. The static port is most often a flush-mounted hole on the fuselage of an aircraft, and is located where it can access the air flow in a relatively undisturbed area. In situations where an aircraft has more than one static port, there is usually one located on each side of the fuselage.

AIRSPEED INDICATOR

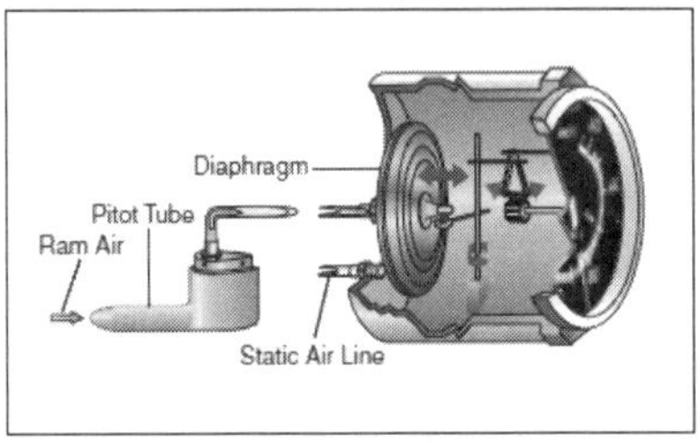

The airspeed indicator is another primary flight instrument that is also a differential pressure gauge. Ram air pressure from the aircraft's pitot tube is gone directly into a diaphragm in an analog airspeed instrument case. Static air pressure from the aircraft static vent is gone directly into the case surrounding the diaphragm. As the speed of the aircraft varies, the ram air pressure varies, expanding or contracting the diaphragm.

ALTIMETER AND ALTITUDE

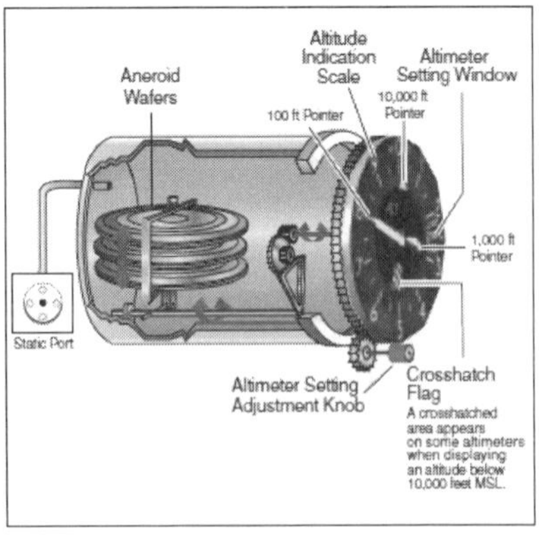

General description

An altimeter indicates the height of the aircraft above a predetermined level, such as sea level or the height of terrain below the aircraft. The most common way to measure this distance is rooted in discoveries made by scientists centuries ago. Seventeenth century work proving

that the air in the atmosphere applied pressure on the things around us led Evangelista Torricelli to the invention of the barometer. Also in that century, using the concept of this first atmospheric air pressure measuring instrument, Blaise Pascal could show that a relationship exists between altitude and air pressure.
As altitude increases, air pressure decreases. The amount that it decreases is measurable and consistent for any given altitude change. Thus, by measuring air pressure, altitude is determined.

VERTICAL SPEED INDICATOR

In aviation, a variometer — also known as a vertical speed indicator(VSI), or vertical velocity indicator(VVI) — is one of the flight instruments in an aircraft used to inform the pilot of the rate of descent or climb.

DIRECTION INDICATING INSTRUMENTS

A lot of techniques and instruments exist to help the pilot in navigation of the aircraft. An indication of direction is part of this navigation. The primary direction indicators are magnetic compass, also called, standby compass, RMI(Radio Magnetic Indicator), HSI(Horizontal Situation Indicator) and ND(Navigation Display).

MAGNETIC COMPASS

The magnetic compass is a direction finding instrument that used for navigation for hundreds of years. It is a simple instrument that takes advantage of the earth's magnetic field. All aircraft are equipped with this magnetic compass as a standby indicator.

ATTITUDE INDICATOR

A gyroscopic flight instrument that gives the pilot an indication of the attitude of the aircraft relative to its pitch and roll axes. The attitude indicator in an autopilot is in the sensing system that detects deviation from a level-flight attitude.

HEADING INDICATOR

A gyroscopic flight instrument that gives the pilot an indication of the heading of the aircraft.

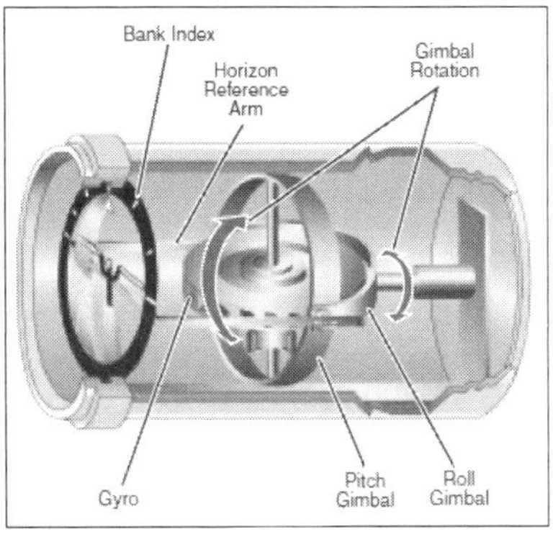

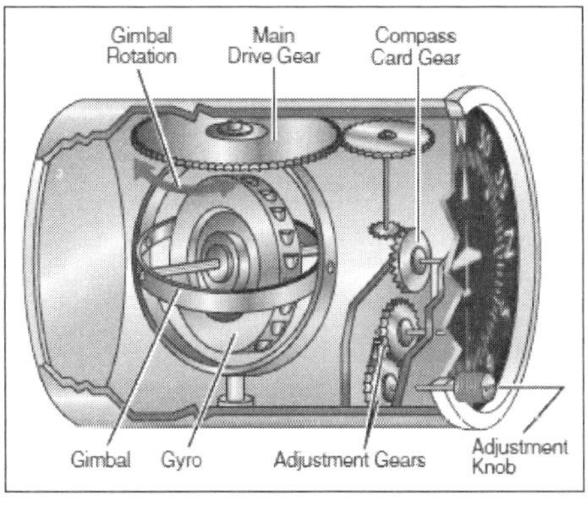

ATA 32 - LANDING GEAR

The landing gear supports the entire weight of an aircraft during landing and ground operation.

General Description
The commercial jet airplanes are equipped with a tricycle type landing gear employing two sets of main wheels and one set of nose wheels. The landing gear is normally retracted and extended using hydraulic power. Brakes are installed on the main wheels.

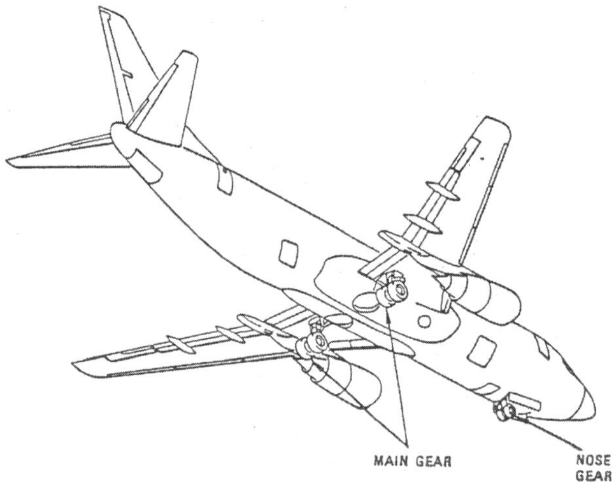

TIRE AND WHEEL PROTECTION

Purpose
Tires and wheels are protected by over inflation valves and thermal plugs.

Physical Description
To prevent tire burst or wheel separation due to tire over inflation, wheels are provided with pressure relief valves. Depending on the tire and wheel combination, the pressure relief point will be in the range of 375 to 450 psi. Thermal plugs installed in the wheels are designed to melt if the wheel temperature rises to a dangerous level due to brake typical release temperatures 350°F to 400°F.

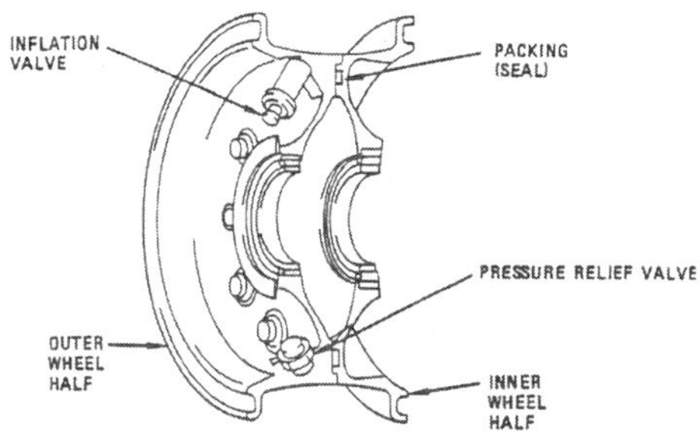

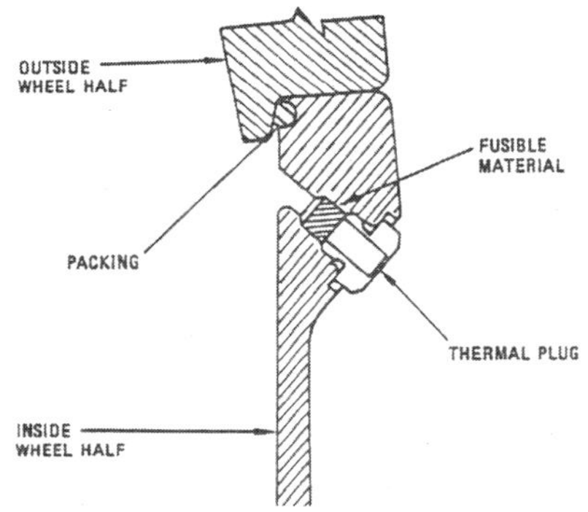

WHEEL SPEED TRANSDUCER

Purpose

The wheel speed transducer is used to provide an electrical signal proportional to wheel speed.

Physical Description

The transducer consists of a rotor atached to the wheel and a stator attached to the wheel axle.

Operation

As the wheel turns, the rotating rotor cuts the magnetic field between the inner and outer stator. This change in the magnetic field induces a voltage in the coils of the rotor which varies in frequency directly proportional to the rotational speed of the wheel. The resulting signal is used for brake anti-skid and other control purposes.

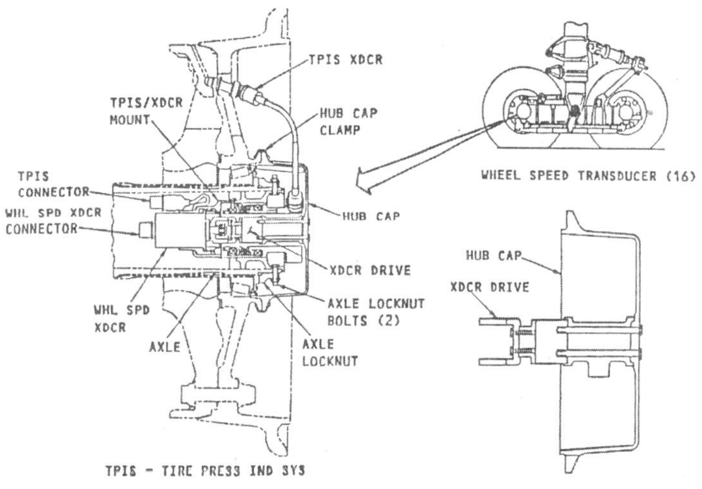

PROXIMITY SENSORS

Purpose

Proximity sensors are used to electrically signal that an action has or has not taken place.

Physical Description

Proximity sensors are round or rectangular in shape. The sensor unit consists of an inductor (coil) and resistance. The actuator, or target, is a ferrous metal plate or other ferrous metal object. Proximity sensors are used for such purposes as determining thrust reverser operation, landing gear status, door status, air/ground status and similar operations.

Operation

When the ferrous metal actuator moves into the proximity of the sensor, the metal disturbs the magnetic lines of force around the sensor coil. The resulting change in current flow through the coil is then detected and used to provide a high or low logic signal or to open or close a solid state switch.

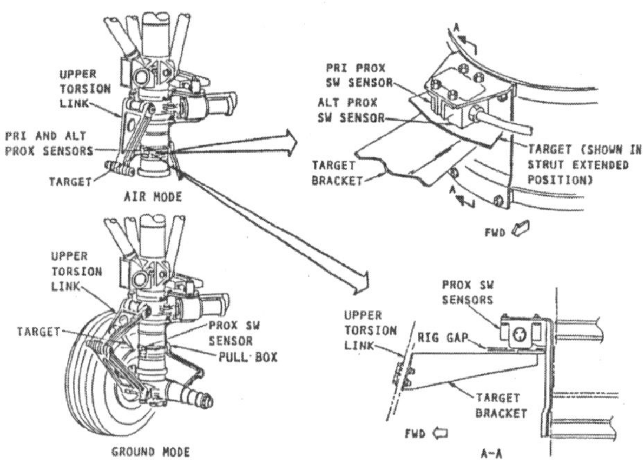

NOSE GEAR COMPRESSED SENSOR INSTALLATION

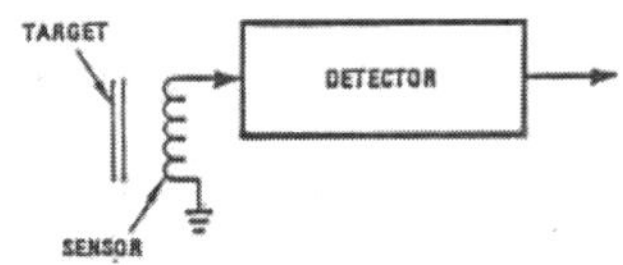

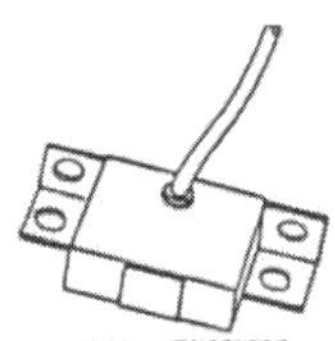

PROXIMITY SENSOR (RECTANGULAR)

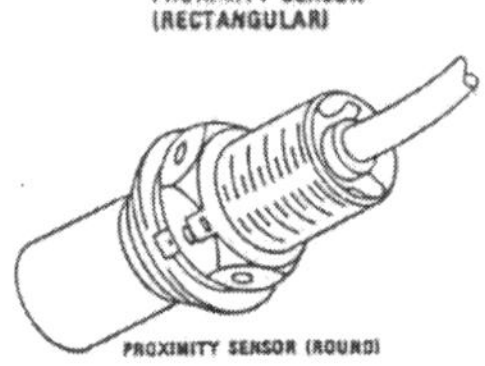

PROXIMITY SENSOR (ROUND)

ATA 34 - NAVIGATION

Moving of an aircraft from one point to another is the most important part for any kind of mission. Plotting on the paper or on the map a course towards a specific area of the earth was a task assigned to a specialized member of the aircraft's crew such a navigator. Such a task was quite complicated and not always accurate because it depended on the observation by using simple maps and geometrical instruments for calculations. Today, aerial navigation became an art which nears to perfection. Both external Navaids (Navigational Aids) and on-board systems help navigate any aircraft over thousands of miles with such accuracy that only imagined a few decades ago.

THE METHOD OF NAVIGATION

There were three basic methods of air navigation and two new methods in recent with technology development.

PILOTAGE OR PILOTING

General description

It is the most common method of air navigation. This method, the pilot keeps on course by following a series of landmarks on the ground. Usually before take-off, pilot will making pre-flight planning, the pilot will draw a line on the aeronautical map to indicate the desired course.

Pilot watches various landmarks, such as:

- Highways
- Railroad tracks
- River
- Bridges
- Mountains
- Villages

When the pilot flies over each of landmark, pilot will checks it off on the chart or map. If the plane does not pass directly over the landmark, the pilot will know that he must correct the course.

DEAD RECKONING

It is the primary navigation method used in the early days of flying. It is the method on which Lindberg relied on his first trans-Atlantic flight. A pilot used this method when flying over large bodies of water, forest, deserts. It demands more skill and experience than pilotage does. It is based on time, distance, and direction only. The pilot must know the distance from one point to the next, the magnetic heading to be flown.

RADIO NAVIGATION

General description

Most of all pilots use the radio navigation. Pilots can find out from an aeronautical chart what radio station they must tune to in a particular area. They can then tune their radio navigation equipment to a signal from this station. A needle on the navigation equipment tells the pilot where they are flying to or from station, on course or not. The typical radio navigation systems are:

- ADF/NDB(Automatic Direction Finder/Non-Directional Beacon)
- VOR (VHF Omni-directional Range)
- DME (Distance Measuring Equipment)
- ILS (Instrument Landing System)

INERTIAL NAVIGATION SYSTEM (INS)

An inertial navigation system(INS) is a navigation device that uses a computer, motion sensors (accelerometers) and rotation sensors (gyroscopes) to continuously calculate by dead reckoning the position, the orientation, and the velocity(direction and speed of movement) of a moving object without the need for external references. Often the inertial sensors are supplemented by a barometric altimeter and sometimes by magnetic sensors (magnetometers) and/or speed measuring devices. INSs are used on mobile robots and on vehicles such as ships, aircraft, submarines, guided missiles, and spacecraft.

The inertial reference system (IRS) constantly supplies safety critical position and attitude data to the aircraft flight controls, avionics and mechanical systems. Inertial reference systems use gyroscopes, accelerometers and electronics to provide precise attitude, velocity and navigation information so an aircraft can determine how its moving through the airspace. More specifically, these systems provide data on linear and angular acceleration, linear and angular velocity, position, attitude (roll, pitch), platform azimuth, magnetic and true heading, altitude, body angular rates and more.

GLOBAL POSITIONING SYSTEM (GPS)

GPS was introduced to reduce and correct these INS errors. The GPS is a space-based radio-navigation system. It is a navigation system based on the transmission of signals from satellites provided and maintained by the USA. It consists of:

- A constellation of satellites broadcasting navigation signals
- A network of ground stations
- Satellite control stations used for monitoring and control

Currently 31 GPS satellites orbit the Earth at an altitude of approximately 11,000 miles providing users with accurate information on position, velocity, and time anywhere in the world and in all weather conditions.

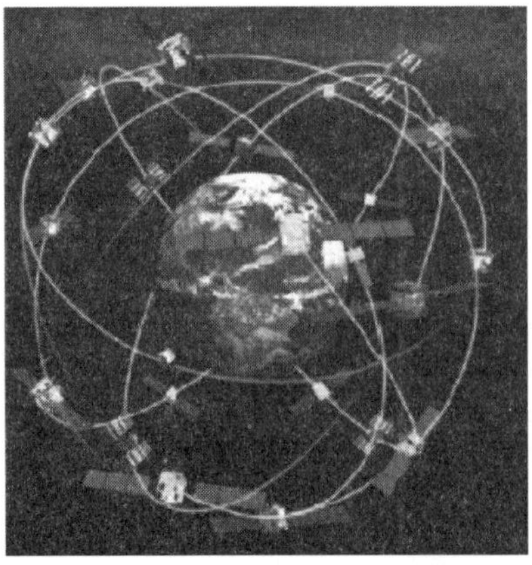

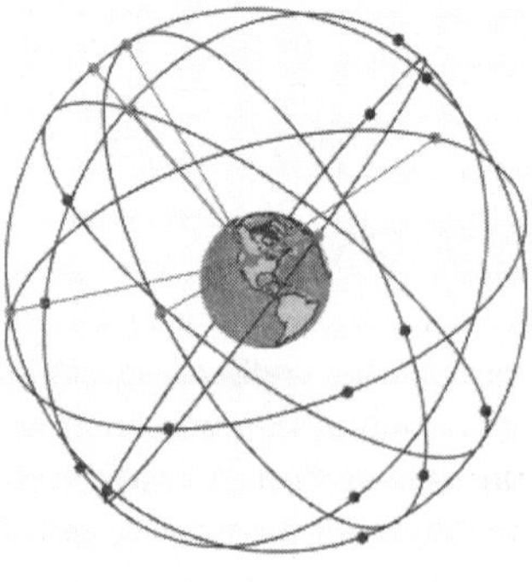

AUTOMATIC DIRECTION FINDER (ADF)

It is an aircraft radio navigation which senses and indicates the direction to a Low/Medium Frequency non-directional radio beacon (NDB) ground transmitter. It is the first navigation system developed and used among radio navigation systems. Currently, it is partially operated as a backup.

VERY HIGH FREQUENCY OMNIDIRECTIONAL RANGE (VOR)

It is a radio navigation aid operating in the 108-118 MHz band. A VOR ground station transmits a two-phase directional signal through 360 degrees. The aircraft's VOR receiver lets a pilot to identify his radial or bearing From/To the ground station. VOR is the most commonly used radio navigation aid in private flying.
VORTAC is a special VOR which combines VOR and DME for civil and military used. System provides information about the bearing and distance from the ground station we have tuned into.

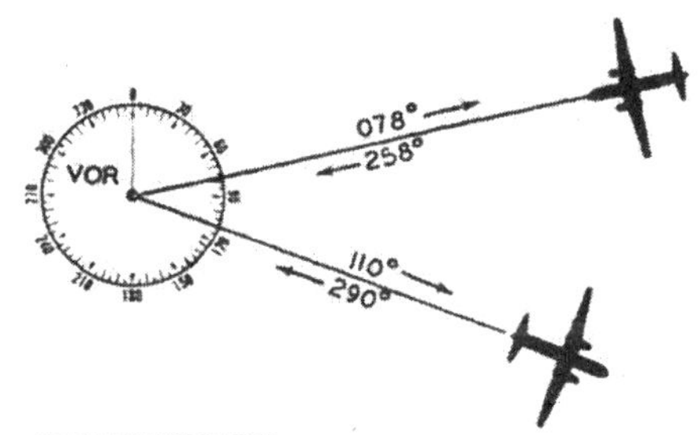

TO OR FROM THE STATION

DISTANCE MEASURING EQUIPMENT(DME)

DME is ground and aircraft equipment which provide distance information and primary serve operational needs of en-route or airport terminal area navigation. The distance information is the slant distance information from the airplane to the DME station on the ground. Based on this information, the system provides important navigation information such as ground speed, distance, elapsed time, and remaining distance and time to the destination.

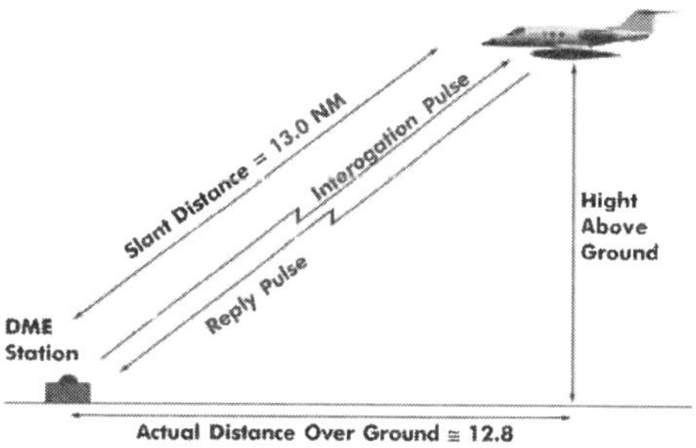

INSTRUMENT LANDING SYSTEM (ILS)

ILS is a standard ICAO precision-landing-aid. It is used to provide accurate azimuth and descent guidance signals to approaching aircraft for landing on the runway under normal or adverse weather conditions. ILS facility is a highly accurate and dependable means of navigating to the runway in IFR conditions. When all components of the ILS system are available, including the approved approach procedure, the pilot can execute a precision approach. ILS has the localizer, the glide slope and marker radio beacons.

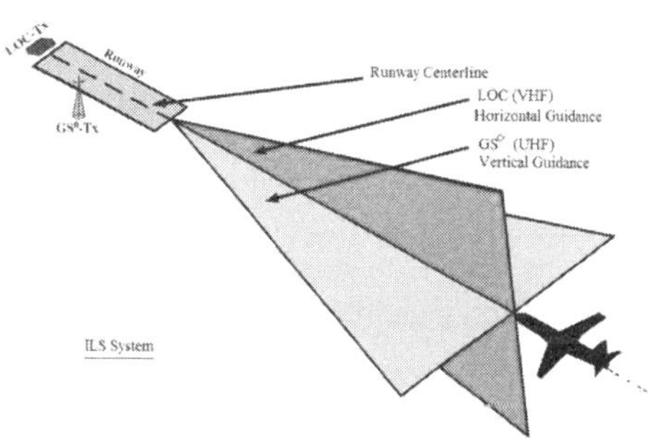

ILS System

BASIC TERMINOLOGY

True north

It is the direction that points directly towards the geographic North Pole. True north is the direction along the surface of the earth that ends in the location of the North Pole. This is a fixed point on the Earth's globe.

Magnetic north

It is the location where the Earth's magnetic field points vertically downward and the direction of compass points is magnetic north. True north is a fixed point on the Earth's globe but Magnetic north is quite different. Magnetic north is the direction that a compass needle points to as it aligns with the Earth's magnetic field.

Heading

It is just the direction the nose of airplane is pointed. This does not factor for wind, or the actual movement of the airplane across the ground. It only refers to what the compass reads based on where the nose is pointed.

Track

It is the actual direction of the airplane tracking across the ground. When navigating in the air, your track is really all that matters in terms of getting to where you want to go. If you need to go northeast to your destination, and have a significant wind from the west, your heading might be to the north in order to achieve a track to the northeast.

Bearing

It is the angle between any two points(location or destination), Bearing can be confusing sometimes because has some overlap with course. Bearing is simply the angle or direction between two points. A practical application of this is in VOR navigation. It's a common thing to hear someone say "we are bearing 090 from the station". This simply means that off of the VOR they are tracking on the 090 radial outbound from the station. In relation to the VOR they are bearing 090.

Course

It is your intended path of travel to your destination. Course is very similar to bearing in that it's the desired direction for your route of flight. If you are going directly from one airport to the other, your course and bearing will be the same along the route of flight. If you are flying from an airport to a VOR to another airport, your course will change in each leg, as will your bearing.

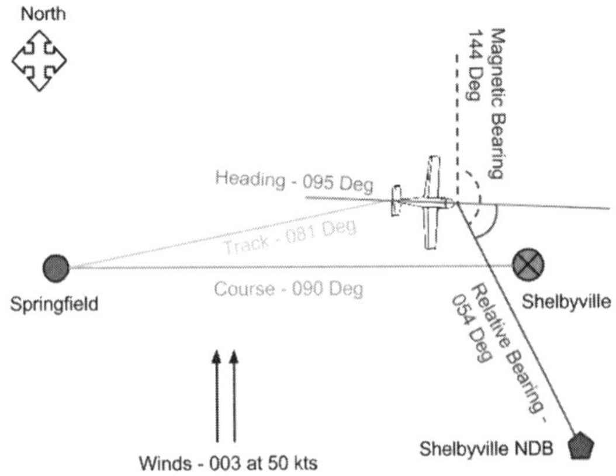

ATA 36 - PNEUMATIC

The purpose of pneumatics is to extract compressed air from the engines or Auxiliary Power Unit (APU) for air conditioning, pressurization, wing anti-icing and engine starting.

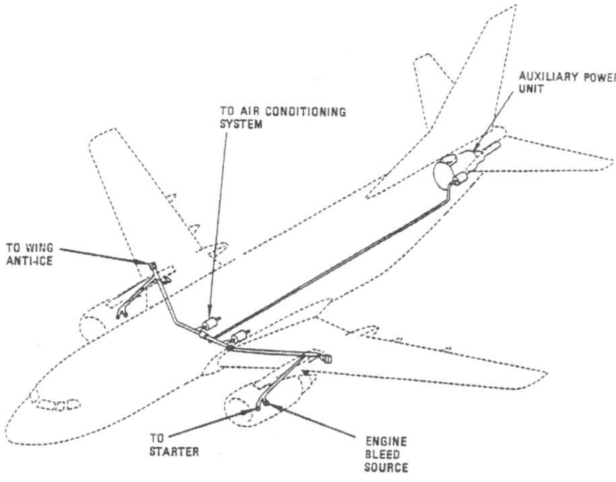

Purpose
The purpose of pneumatics is to supply compressed air to the user systems during all phases of flight and ground operation. Air is obtained from the engine bleed or auxiliary power unit, and distributed through ducts to the user systems.

General Description
Air extracted from the engines or APU is controlled by different types of valves. Electric power is also required for control, warning functions.

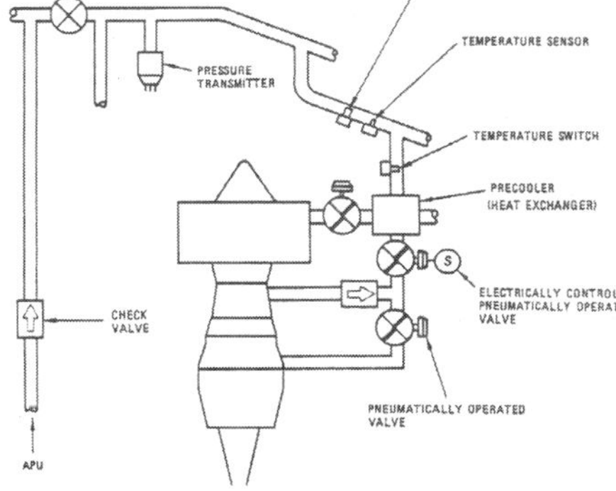

ELECTRICALLY CONTROLLED PNEUMATICALLY OPERATED VALVE

Purpose

The electrically controlled pneumatically operated valve allows airflow through the pneumatic manifold to the user systems.

Physical Description

The valve is electrically controlled by a constant duty solenoid. The latch-type solenoid plunger will remain in the last energized position when power is removed. The plunger has a manual override feature for opening or closing the switch or valve. The pneumatically operated valve is spring-loaded to the "closed" position.

The valve consists of a spring-loaded closed butterfly assembly, pneumatic operated diaphragm actuator which is mechanically coupled through a lever arm to the butterfly shaft; a single-ball switcher valve in the upstream pressure supply line and a latching solenoid assembly. The "open" solenoid, when energized, will position the switcher valve to the open position.

Operation

With the "close" solenoid energized, upstream air pressure is shut off. Chamber A is vented to ambient and the valve will be held closed by spring pressure.

Valve opening is initiated by energizing the "open" solenoid. The switcher valve is moved open and upstream air passes through the switcher valve into Chamber A. The ambient vent for Chamber A is now closed off. Pressure in Chamber A, acting on the diaphragm moves the mechanical linkage, causing the butterfly to move toward the open position.

Shutoff occurs when electrical power is applied to the solenoid "close" winding. The solenoid then moves the switcher valve closed opening Chamber A to ambient. The pressure differential across the diaphragm, in conjunction with the spring forces the butterfly valve closed.

Manual operation may be obtained by the manual override plunger. The plunger can be pushed to close the valve and pulled to open the valve.

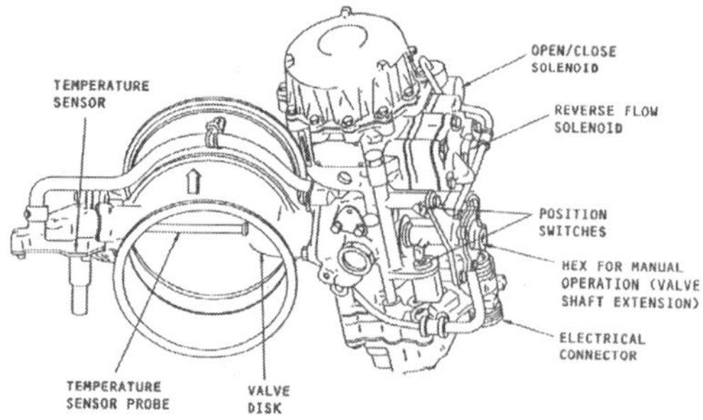

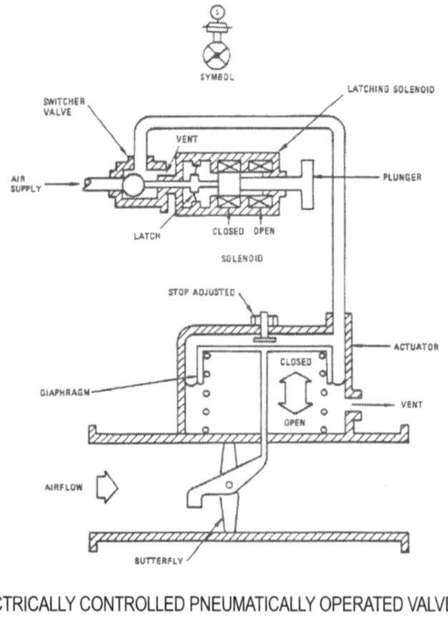

ELECTRICALLY CONTROLLED PNEUMATICALLY OPERATED VALVE

PNEUMATICALLY OPERATED VALVE

Purpose

The pneumatically operated valve allows airflow through the pneumatic manifold to the user systems.

Physical Description

The valve is pneumatically controlled by the pressure regulator and is spring-loaded to the closed position. Pneumatic pressure is required for valve actuation.

The valve consists of a butterfly assembly, a pneumatically operated diaphragm actuator which is mechanically coupled through a lever arm to the butterfly shaft and a pressure regulator. The bleed-off regulator will limit opening chamber pressure to a preset value.

Operation

Chamber A is vented through the orifice to ambient and the valve will be held closed by the spring pressure.

Upstream air passes through the regulator, into Chamber A. Pressure in Chamber A, acting on the piston and mechanical linkage, moves the butterfly against the closing spring, toward the open position. The rate of opening is controlled by the rate at which upstream air can pass through the regulator and orifice.

As upstream pressure increases, the poppet in the regulator moves to restrict the air flow to Chamber A.

The valve closes when air pressure is stopped to the regulator and Chamber A vents to ambient through the orifice. Differential pressure across the piston, in conjunction with the force of the spring, moves the butterfly valve toward closed.

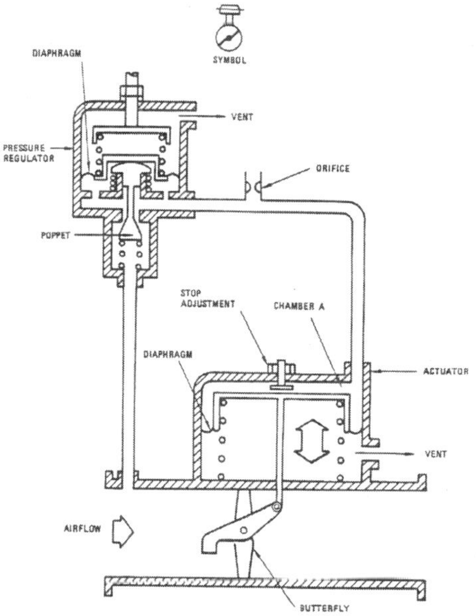

PNEUMATICALLY OPERATED VALVE

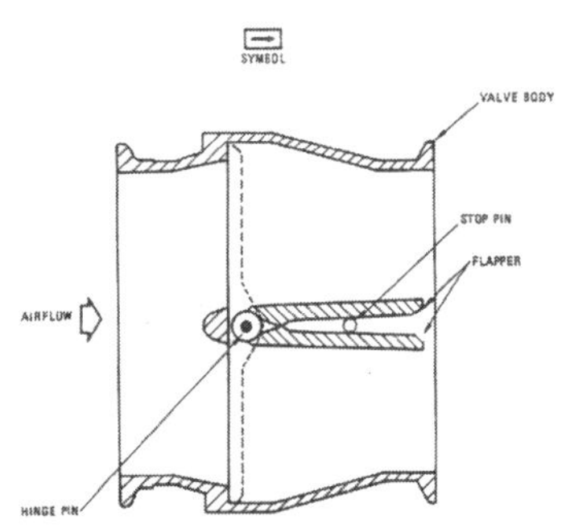

FLAPPER CHECK VALVE

FLAPPER CHECK VALVE

Purpose
A flapper check valve is used to control air flow direction in a pneumatic system.

Physical Description
The valve consists of a valve body containing two semicircular flappers hinged to a pin extending through the center of the valve body. A pin, integral with the valve body, is located downstream of the flapper hinge and provides a stop for the flappers.

Operation
Air pressure on the upstream side of the check valve acts on the surface of the flappers; moving them toward the open position allowing air flow through the valve. Reverse air flow causes the flappers to close, shutting off air flow.

ELECTRIC MOTOR OPERATED VALVE

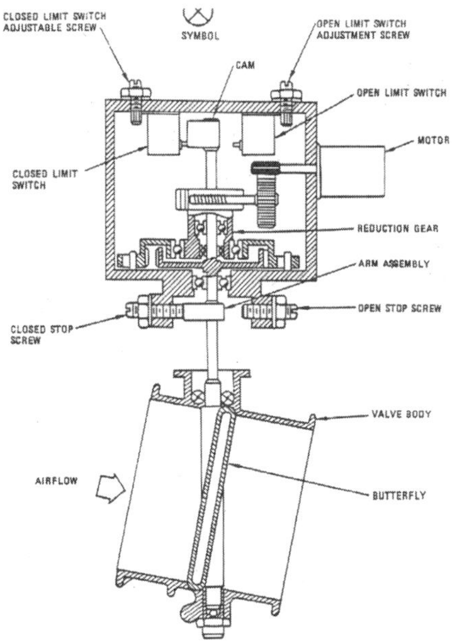

Purpose
The electric motor operated valve allows airflow through the pneumatic manifold to the user systems.

Physical Description
The valve is an electromechanical rotary actuator and valve assembly, consisting of a convolute-type butterfly mounted within the housing airflow duct.
A gear with an offset shaft and limit switch actuating cam are mounted on the drive shaft within the housing. Each limit switch includes a limit switch and a limit position signal switch.

Operation
When the actuator is energized, motor rotation is transmitted through the gear assembly. The drive shaft moves in the direction of the force exerted. Drive shaft rotation is limited in either direction by limit switches actuated by the shaft-mounted cam. If a limit switch fails, rotation will be stopped by mechanical stops provided in the housing.

THERMOSTAT

Purpose
The thermostat is used to reduce control pressure in a pneumatic control valve at a preset temperature.

Physical Description
The thermostat consists of a hermetically sealed fluid, a bellows assembly and an output rod. The ball valve consists of a seat, ball, spring and housing. The ball valve assembly is actuated by the

output rod of the sensing element.

Operation
As temperature increases, the hermetically sealed fluid within the sensing tube expands and acts on the bellows assembly. The force against the bellows causes the output rod to move the ball valve off the seat allowing control pressure through the vent to ambient. The result is a decrease in control pressure in response to an increase in sensed temperature. When the temperature decreases, fluid contraction causes the output rod and ball valve to move in the closing direction. This reduces the control pressure output through the vent.

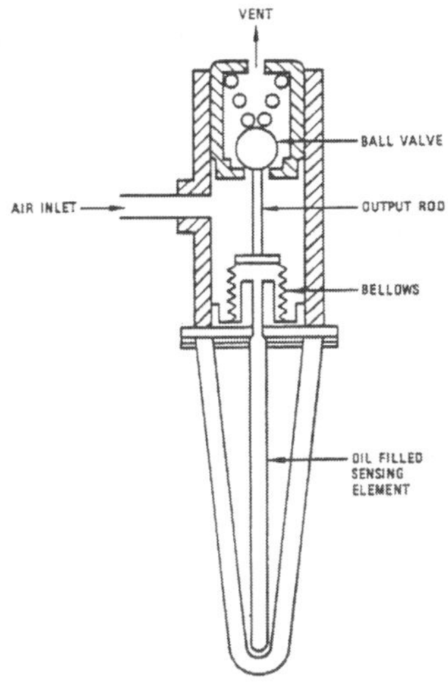

TEMPERATURE SENSOR

Purpose
The temperature sensors are used to measure air, oil and fuel temperatures.

Physical Description
The temperature sensors consist of a thermally-open sensing element, composed of a single thermistor bead mounted in a probe type housing.

Operation
The resistance of the sensing element varies inversely with any change in ambient temperature. The resistance valve is read on an indicator as temperature.

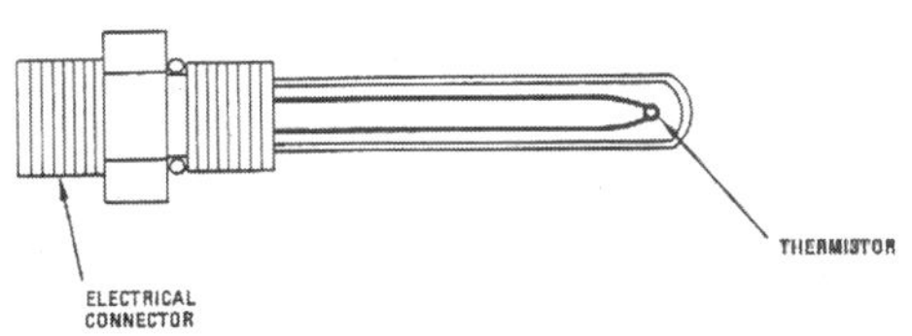

TEMPERATURE SENSOR

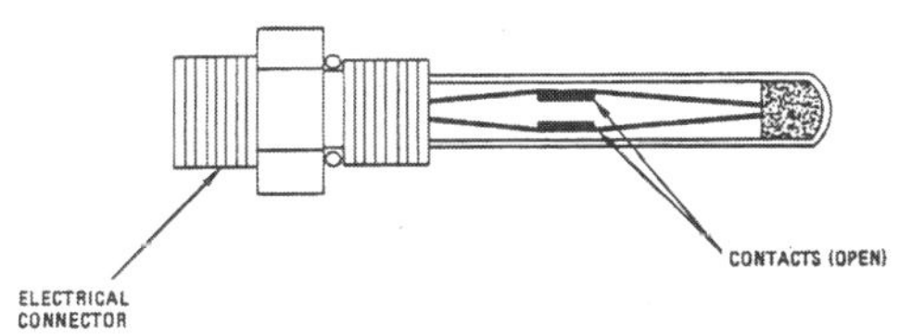

TEMPERATURE SWITCH

TEMPERATURE SWITCH

Purpose
The switch is installed in an air system to provide for system shutdown if the set temperature is exceeded.

Physical Description
The temperature switch consists of a bimetallic element enclosed in a steel probe. Electrical connections are made through a hermetically-sealed connector at the end of the switch.

Operation
The bimetallic switch in the probe remains open as long as the temperature remains below 1ts calibrated level. When the temperature exceeds the calibration setting, the bimetallic switch closes and will remain closed until the temperature drops below its calibration setting. When the contacts are closed, it completes an electrical circuit which will cause the user system to shut down.

ATA 49 - AIRBORNE AUXILIARY POWER

The Auxiliary Power Unit (APU) is used for providing electrical power for the airplane electrical system on the ground or in-flight, providing compressed air for engine starting and providing compressed air for air conditioning system on the ground or in-flight.

Purpose
The auxiliary power unit provides electrical and pneumatic power for the airplane systems.

General Description
The auxiliary power unit consists of a gas turbine engine driving a generator through a gearbox and supplying air from the compressor through a bleed valve.
The fuel supply passes through the fuel heater controlled by a valve. Fuel is scheduled to the fuel atomizer in the combustion chamber by a governor. A thermostat is used for controlling the bleed valve. Solenoid valves are used in the fuel and pneumatic lines of the APU.

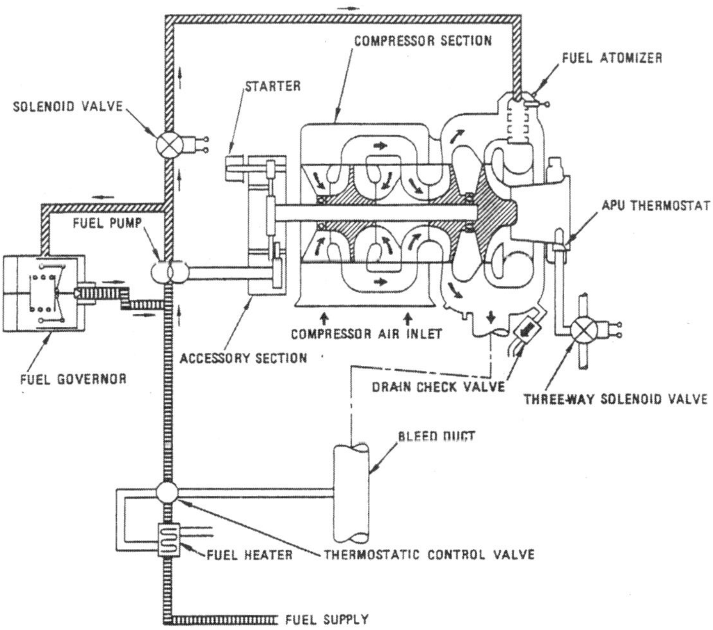

THERMOSTATIC CONTROL VALVE

Purpose
The purpose of the thermostatic control valve is to provide air flow to the fuel heater as a function of fuel temperature.

General Description
The valve consists of a poppet connected to a rod made from a thermostatic material enclosed in a housing. Fuel and air lines are connected to the housing.

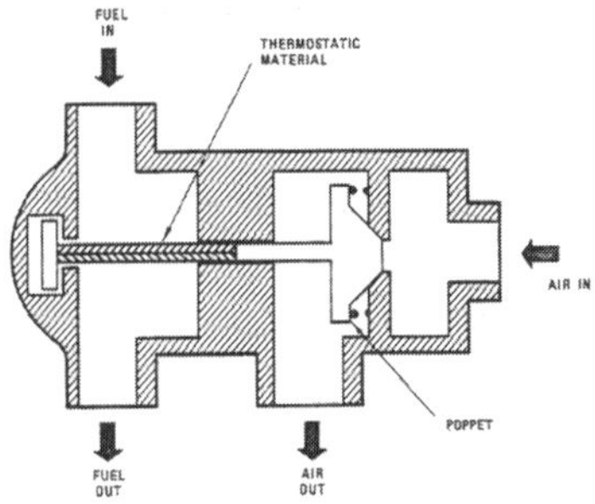

Operation
When cold fuel flows past the thermostatic rod, the rod contracts moving the poppet. This allows air to flow past the poppet through the valve. As fuel temperature - increases, the thermostatic rod expands causing the poppet to close and stop the airflow through the valve.

GOVERNOR

Purpose
The governor is used for scheduling fuel flow as a function of the rotational speed of the engine.

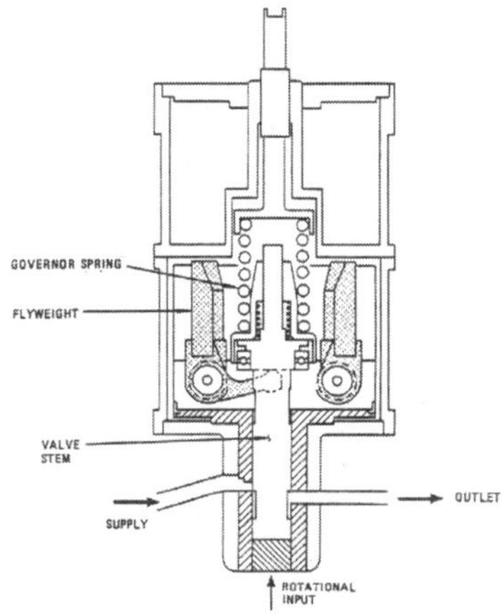

Physical Description
The governor consists of a speed input, two flyweights, springs and a spool valve.

Operation
During steady state condition, a speed input to the governor positions the spool valve by the flyweights. When speed input changes, the flyweights change position, moving the spool valve, hence the fuel flow changes.

FUEL ATOMIZER

Purpose
The fuel atomizer or nozzle is necessary to break up the fuel flowing into the flame tube into very small droplets (atomize the fuel), so that it will vaporize more readily, since only a vapor form can be combined with the air and burned.

Physical Description
The fuel atomizer consists of a concentric nozzle, a screen and flow divider valve.

Operation
Fuel enters the nozzles through a screen and flows to the small primary orifice, which forms the spray pattern for light-off. As the engine accelerates and fuel pump speed increases, an increasing fuel flow will be forced through the nozzle, increasing the nozzle pressure. At a predetermined pressure the flow divider valve will open, allowing additional fuel to flow through the secondary orifice which surrounds the primary orifice.

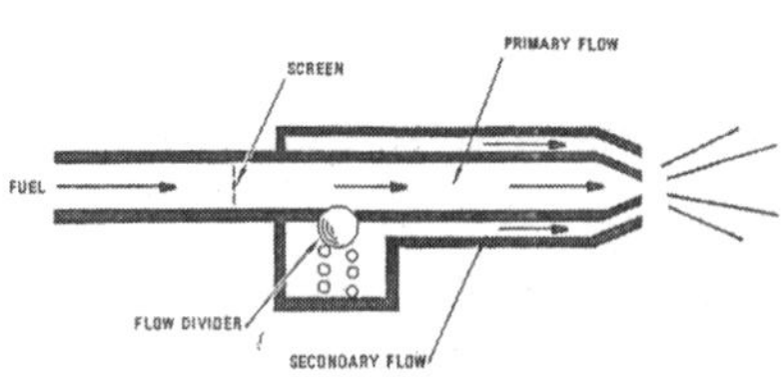

FUEL ATOMIZER

DRAIN CHECK VALVE

Purpose

A drain is provided to remove liquid fuel when the engine is not running. This is necessary after a false start when fuel is admitted to the combustion chamber but is not burned, as the fuel could be a hazard during the next start.

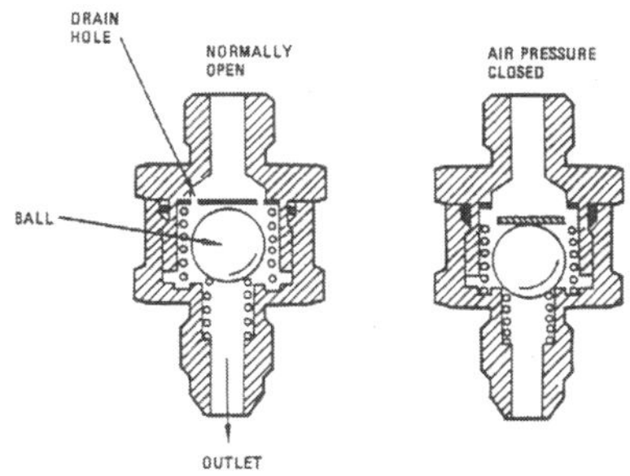

Physical Description

The valve consists of a housing enclosing a ball and spring.

Operation

The valve is spring-loaded open to allow free passage of liquid when the unit is stopped. When plenum pressure rises, the flow of air through the drain line will force the valve onto its seat to stop the leakage:

2 WAY SOLENOID VALVE

Purpose

The solenoid valve is used for connecting fluid or air flow.

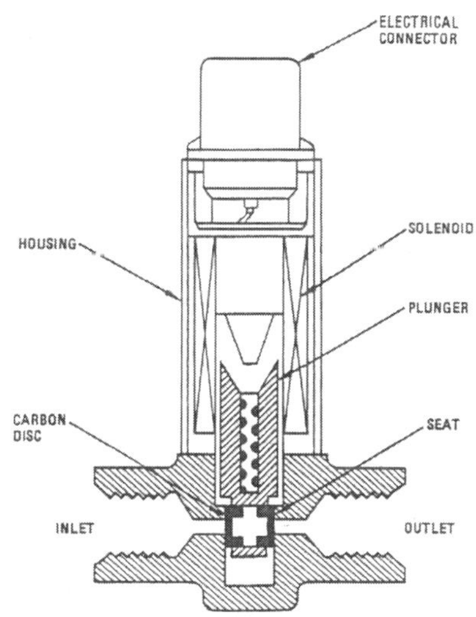

Physical Description

The solenoid valve is a normally closed, straight-through-flow type electrically actuated shutoff valve. It consists of a solenoid coil, movable plunger and valve body. The lower end of the plunger has a hole in which are set two spring loaded carbon discs.

Operation

With the solenoid de-energized the plunger is positioned by its spring to interpose the discs in line with the holes on the seats. Incoming fluid or air pressure, entering through either port, forces its way into the area between the carbon discs and forces the downstream disc tightly against the seat. When the solenoid is energized, the plunger is magnetically attracted upward by the core, sliding the discs over the seat faces until they clear the holes in the seats. The fluid or air then passes through the valve body.

3 WAY SOLENOID VALVE

Purpose
The solenoid valve is used for connecting fluid or air flow.

Physical Description
The solenoid valve is an electrically actuated shutoff valve. It consists of a solenoid coil, movable plunger and valve body. The lower end of the plunger has a ball.

Operation
With the solenoid de-energized the plunger is positioned by its spring to connect port 1 with port 3 and port 2 is blocked. When the solenoid is energized, the plunger is magnetically attracted upward by the core, closing port 1 and connecting port 2 with port 3.

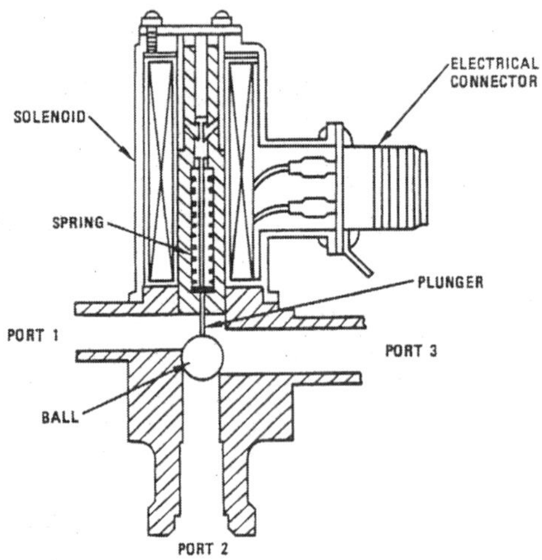

ATA 51 - STANDARD PRACTICES-STRUCTURES

The structure of an airplane is designed to provide maximum strength with minimum weight. This provides maximum payload capability with long life and maximum safety.
This is achieved by designing alternate load paths into the structure so that a failure of one segment cannot endanger the airplane. A semi-monocoque fuselage type construction is used with cantilever wings and integral tanks to provide the maximum strength to weight ratios possible. Materials are appropriately selected with the use of advanced composite replacing the heavier metal components whenever practicable.

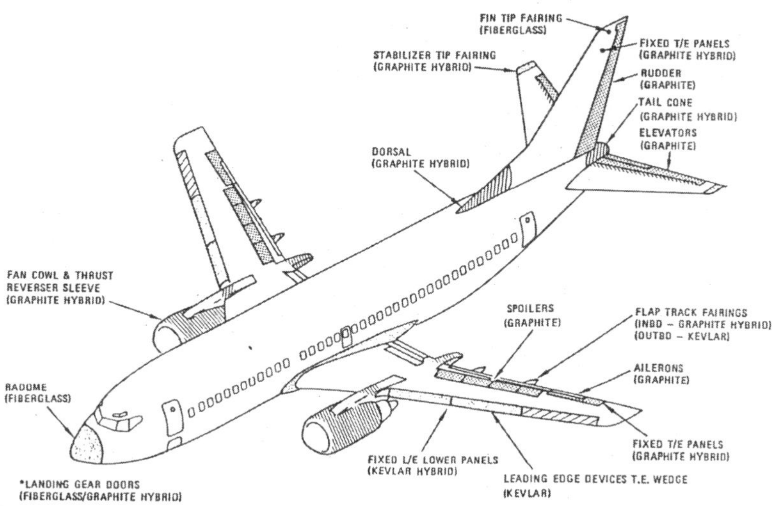

ADVANCED COMPOSITE APPLICATIONS –

SEMIMONOCOQUE FUSELAGE CONSTRUCTION

Purpose

The semi-monocoque fuselage does not depend on a few members for strength and rigidity, therefore it may withstand considerable damage and still be strong enough to hold together. Also the technique facilitates the design and construction of a streamlined fuselage.

Physical Description

The semi-monocoque fuselage relies on a stringer reinforced skin to carry the primary stresses. Vertical structural members called frames and bulkheads provide the shape of the fuselage. The heaviest of these are located to carry concentrated loads and at points of attachment to other units. Transverse beams support the floor panel s except over the wheel wells and center wing where longitudinal beams are used. A forward and an aft bulkhead are specially designed to withstand pressurization loads.
Non-structural fairings provide aerodynamic streamlining at major assembly joints such as wing to body.

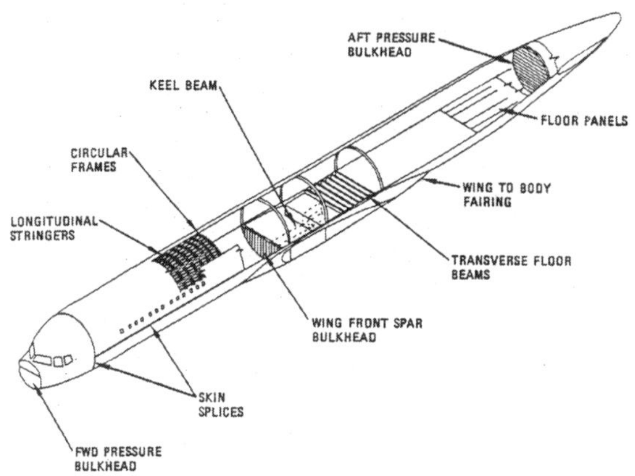

KEEL BEAM

Purpose

The keel beam carries the bending loads acting across the cavities for the center wing and wheel wells.

Physical Description

The keel beam extends from aft of the wheel wells to forward of the center wing section.

The lower portion of this beam is usually a box type beam while the upper part may be a single reinforced web between the left and right wheel wells.

The area below the center wing section is normally enclosed with fairing and used to house the air conditioning equipment.

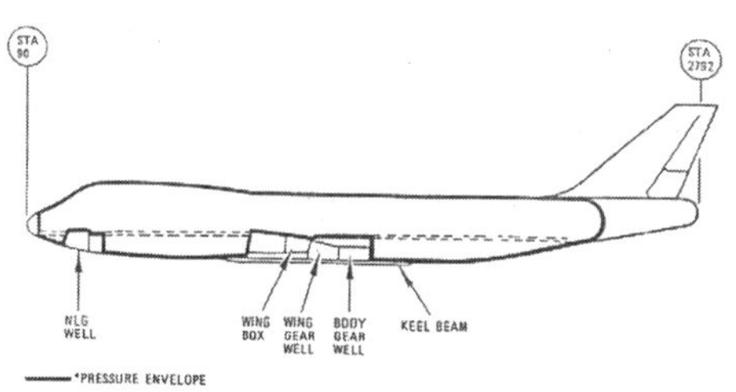

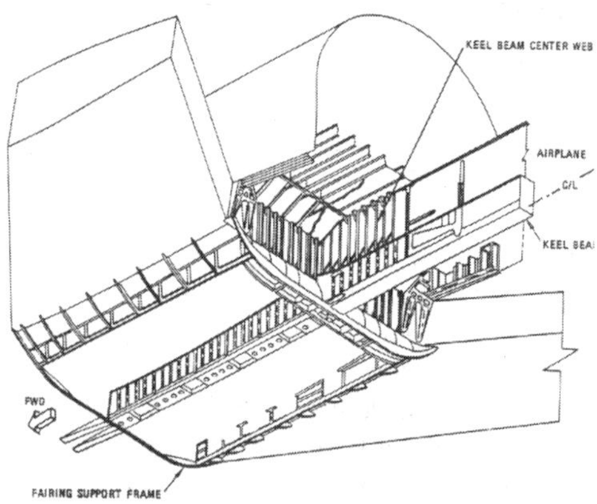

KEEL BEAM

FRAME STRINGER ASSEMBLY

Purpose

The purpose of the frames and stringers is to stabilize the skin and to distribute loads.

Physical Description

The frames are Z-section vertical members, generally spaced at regular intervals along the length of the fuselage.

A heavy frame may be reinforced with an angle and is sometimes called a bulkhead.

Stringer-longitudinal member which reinforces the skin and distributes the skin loads.

Doublers and/or doubler straps are used at intervals and around openings to reinforce the skin and act as a tear stopper (dual load path). They are bonded and riveted to the skin.

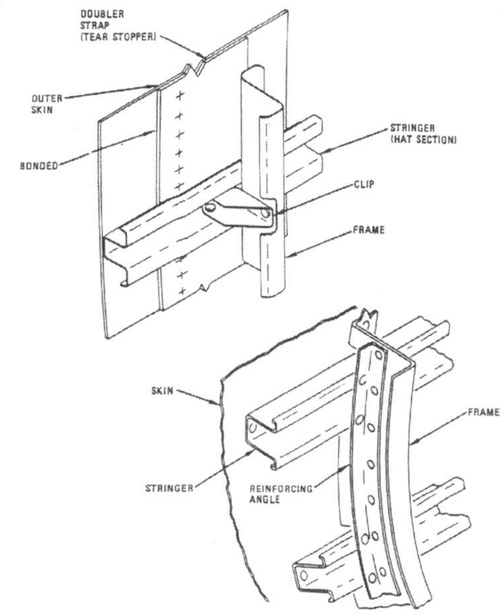

FRAME STRINGER ASSEMBLY

BONDED WAFFLE DOUBLER

Purpose

Purpose of this type doubler is the same as the tear stop strap type to reinforce the skin. This type construction allows use of a lighter weight (thinner) skin with the same strength and safety factors.

Physical Description

The doubler consists of an additional skin section bonded on the inside of the fuselage skin.

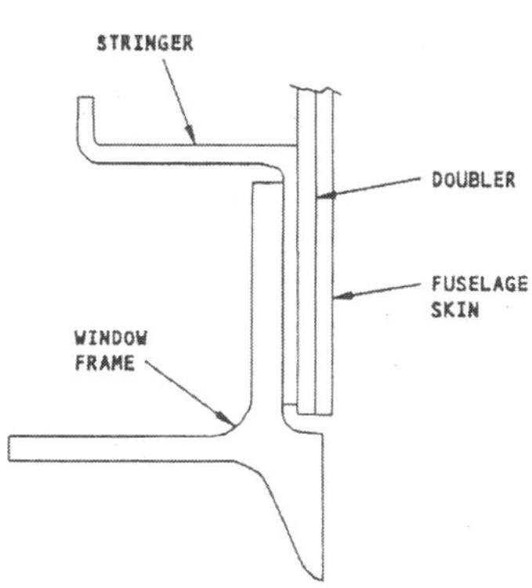

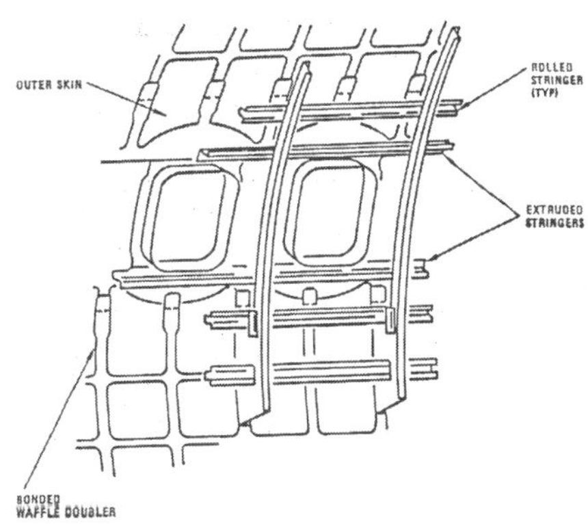

FUSELAGE SKIN JOINTS

Purpose

The skins are joined to provide a continuous and aerodynamically cl external fuselage surface.

Physical Description

Two types of skin joints are normally used.

Longitudinal lap joints are used to join upper and lower skin panels.

The upper panel will be lapped over the lower panel and in addition to riveting will be bonded and sealed to prevent water entry.

Circumferential skin splices are normally used where sections of the fuselage are joined (production breaks). In this case each skin section is riveted to a bulkhead and the joint sealed.

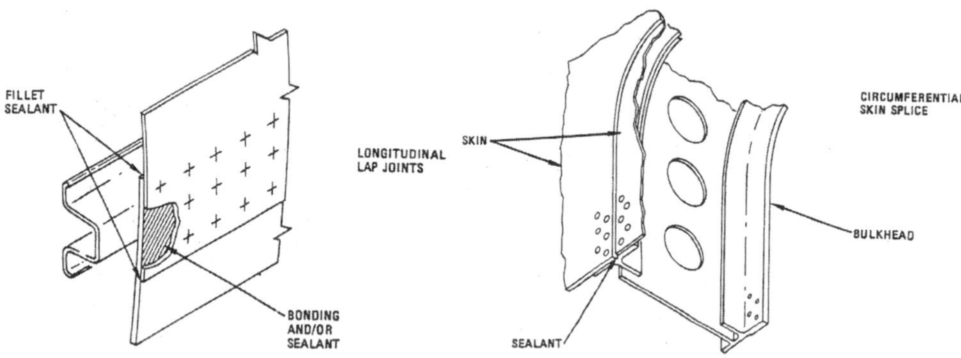

BILGE DRAINS

Purpose

The bilge drains are used along the lower portion of the fuselage to allow liquids to drain when the airplane is not pressurized.

Physical Description

Two types are used. Both work the same way. A spring opens the valve portion and allows any liquid accumulation to drain. When the airplane is pressurized the pressure overcomes the spring force and closes the valve, thus not effecting pressurization control.

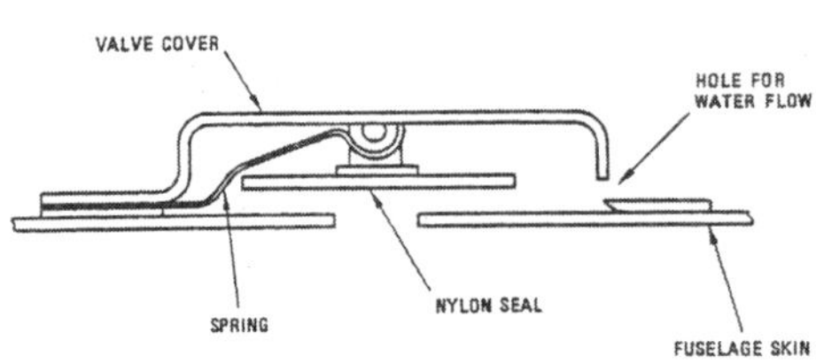

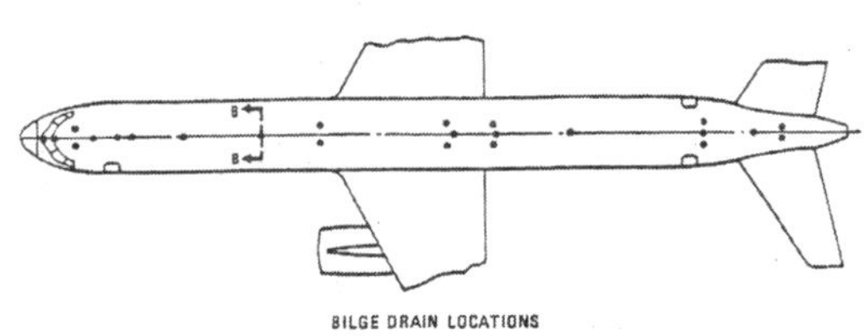

VORTEX GENERATORS

Purpose

The purpose of the vortex generators is to increase the energy of boundary layer air and delay the airflow separation from the wing. This aids in maintaining aileron effectiveness at high speeds.

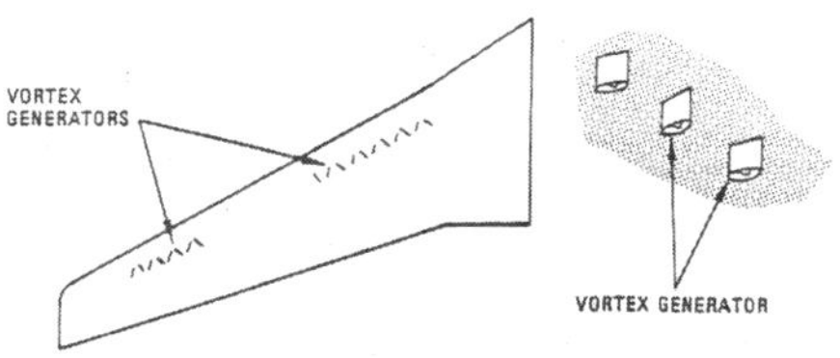

Physical Description

A vortex generator is a pair of small airfoils mounted at opposite angles of attack to each other and perpendicular to the surface they serve.

Like any airfoil the vortex generators develop lift and tip vortices. The tip vortices draw high-energy air from outside the boundary layer into the slower moving air next to the skin thus delaying the unwanted airflow separation.

ADVANCED COMPOSITE APPLICATIONS (TYPICAL)

Purpose

The airplane has been updated to include the latest in structural technology. Substantial amounts of advanced composite material are used structurally as a lightweight replacement for aluminum. A weight savings is realized compared to the conventional metal construction.

General Description

Composite components of graphite, Kevlar, fiberglass, and hybrids of these are used in the secondary flight control surfaces, fairings, and interior.

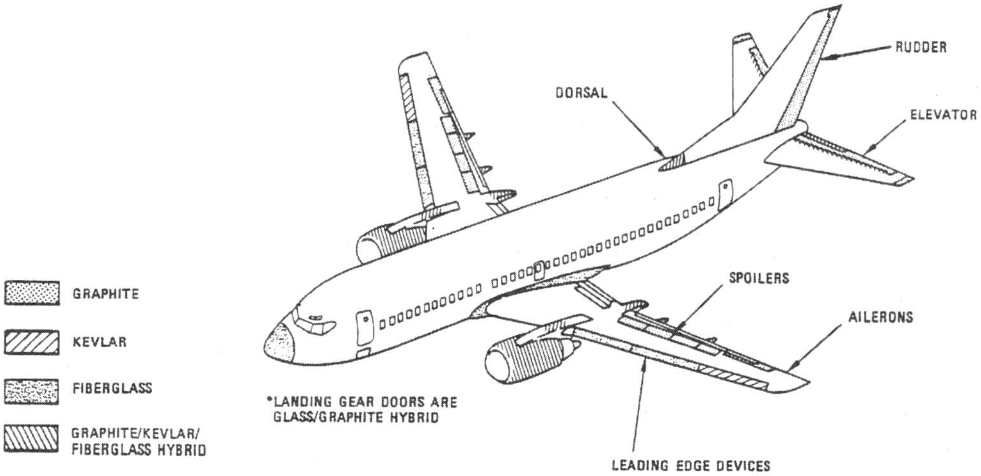

COMPOSITE MATERIALS

Physical Description

The term "Advanced Composites" covers a variety of fiber and resin systems.

Composites differ from metals. Composites are combinations of materials differing in composition or form. Although they act together, composite materials do not dissolve or merge as do metals when they are alloyed.

Advanced composite materials consist of high strength fibers (graphite or Kevlar) embedded in a resin matrix. Composites offer the potential for major weight savings in airplane design. They have high strength to weight ratios. When replacing aluminum structure with graphite/epoxy composite structure, weight reductions of 20% or better are common. Weight reduction is the greatest advantage of composite material and is one of the key items in decisions regarding its selection.

MATERIAL FORMS

Physical Description

Tape and Fabric

The common forms of materials used in airplane structure are tape and fabric. In the tape form, the filaments are laid in a single direction within a resin matrix. The fabric form is a material constructed of interlaced yarns, fibers or filaments, usually in a planar structure. The fabric form can be described as a woven material.

Warp and Fill

Warp and fill refer to the orientation of yarn in a fabric material. The warp is the longitudinally oriented yarn in a group of yarns in long lengths and approximately parallel. Fill refers to yarn oriented at right angles to the warp in a woven fabric.

To aid in distinguishing warp and fill yams in graphite fabrics, Kevlar tracer yarns are woven every 2 inches in warp direction and every 6 inches in fill direction.

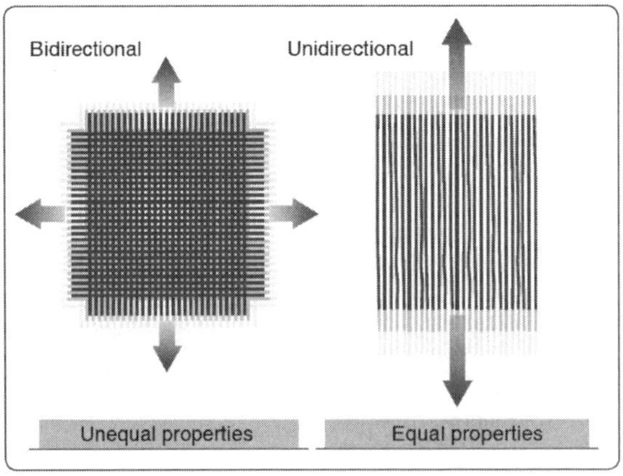

Bidirectional and unidirectional material properties.

HONEYCOMB SANDWICH

Purpose

Honeycomb sandwich construction is used to increase thickness and strength with a very small increase in weight.

Physical Description

A laminate is bonded on either side of a lightweight core "sandwich fashion". The laminate is made of layers of tape and/or fabric bonded together with an epoxy resin. The fiber direction (ply orientation) is determined by the strength requirements. The laminate need be only one material such as graphite, Kevlar or fiberglass, sometimes a hybrid is used like Kevlar/graphite/epoxy. This hybrid is used for certain areas where high impact resistance is needed.
The core is normally of a lightweight material called "nomex", but in some cases it may be very lightweight aluminum.

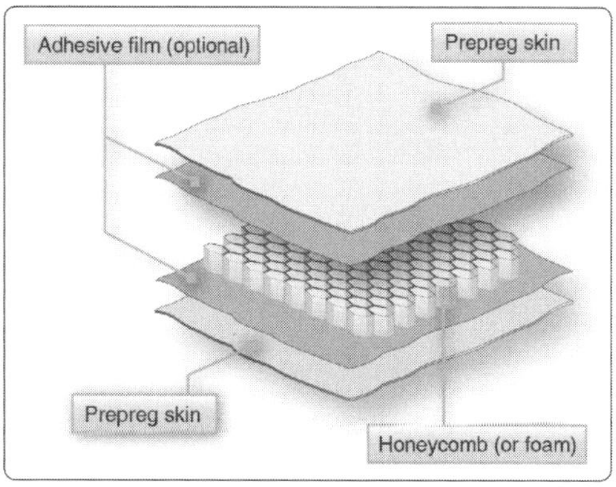

Honeycomb sandwich construction.

Honeycomb core materials.

ATA 71 - POWER PLANT

High bypass ratio turbofan engines provide thrust for the airplanes. They also supply electric, hydraulic, and pneumatic power for the airplane systems.

General Description
Power plant cowling consists of the inlet, fan, fan thrust reverser, and core cowls. Mounts attach the engine to the strut. The fan, fan reverser and core cowls are hinged to the strut and remain attached to the strut during an engine change. Latches at the bottom center pull the cowls together.

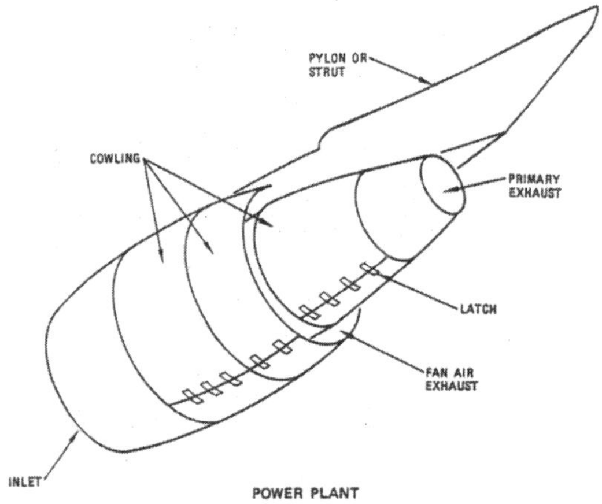

POWER PLANT

COWL HOLD OPEN ROD

Purpose
The rods support the cowl in the open position while maintenance is being performed.

Physical Description
The cowl hold open rod locks into place on the engine flange bracket with a spring loaded interference pin. It is removed by sliding back the lower collar. When fully extended and locked the upper collar will cover the word UNLOCKED.

COWL LATCHES

Purpose
The Cowl Latches pull the two cowl halves together at the bottom of the engine to secure the cowl.

Physical Description
The latch consists of a handle with a latch hook on one cowling half and a keeper on the other cowling half. A tool slot is incorporated in the handle.

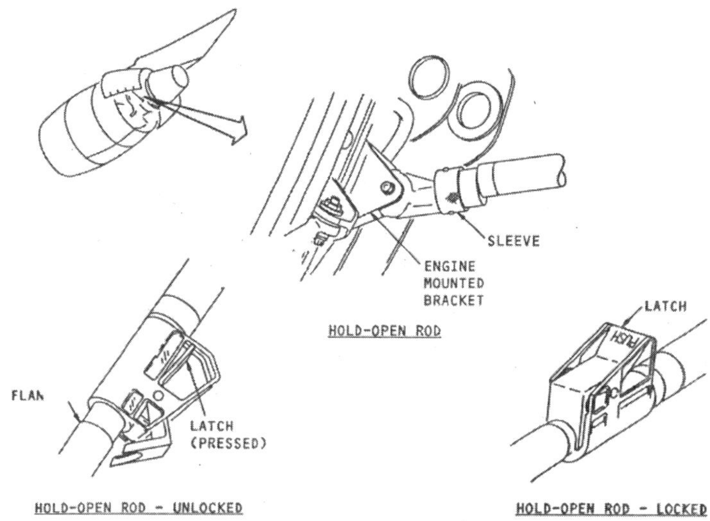

CORE COWL PANELS (OPENING/CLOSING)

Operation

Open Latch

Push trigger to release handle. If the latch is tight, a tool (flat bladed screw driver) may be inserted in a slot provided for the purpose and used to operate the trigger, on some latches a screwdriver slot is designed to operate the trigger and also lever the latch open.

Pull the handle down to release the latch.

Close latch

Engage latch hook into keeper.

Press the handle until it is flush with the skin and the trigger engages.

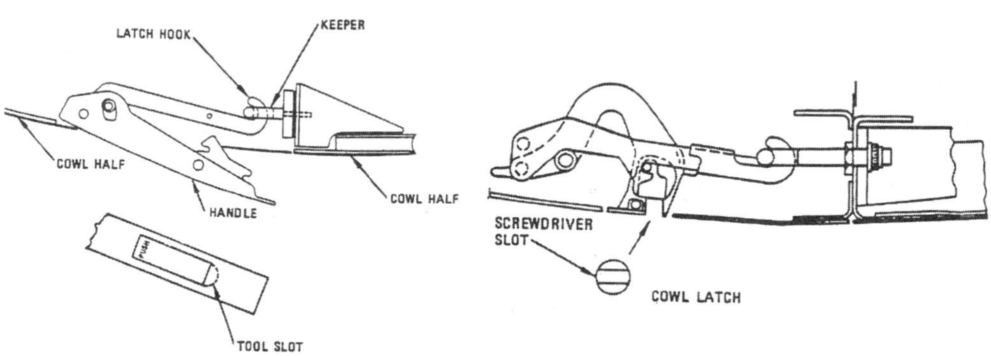

TURBOFAN ENGINE

Purpose

The powerplant provides the thrust required for flight and also supplies power to aid in stopping the airplane (thrust reversers), provides bleed air for air conditioning, pressurization, and anti-icing and drives a gearbox used for electrical and hydraulic power.

General Description

The modern turbofan engine is a multi-spool, axial flow, high bypass ratio engine. The low pressure Compressor (N1) has multi pie stages including a fan and is driven by a multiple stage low pressure turbine. The high pressure compressor (N2) also has multiple stages and is driven by the high pressure turbine. A singular annular combustion chamber separates the compressor sections from the turbine sections. Bearings are used to support each rotating assembly.

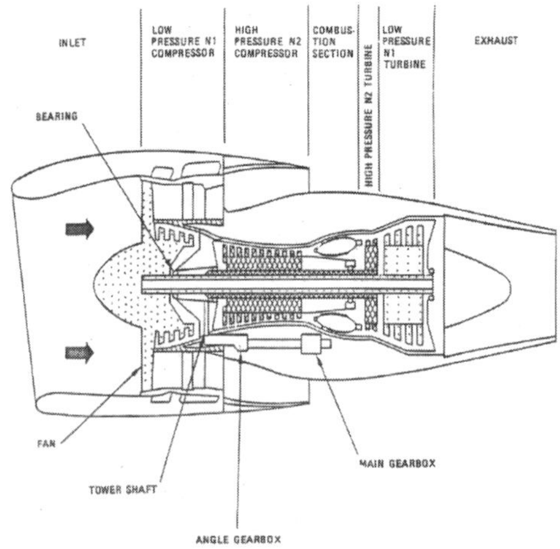

TURBOFAN ENGINE

The accessories are mounted on and driven by the accessory gearbox at the bottom of the engine. The turbofan engine produces approximately 75% of its thrust by moving a large amount of air at relatively low velocity. This is accomplished by use of the large fan stage of the N1 compressor.

LOW PRESSURE SECTION

Purpose

The Low Pressure Section contains the fan and the first few stages of the compressor. It is driven by the multi-stage low pressure turbine.

Physical Description

An axial-flow compressor has two main elements, a rotor, and a stator. The rotor has airfoils or blades which impel air rearward in the same manner as a propeller. The stator vanes act as diffusers at each stage, partially converting velocity to pressure and also directing the air to the next rotor stage for maximum efficiency. Each pair of rotor and stator blades constitute a pressure stage.

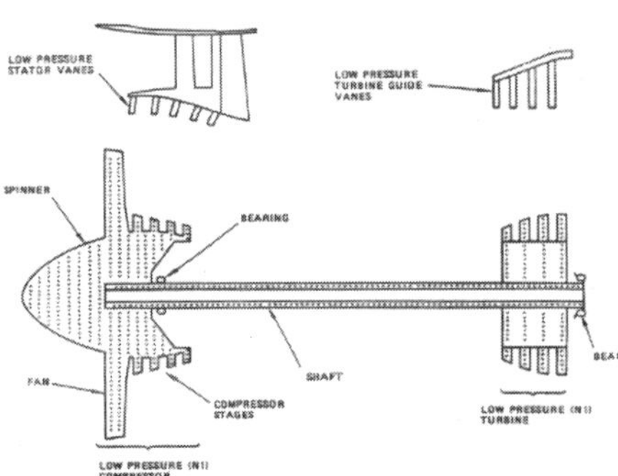

LOW PRESSURE SECTION

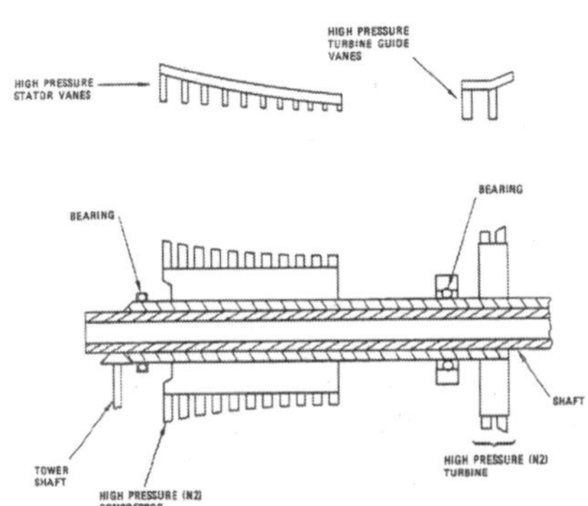

HIGH PRESSURE SECTION

The low pressure turbine consists of two basic elements, the stator and the rotor. The stator element is known by a variety of names. The most common are turbine guide vanes, turbine nozzle vanes and nozzle diaphragm. The purpose of the turbine guide vanes is to increase the velocity of the exhaust gases and guide the gases onto the turbine blades properly. A shaft from the turbines drives the low pressure compressor. Bearings support the Low Pressure Rotating Assembly.

HIGH PRESSURE SECTION

Purpose

The High Pressure Compressor (N2) provides the final compression of the air for the combustion section. It also drives the engine accessory section.

Physical Description

The High Pressure Compressor (N2) like N1 consists of a rotor and a stator and works in the same manner. One difference is, in many engines the first few rows (stages) of stators are variable for airflow control.

The multi-stage turbine drives the N2 compressor by means of a large hollow shaft. This shaft is supported by bearings. A bevel gear attached to the front of the shaft provides a drive for the accessory gear box. The N1 shaft passes through the hollow N2 shaft, however there is no mechanical interconnection.

COMBUSTION SECTION

Purpose

The purpose of the Combustion Section is to burn the fuel/air mixture, thereby adding heat energy to the air.

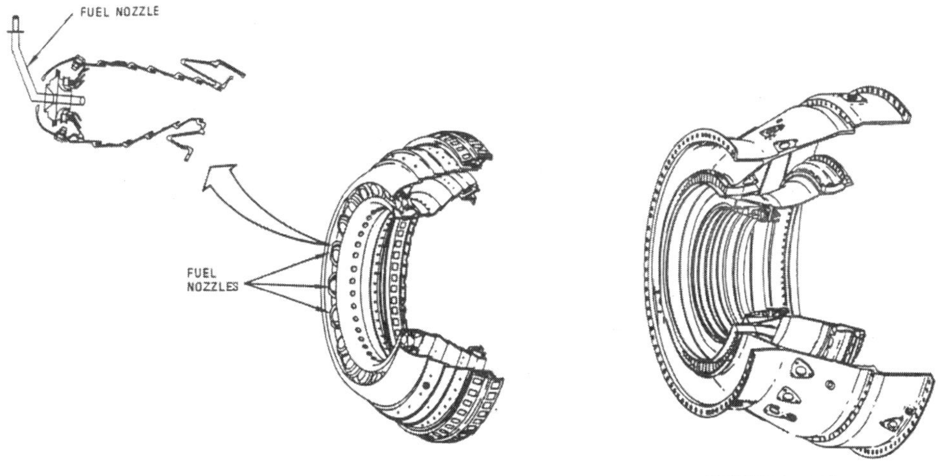

COMBUSTION SECTION

Physical Description

There are three basic types of combustion chambers, variations within these types being in details only. These types are:

1. multiple chamber or can type
2. annular or basket type
3. can-annular type

The annular chamber is the type most currently in use on the new turbofan engines and consists of a housing and a liner.

The housing or combustion chamber is an undivided chamber extending all the way around the outside of the turbine shaft housing. Provisions are made for installation of fuel nozzles. The fuel nozzle delivers the fuel into the liner in a finely atomized spray.

The liners have perforations of various sizes and shapes, each having a specific purpose. Some of the air entering the combustion chamber is directed inside the liner at the front end, where it mixes with the fuel and is burned. Cooling air passes between the chamber wall and liner. Louvers or holes are provided along the length of the liner to direct a cooling layer of air along the inside wall of the liner. This layer of air also controls the flame to keep it centered in the liner, preventing burning of the liner walls.

There are usually two spark igniter plugs mounted on bosses on the combustion chamber. These plugs must be long enough to extend into the liner for ignition of the fuel/air mixture.

ACCESSORY DRIVE

Purpose

The accessory drive gearbox provides for mounting and driving the engine driven accessories.

Physical Description

The accessory drive gearbox is mounted to the engine case and is driven by the high pressure compressor (N2) Shaft. A set of bevel gears at the front of the N2 shaft transmits rotary motion outside the engine through a radial shaft(commonly called tower shaft.) An angle gearbox then transmits the power to a drive shaft which operates the accessory gearbox.

The angle gear box normally does not drive any accessories except in some cases its own oil scavenge pump. Magnetic chip detectors may be built into either or both gearboxes.

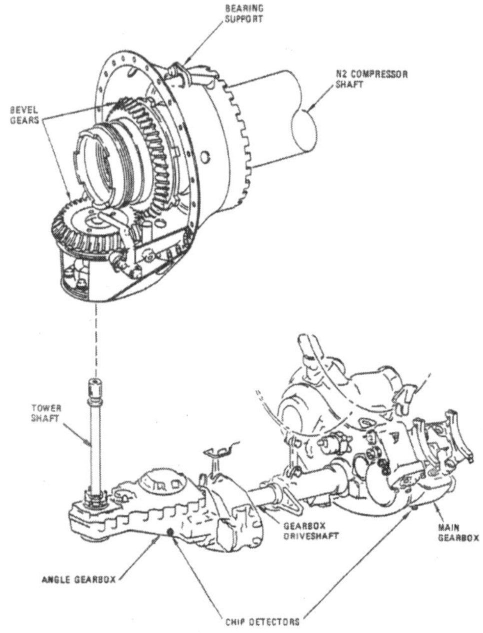

ACCESSORY DRIVE

MAGNETIC CHIP DETECTOR

Purpose
To collect ferrous metal particles for later inspection as an aid in troubleshooting.

Physical Description
The detector is a quick release, quarter-turn type fastener with a magnet on the end of a knob. The magnetic portion has two seals, and fits into a housing with a spring loaded shutoff valve.
Turning the knob one-quarter turn counterclockwise releases the detector from its housing. The spring loaded shutoff valve prevents oil from leaking once the detector is removed.

DO NOT INSTALL A DETECTOR WITH A MISSING OR DAMAGED SEAL.

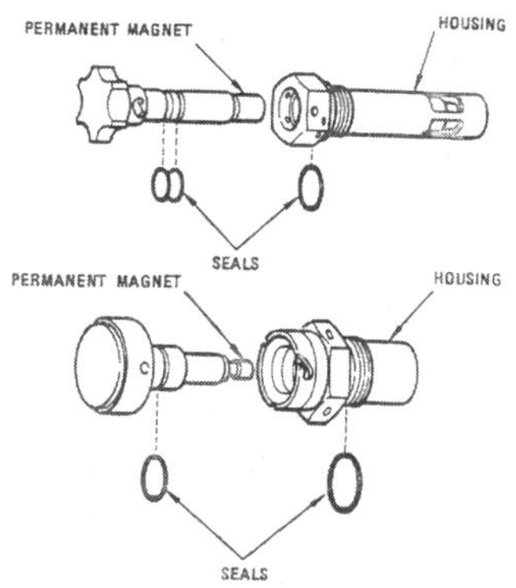

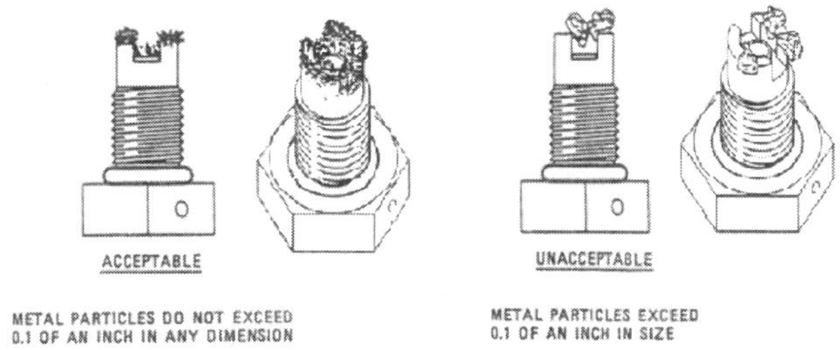

MAGNETIC CHIP DETECTOR INSPECTION

ENGINE FUEL SYSTEM

Purpose
The engine fuel system automatically delivers fuel to combustion system at a flow rate consistent with engine power requirement set by the flight compartment thrust lever. The system maintains the set power regardless of ambient temperature up to specific International Standard Atmosphere (I.S.A.) limits for takeoff, climb and cruise power requirement.

General Description
The engine fuel system receives fuel from the airplane tank fuel boost pumps or by suction flow. The fuel is delivered by a low pressure (L.P.) fuel pump, through a fuel heater and fuel filter to a high pressure (H.P.) fuel pump. The H.P. pump delivers the fuel through a main engine control and fuel manifold to spray nozzles located in the combustion section. In most systems servo fuel (high pressure) goes through the main engine control to operate hydraulic type actuators in the compressor airflow control.

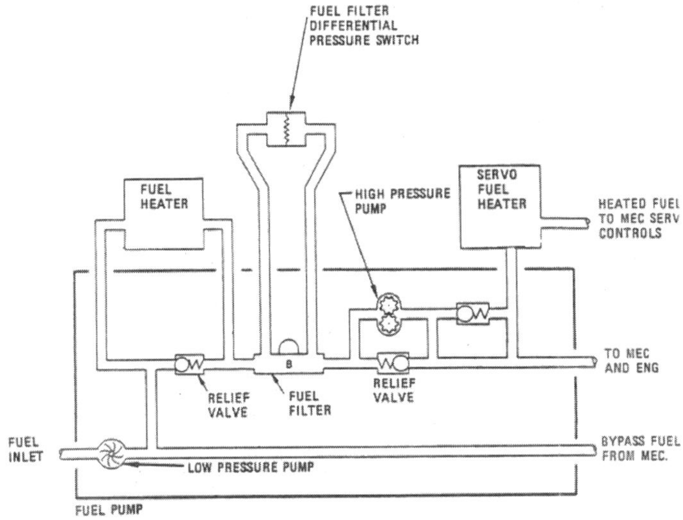

ENGINE FUEL SYSTEM

FUEL PUMP AND FILTER

Purpose
Provides filtered pressurized fuel to the main engine control.

Physical Description
The fuel pump comprises a centrifugal boost stage, a single gear high pressure stage and a filter with bypass valve. A filter delta P switch is mounted to the filter housing. The gear stage pump pressure is limited by gear stage pressure relief valve.

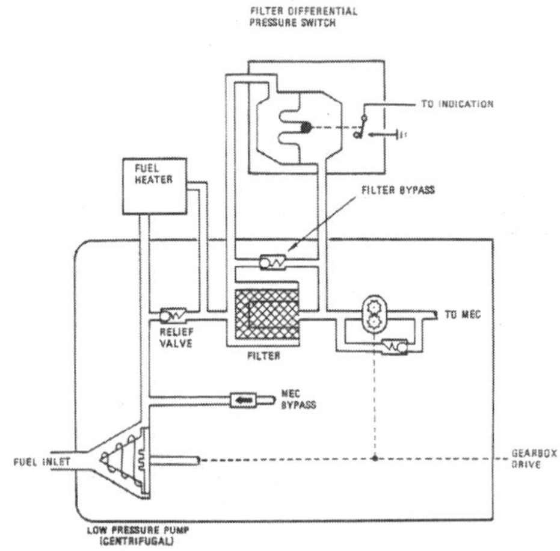

PUMP AND FILTER ASSY

Operation
The fuel pump is supplied with fuel by the fuel tank boost pumps, through engine fuel supply lines. Fuel leaving the boost stage mixes with bypass return fuel from the Main Engine Control and then is directed through the fuel heater before returning to pass through the filter, immediately upstream of the gear stage (high pressure). The filter incorporates a bypass feature to ensure against fuel starvation under abnormal contamination conditions, the fuel filter bypass warning system indicates an impending bypass of the fuel pump filter. When the pressure drop across the fuel pump filter exceeds a given value, the filter delta P switch closes a circuit to ground to provide an indication in the flight compartment. On some engines the fuel heater is replaced by a fuel cooled oil cooler.

ENGINE FUEL HEATER

Purpose

The engine fuel heater is designed to melt entrained ice in the fuel, caused by low fuel temperatures.

Physical Description

The fuel heater is a heat exchanger with no moving parts. The compressor bleed air flows through when the fuel heat valve is open. The air exhausts overboard.

Fuel is utilized to cool engine oil. Thus, when fuel heat is "ON" it is normal to see an increase in oil temperature.

The bleed air extraction causes a slight decrease in engine pressure ratio.

IGNITION EXCITER

Purpose

The purpose of the ignition exciter is to provide high energy ignition for initial combustion and to sustain combustion during certain engine operating conditions.

Physical Description

The ignition exciter is a sealed unit with a power input connector and an igniter output connector. Normally two are installed on each engine. One is sufficient to start the engine. Each exciter is of capacitance discharge type, converting electrical power supply to 24,000 volts. A power transformer inside the exciter steps up the voltage. It is converted to DC and used to charge a storage capacitor. When the capacitor is discharged, an arc is formed across the igniter plug. A bleed-down resistor will dissipate voltage to prevent injury.

WARNING. IGNITION SYSTEM VOLTAGE IS DANGEROUSLY HIGH. MAINTENANCE MANUAL PROCEDURES CONCERNING BLEED DOWN TIME MUST BE FOLLOWED AFTER OPERATION OF SYSTEM PRIOR TO MAINTENANCE ON THE SYSTEM.

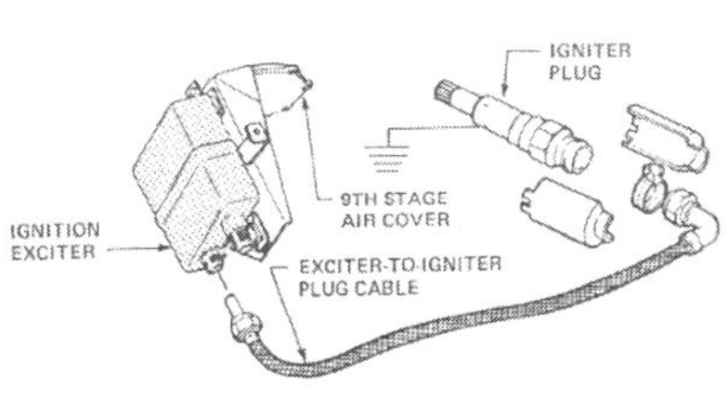

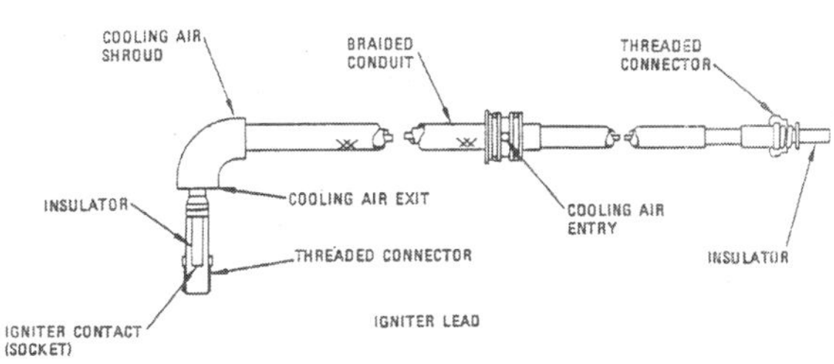

IGNITER LEAD AND PLUG

Purpose
The cables carry the high voltage DC from the exciters to the igniters. The igniter plugs develop a high voltage arc to ignite the fuel/air mixture.

Physical Description
The exciter to igniter plug cable consists of an insulated high tension lead enclosed in an air-cooled flexible steel conduit. The steel conduit is fed with fan discharge air and provides shielding for the high tension lead.
Each plug consists of a center electrode within a metal body. The electrode is insulated from the body and sealed to form a gas tight joint. Provisions are made for spacer gaskets to control the immersion depth of the plug, on some engine types the spacers are placed on the plug, on others the spacers are under the mounting boss.

AIRFLOW CONTROL

Purpose
The Airflow Control System permits optimum compressor performance to achieve thrust output and fuel consumption objectives with adequate stall margin.

Physical Description
The Airflow Control System consists of:
(a) the variable bleed valves
(b) the variable stator vanes
(c) some installations also utilize high stage compressor bleeds

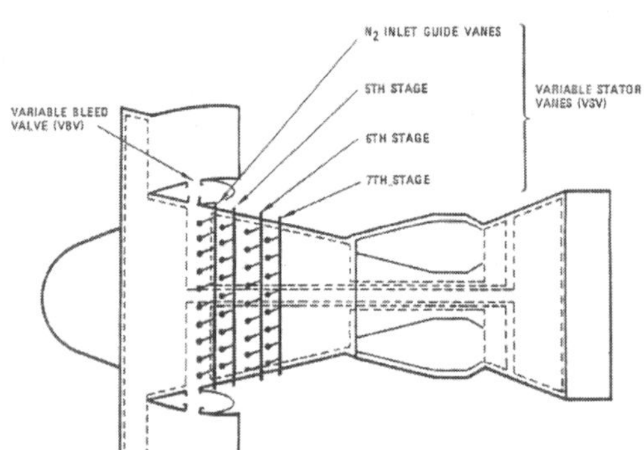

AIRFLOW CONTROL

The modern turbofan engine is designed to be most efficient and operate the best in the power area where it will spend most of its running time. The two rotating assemblies (compressors) turn at different speeds. At low powers and/or rapid changes, (acceleration, deceleration) it is possible that the low pressure compressor may provide more air than the high pressure compressor can accept or handle, this will result in a compressor surge or stall. To prevent a surge or stall in this condition it is necessary to unload the compressor by venting some of the low compressor air overboard (variable bleed valves-VBV) or to restrict the airflow into the high speed compressor (variable stator vanes-YSV) until the high speed compressor can increase or stabilize its speed. Normally both of these systems work together.

VARIABLE BLEED VALVES

Purpose

The purpose of the variable bleed valves is to bleed some of the low pressure compressor air overboard to help prevent engine stalls or surges during starting, low power settings, rapid power changes and reverse thrust.

General Description

Two types of valves may be used depending on the engine make and model. One is a 360° ring that circles the engine and is moved forward or aft by an actuator system to open or close overboard ports.

The other type has several flap type valves that are opened and closed by art actuator and drive mechanism.

The engine vane and bleed control or main engine control, depending on engine type, control the actuators by directing servo fuel pressure to the actuators. The servo fuel pressure will act as a hydraulic medium to position the bleed valve actuators to the open, close or any intermediate position as scheduled by the controlling unit.

VARIABLE STATOR VALVES

Purpose

The purpose of the variable stator vanes is to improve the compressor stall margin for engine starting, acceleration and part power operation.

Physical Description

The first few rows of the high pressure compressor stators (including inlet guide vanes) are constructed so that their angle relative to the airflow path can be varied.

The stator vanes are mounted so they can pivot, a link or lever arm is attached to the outer end. A unison or actuation ring circles the engine for each stage of variable stator vanes, all lever arms for that stage are attached to the actuation ring. A servo fuel pressure operated actuator moves all actuation rings through a bell crank and linkage assembly. Servo fuel pressure is directed to the actuator by the engine vane and bleed control or main engine control (depending on type engine).

Operation

The variable stator vanes are rotated to partially restrict airflow through the front stages of the N2 compressor during low RPM operation.

During low RPM operation, idle for example, the vanes are closed or near the closed stop restricting airflow into the compressor. As the RPM increases the vanes start to move open and modulate toward their full open limit or stop, at a rate schedule to give a good front-to-back compressor match as the N2 compressor can accept more air from the N1 compressor due to the increases in RPM.

POWER PLANT CONTROL

Purpose

The power plant control system provides the means for controlling fuel and ignition for starting, operating and shutting down the engines and for controlling engine thrust level in forward and reverse thrust ranges.

General Description

The power plant control system consists of forward and reverse thrust levers, thrust control cables, engine strut control box and a push-pull cable to the fuel control.

A start lever controls fuel supply to the engine and ignition. Switches operated by the start 1 ever operate to open the fuel valves allowing fuel from the tank to the engine. Ignition is enabled at the same time. A cable system from the start lever mechanically operates a final fuel shutoff at the fuel control, or this is accomplished electrically.

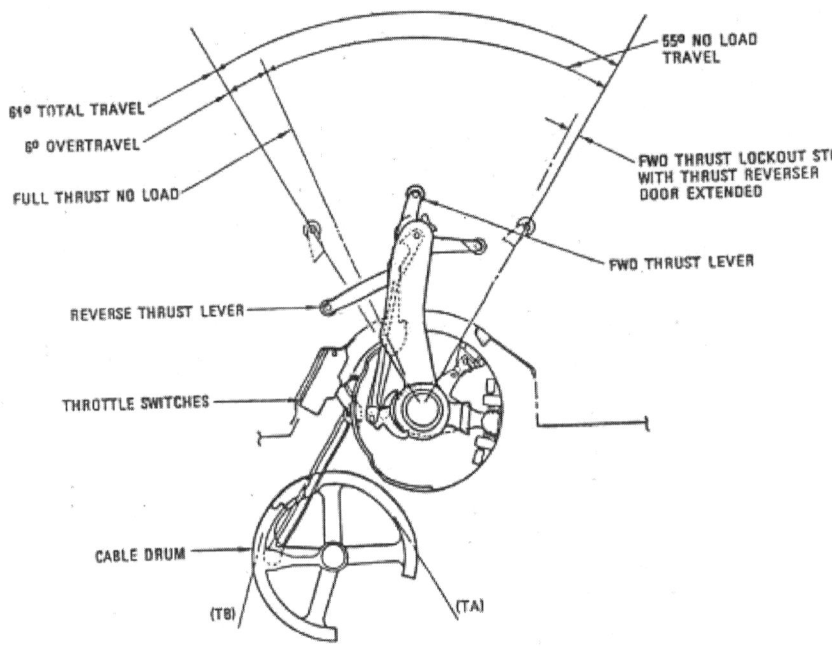

ENGINE INDICATING

Purpose

To display engine performance parameters (EPR, RPM, EGT and vibration) in the flight compartment.

Engine Pressure Ratio

Engine Pressure Ratio (EPR) indicating system provides a ratio of inlet to exhaust pressure to the indicators in the flight compartment. EPR is used for selecting thrust required.

Engine Tachometers

The engine tachometer systems determine shaft rotational speeds and provides each speed as a percentage equivalent.

Engine Exhaust Gas Temperature
Exhaust Gas Temperature (EGT) indicating system provides total temperature of the exhaust gases to the indicators in the flight compartment.
Airborne Vibration Monitor
Airborne Vibration Monitor (AVM) provides engine vibration levels to the indicators in the flight compartments.

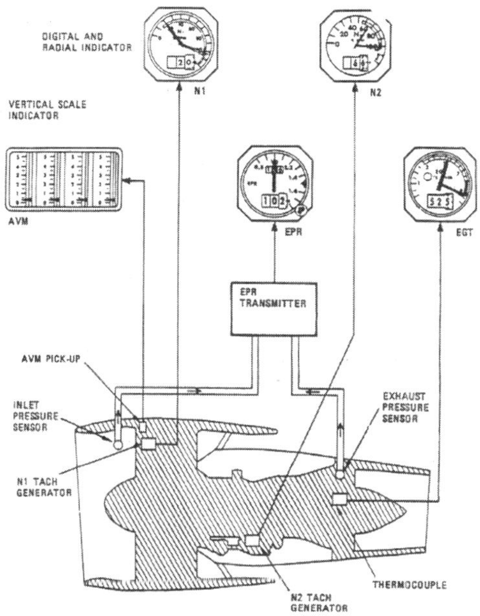

ENGINE INDICATING

FAN THRUST REVERSER

Purpose
The fan thrust reverser provides the means to reverse the direction of the fan exhaust.

General Description
The Thrust Reverser is an integral part of the fan stream duct and comprises two "C" sections, hinged at the top to the airplane strut. They can be opened to provide access to the engine for maintenance.
Stowed, the thrust reverser provides a smooth surface that enables the fan exhaust to produce forward thrust.
Deployed, the thrust reverser redirects the fan exhaust to produce reverse thrust.
The thrust reverser is controlled from the flight compartment by the reverse thrust levers, one for each engine. The thrust reverser is actuated utilizing the airplane hydraulic system pressure or engine pneumatic pressure. Operation of the thrust reverser is indicated in the flight compartment.

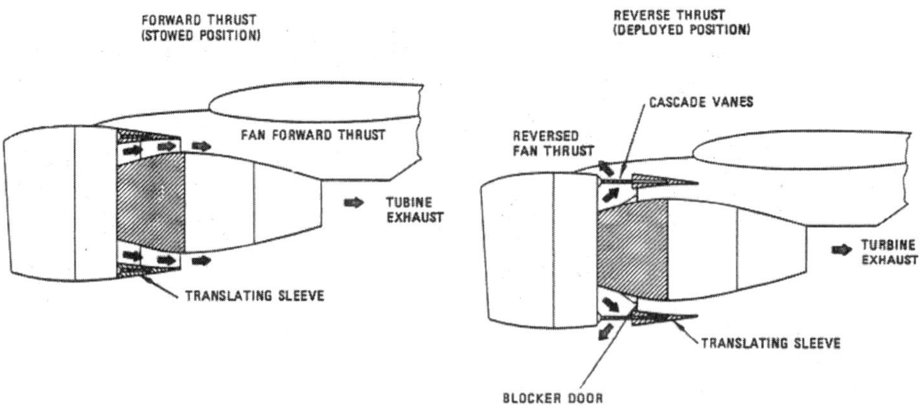

FAN THRUST REVERSER

BLOCKER DOORS

Purpose

The purpose of the blocker doors is to block the inner duct when the thrust reverser sleeve is aft and direct the fan air exhaust through the cascade vanes to produce reverse thrust.

Physical Description

The blocker doors are attached to the translating sleeve by two hinges on each blocker door's forward side. A drag link attaches to an attachment fitting on each door. The other end of the drag link attaches to the inner duct.

When the reverser is stowed the doors are flush with the inner side of the translating sleeve. When the thrust reverser is deployed the sleeve moves aft with the blocker doors, the drag link pulls the doors across and blocks the fan duct.

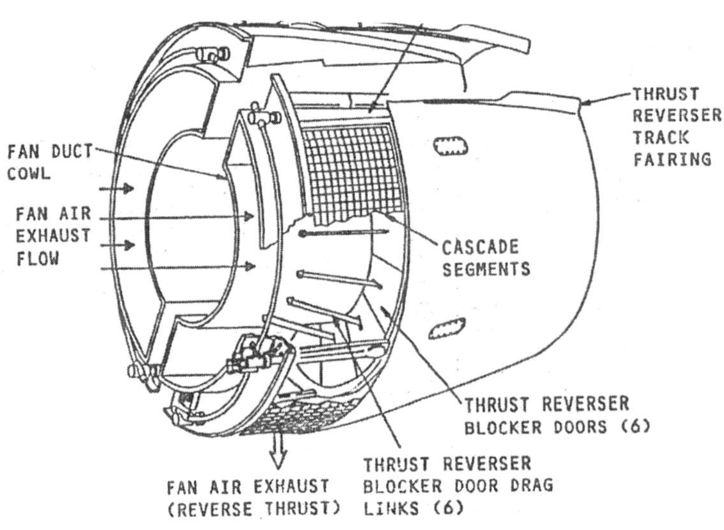

CASCADE VANES

Purpose
To direct the fan exhaust into the desired external flow pattern.

Physical Description
The cascade vanes are bolted on their leading edge to the thrust reverser torque box and on their aft end to the cascade support ring.
Each vane segment contains axial vanes (cascades) and circumferential turning vanes of various arrangements. The arrangements of the segments are different for right and left engines and are not interchangeable. An identification plate giving the positions at which a segment may be fitted is located on the front face of each segment. The segments are numerically identified clockwise viewed from the rear.

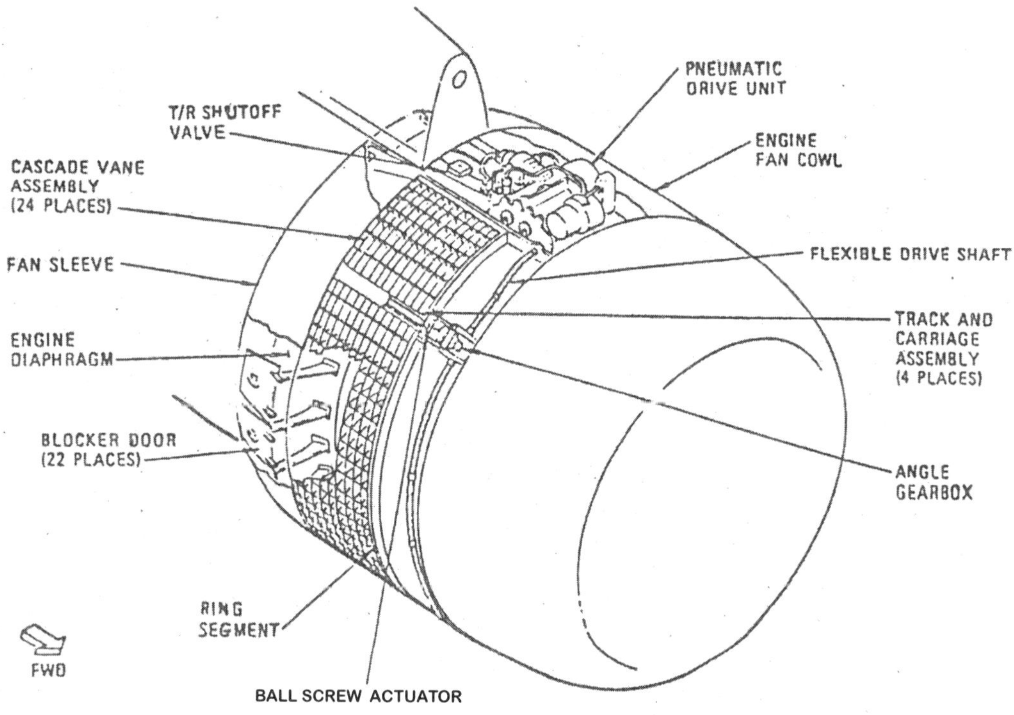

Blank Page

PART 3 - DICTIONARY

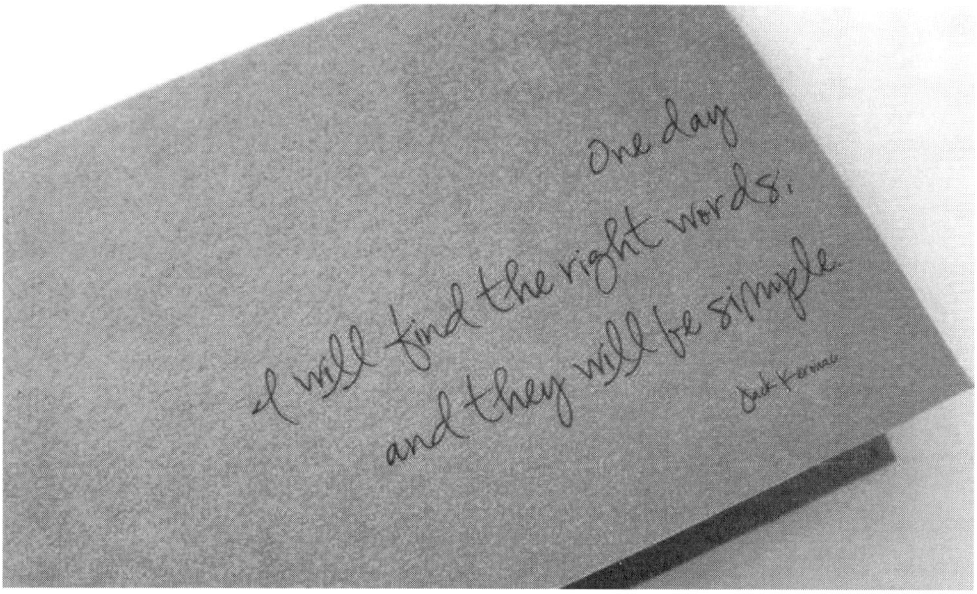

Blank Page

Introduction

General

The dictionary gives all the general words that are approved in STE and the examples that show how to use each word correctly. There is also a selection of unapproved words, which refer to the equivalent words that are approved in STE. Their examples show you how to use the approved words to express the same idea (sometimes with different sentence constructions.)

The dictionary does not include technical names (TN) or technical verbs (TV) in the word list. But some of them are referred to as alternatives to unapproved words.

How to use the dictionary

The dictionary has four columns:

- The word and its part of speech
- The approved meaning or alternatives for each word
- The approved examples
- The examples that are not approved.

Example:

Word (part of speech)	Approved meaning/ ALTERNATIVES	APPROVED EXAMPLE	Not approved example
accuracy (n)	PRECISION (n)	THE PRECISION OF THE ADJUSTMENT CAN CHANGE.	The accuracy of the adjustment can vary.

Column 1: Word (part of speech)

All words are in **bold** typeface.

A word in **UPPERCASE** letters shows that you can use this word (it is approved.)

Examples:

Word (part of speech)	Approved meaning/ ALTERNATIVES	APPROVED EXAMPLE	Not approved example
ABRASIVE (adj)	That can remove material by friction	DUST, WHEN MIXED WITH OIL, HAS AN ABRASIVE EFFECT.	
AID (n)	Help that is given	IF YOU GET THE SOLUTION IN YOUR EYES, GET MEDICAL AID IMMEDIATELY.	

A word in **lowercase** letters shows that you must use another word or a different construction because the word is not approved in STE.

Examples:

Word (part of speech)	Approved meaning/ ALTERNATIVES	APPROVED EXAMPLE	Not approved example
main (adj)	PRIMARY (adj)	THE PRIMARY CAUSE OF VALVE FAILURE IS CONTAMINATION OF THE HYDRAULIC FLUID.	The main cause of valve failure is contamination of hydraulic fluid.
build (v)	ASSEMBLE (v)	ASSEMBLE THE UNIT.	Build the unit.

The **part of speech** for each word is given in parentheses. Use an approved word only as the part of speech given (refer to part 1, section 1.)

In the dictionary, there are eight parts of speech:

1. **noun (n):** A word that is the name of a person, place, object, idea, quality, or activity.
2. **verb (v):** A word that describes a state of being or an action. Its tense (present, past, future) tells you when the action occurs.
3. **adjective (adj):** A word that modifies a noun or noun phrase. It describes, for example, the type, size, color, or number of a noun or noun phrase.
4. **adverb (adv):** A word that modifies a verb, an adjective, or another adverb. It answers the questions, "how?", "where?", "when?", "how often?", and "how much?".
5. **pronoun (pron):** A word that replaces a noun or noun phrase.
6. **article (art):** A word that tells you whether a noun phrase is new (indefinite = a, an) or one that you already know (definite = the).
7. **preposition (prep):** A word that shows how a noun, noun phrase or pronoun is related to other parts of the sentence.
8. **conjunction (conj):** A word or phrase that connects words, phrases, and clauses.

Forms of approved words

Nouns: Nouns are given only in their singular form, but the plural form of countable nouns is permitted unless the help tells you differently.

Example:

Word (part of speech)	Approved meaning/ ALTERNATIVES	APPROVED EXAMPLE	Not approved example
AGENT (n)	One of a group of materials made to do a specified task	DO NOT USE THESE CLEANING AGENTS ON HOT SURFACES.	

Verbs: Verbs are given in the forms that are permitted (refer to part 1, section 3.) Do not use verb forms that are not listed. The meaning of the verb can be different if it has an object (transitive) or no object (intransitive). Such verbs have more than one approved meaning and approved example.

Example:

Word (part of speech)	Approved meaning/ ALTERNATIVES	APPROVED EXAMPLE	Not approved example
ADAPT (v), ADAPTS, ADAPTED, ADAPTED	To change or adjust to that which is necessary	ADAPT THE PRESSURE CONNECTION TO THE PITOT HEAD.	
		THE SYSTEM INTERFACE CIRCUITS ADAPT TO THE PHYSICAL PROPERTIES OF THE CONNECTED SYSTEMS.	

Adjectives: Adjectives are given in their basic form, with their comparative and superlative forms in parentheses. Adjectives that make their comparative and superlative forms with "more" and "most" do not have these forms listed in the dictionary, because "more" and "most" are also listed as approved words.

Example:

Word (part of speech)	Approved meaning/ ALTERNATIVES	APPROVED EXAMPLE	Not approved example
SLOW (adj) (SLOWER, SLOWEST)	At low speed	MAKE SURE THAT THE MOVEMENT OF THE ELEVATORS IS SLOW.	

Adverbs: Frequently (but not always), you can make an adverb from an adjective when you attach an "-ly" ending to it. The comparative and superlative forms of adverbs are also made with "more" and "most". Thus, they are not given in the dictionary.

Examples:

Word (part of speech)	Approved meaning/ ALTERNATIVES	APPROVED EXAMPLE	Not approved example
SLOWLY (adv)	In a slow manner	TURN THE KNOB SLOWLY.	
briskly (adv)	QUICKLY (adv)	RUB THE SURFACE QUICKLY WITH A SOFT, DRY CLOTH.	Rub the surface briskly with a soft, dry cloth.

Column 2: Approved meaning/ALTERNATIVES

Approved words

This column gives the approved meaning (or definition) of an approved word in STE. The text in these definitions is not written in STE. If a meaning is not given in the dictionary, you cannot use the word in that meaning. Use an alternative word.

Example:

Word (part of speech)	Approved meaning/ ALTERNATIVES	APPROVED EXAMPLE	Not approved example
BEHIND (prep)	In a position at the rear of	THE PUMP IS INSTALLED BEHIND THE HYDRAULIC MOTOR.	

Unapproved words

For unapproved words, this column gives approved alternatives that you can use to replace the unapproved words. These alternatives are in uppercase letters and are only suggestions to help you.

Example:

Word (part of speech)	Approved meaning/ ALTERNATIVES	APPROVED EXAMPLE	Not approved example
addition (n)	ADD (v)	TO GET THE CORRECT CLEARANCE, ADD SPECIAL SHIMS, AS NECESSARY.	Adjust the clearance by the addition of special shims, as necessary.

It is possible that the suggested alternative for an unapproved word has a different part of speech. Usually, the first suggested alternative has the same part of speech as the unapproved word.

Example:

Word (part of speech)	Approved meaning/ ALTERNATIVES	APPROVED EXAMPLE	Not approved example
maintain (v)	KEEP (v)	KEEP THE FLUID TEMPERATURE AT 70 °F.	Maintain the fluid temperature at 70 °F.
	HOLD (v)	TURN THE ROTATING TUBE 75 DEGREES. THEN HOLD THE TUBE IN THIS POSITION.	Turn the rotating tube 75 degrees. Then maintain the tube in this position.
	MAINTENANCE (n)	DO MAINTENANCE ON THE FUEL SYSTEM.	Maintain the fuel system.

The dictionary does not include technical names or technical verbs. But, if a technical name or a technical verb is used in the suggested alternatives, this word is identified as (TN) or (TV). Antonyms are not included in the dictionary.

Example:

Word (part of speech)	Approved meaning/ ALTERNATIVES	APPROVED EXAMPLE	Not approved example
uncovered (v)	COVER (TN)	DO NOT PUT A COVER ON THE CONTAINER.	Leave the container uncovered.

The approved alternative for an unapproved word can also be more than one word. For these alternatives, the parts of speech are not given.

Example:

Word (part of speech)	Approved meaning/ ALTERNATIVES	APPROVED EXAMPLE	Not approved example
simultaneously (adv)	AT THE SAME TIME	DO THESE TWO STEPS AT THE SAME TIME.	Do these two steps simultaneously.

Column 2: Approved meaning/ALTERNATIVES

Help

For some words, there is a graphic symbol and related text that tells you to use another approved alternative or a different construction. This graphic symbol and its related text is referred to as the "help" and replaces the old word "note" in the dictionary.

There are four categories of help and they are explained in the examples that follow.

Examples:

Category 1

Recommendations that give you more information or instructions about how to use the approved word correctly.

Word (part of speech)	Approved meaning/ ALTERNATIVES	APPROVED EXAMPLE	Not approved example
PUSH (v) PUSHES, PUSHED, PUSHED	1. To apply a force to something to move it away from the source of the force	PUSH THE INSERT DOWN UNTIL IT TOUCHES THE BOTTOM OF THE HOLE.	
	2. To move with a force against something Use this word together with a preposition or an adverb to show direction.	THE SPRING UNIT PUSHES AGAINST THE BRAKE PEDAL.	

Category 2

This category of help tells you that the meanings of some approved words, for example ABOUT (prep), are restricted. Thus, you must use the suggested alternatives. When the help restricts the meaning of approved words, the dictionary gives "approved" and "not approved" examples in columns 3 and 4.

Word (part of speech)	Approved meaning/ ALTERNATIVES	APPROVED EXAMPLE	Not approved example
ABOUT (prep)	Concerned with For other meanings, use:	FOR DATA ABOUT THE LOCATION OF CIRCUIT BREAKERS, REFER TO THE WIRING LIST.	
	APPROXIMATELY (adv)	DRAIN APPROXIMATELY 2 LITERS OF FUEL FROM THE TANK.	Drain about 2 liters of fuel from the tank.
	AROUND (prep)	TURN THE SHAFT AROUND ITS AXIS.	Rotate the shaft about its axis.

Category 3

This category of help tells you to be careful with the use of the words that are approved for one context only. For example, the word SWALLOW (v), is only approved for use in safety instructions. You must not use words like SWALLOW (v) in other different contexts.

Word (part of speech)	Approved meaning/ ALTERNATIVES	APPROVED EXAMPLE	Not approved example
SWALLOW (v) SWALLOWS, SWALLOWED, SWALLOWED	To take through the mouth and esophagus into the stomach Use this word for safety instructions only.	IF YOU SWALLOW NITRIC ACID, GET MEDICAL AID IMMEDIATELY.	

Category 4

This category of help tells you to think about the given information. For example, you can only use the forms of BE (v) that are listed in the dictionary. No other forms are permitted.

Word (part of speech)	Approved meaning/ ALTERNATIVES	APPROVED EXAMPLE	Not approved example
BE (v), IS, WAS (also ARE, WERE) No other forms of this verb.	1. To occur, exist	IF THERE IS CORROSION ON THE PUMP VANES, REPLACE THE PUMP.	
	2. To have a property to be equal to	ACID SOLUTIONS ARE DANGEROUS.	

Column 3: APPROVED EXAMPLE

This column shows you:

- How to use the approved word, or
- How to use the approved alternative (usually a word-for-word replacement), or
- How to keep the same meaning with a different construction.

The wording given in the approved examples is not mandatory. It shows only one method to write the same information with approved words. You can frequently use different constructions with other approved words to say the same thing.

Examples:

Word (part of speech)	Approved meaning/ ALTERNATIVES	APPROVED EXAMPLE	Not approved example
A (art)	Function word: indefinite article	A FUEL PUMP IS INSTALLED IN ZONE 10.	
main (adj)	PRIMARY (adj)	THE PRIMARY CAUSE OF VALVE FAILURE IS CONTAMINATION OF THE HYDRAULIC FLUID.	The main cause of valve failure is contamination of hydraulic fluid.
manufacture (v)	MAKE (v)	YOU CAN MAKE THE CLEARING TOOL LOCALLY.	The clearing tool can be manufactured locally.

Column 4: Not approved example

This column gives examples that show you how the unapproved word is used in standard technical English. The examples also help you to understand how you can use the approved alternatives and/or different constructions to give the same information.

In an unapproved example, it is possible that you will find more than one unapproved word or a non-STE construction. Thus, the approved example shows you how to replace all the unapproved words or use a correct STE construction.

For approved words, column 4 is empty, unless there is help related to other meanings or restrictions. You can see that for ABOVE (prep) column 4 has an unapproved example.

Examples:

Word (part of speech)	Approved meaning/ ALTERNATIVES	APPROVED EXAMPLE	Not approved example
main (adj)	PRIMARY (adj)	THE PRIMARY CAUSE OF THE VALVE FAILURE IS CONTAMINATION OF THE HYDRAULIC FLUID.	The main cause of valve failure is contamination of hydraulic fluid.
previous (adj)	BEFORE (conj)	REMOVE THE PLATE BEFORE YOU ADJUST THE CABLE.	Adjustment of the cable requires previous removal of the plate.
A (art)	Function word: indefinite article	A FUEL PUMP IS INSTALLED IN ZONE 10.	
ABOVE (prep)	In (or to) a position farther up than something For other meanings, use:	LIFT THE CYLINDER ABOVE ITS INSTALLED POSITION.	
	MORE THAN	THE PRESSURE VALUE MUST BE MORE THAN 800 kPa.	The pressure value must be above 800 kPa.

Correct use of the words

STE is a controlled language with a restricted vocabulary.

As a result, it is not possible to use all words that you want. If you are not sure about a word that you want to use, refer to the flow chart that follows.

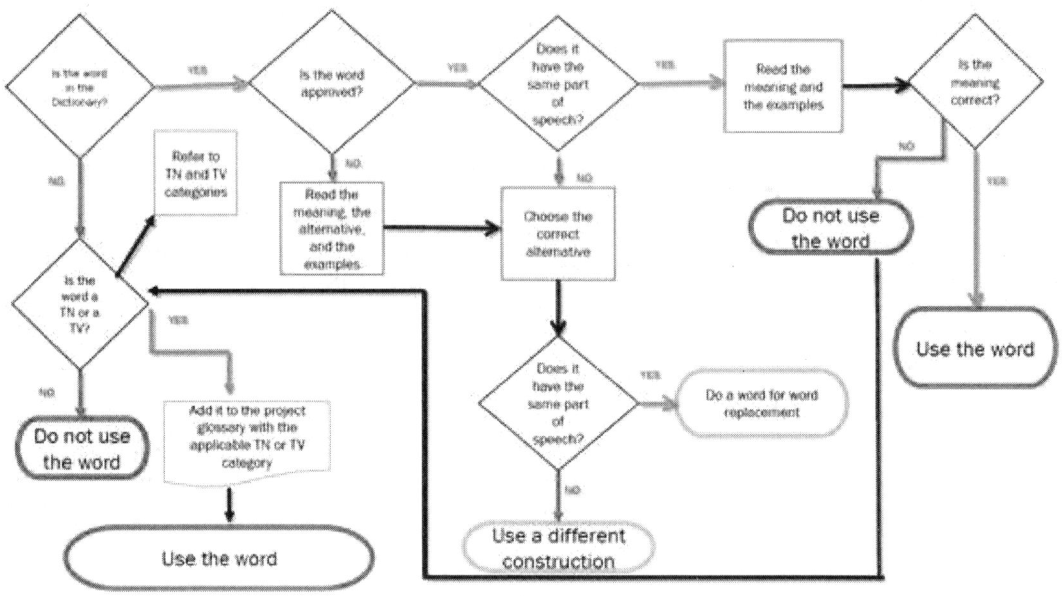

The table that follows gives you a list of the most "popular" recurring errors that writers do when they use STE. If a word is not approved in the dictionary, do not use it.

Non-STE	STE
acceptable (adj)	PERMITTED (adj)
alternate (adj)	ALTERNATIVE (adj)
any (adj)	None or a different construction
avoid (v)	PREVENT (v)
both (adj)	THE TWO (TN)
check (v)	CHECK (n)
cover (v)	COVER (TN)
damage (v)	DAMAGE (n)
ensure (v)	MAKE SURE (v)
fit (v)	INSTALL (v)
follow (v)	OBEY (v)
further (adj)	MORE (adj)
further (adv)	MORE (adv)
have to (v)	MUST (v)
insert (v)	PUT (v)
main (adj)	PRIMARY (adj)
may (v)	CAN (v)
need (v)	NECESSARY (adj)
now (adv)	AT THIS TIME
over (prep)	ABOVE (prep), ON (prep), ALONG (prep)
perform (v)	DO (v)
press (v)	PUSH (v)
reach (v)	GET (v)
repeat (v)	DO (v) ... AGAIN
required (v)	NECESSARY (adj)
rotate (v)	TURN (v)
secure (v)	ATTACH (v), SAFETY (v)
shall (v)	MUST (v)
should (v)	MUST (v)
since (conj)	BECAUSE (conj)
test (v)	TEST (n)
therefore (adv)	THUS (adv), AS A RESULT

Blank Page

Word (part of speech)	Approved meaning/ ALTERNATIVES	APPROVED EXAMPLE	Not approved example
A (art)	Function word: indefinite article	A FUEL PUMP IS INSTALLED IN ZONE 10.	
abaft (prep)	AFT OF (prep)	THE CONTROL UNIT IS INSTALLED AFT OF THE FLIGHT COMPARTMENT.	The control unit is installed abaft the flight compartment.
abandon (v)	STOP (v)	STOP THE ENGINE START PROCEDURE.	Abandon the engine start procedure.
	GO (v)	IF THERE IS A FIRE, IMMEDIATELY GO TO A SAFE AREA.	If there is a fire, immediately abandon the area.
abate (v)	DECREASE (v)	WHEN THE WIND SPEED DECREASES TO LESS THAN 30 KNOTS, YOU CAN OPEN THE CARGO DOOR.	When the wind speed abates to less than 30 knots, you can open the cargo door.
ability (n)	CAN (v)	ONE GENERATOR CAN SUPPLY POWER FOR ALL THE SYSTEMS.	One generator has the ability to supply power for all the systems.
able (adj)	CAN (v)	IF YOU CAN START THE ENGINE, DO THE APPLICABLE TESTS.	If you are able to start the engine, do the applicable tests.
abnormal (adj)	UNUSUAL (adj)	LISTEN FOR UNUSUAL NOISES.	Listen for abnormal noises.
	INCORRECT (adj)	IF YOU FIND AN INCORRECT QUANTITY OF AIR FROM THE VENT MAST, DO A SYSTEM TEST.	If you find an abnormal quantity of air from the vent mast, do a system test.
abnormality (n)	DEFECT (TN)	EXAMINE THE SEAL FOR DEFECTS.	Examine the seal for abnormalities.

Word (part of speech)	Approved meaning/ ALTERNATIVES	APPROVED EXAMPLE	Not approved example
ABOUT (prep)	Concerned with For other meanings, use:	FOR DATA ABOUT THE LOCATION OF CIRCUIT BREAKERS, REFER TO THE WIRING LIST.	
	APPROXIMATELY (adv)	DRAIN APPROXIMATELY 2 LITERS OF FUEL FROM THE TANK.	Drain about 2 liters of fuel from the tank.
	AROUND (prep)	TURN THE SHAFT AROUND ITS AXIS.	Rotate the shaft about its axis.
ABOVE (prep)	In (or to) a position farther up than something For other meanings, use:	LIFT THE CYLINDER ABOVE ITS INSTALLED POSITION.	
	MORE THAN	THE PRESSURE VALUE MUST BE MORE THAN 800 kPa.	The pressure value must be above 800 kPa.
ABRASIVE (adj)	That can remove material by friction	DUST, WHEN MIXED WITH OIL, HAS AN ABRASIVE EFFECT.	
abrupt (adj)	SUDDEN (adj)	THE DAMPER PREVENTS SUDDEN MOVEMENT OF THE CONTROL.	The damper prevents abrupt movement of the control.
	SUDDENLY (adv)	IF THE ROTORS STOP SUDDENLY, EXAMINE THE INTAKE FOR UNWANTED MATERIAL.	If the rotors come to an abrupt stop, examine the intake for unwanted material.

Word (part of speech)	Approved meaning/ ALTERNATIVES	APPROVED EXAMPLE	Not approved example
absence (n)	NONE (pron)	IF NONE OF THE BRACKETS ARE DAMAGED, CONTINUE THE PROCEDURE.	In the absence of damage to the brackets, continue the procedure.
	NOT (adv)	IF THE BRACKETS ARE NOT DAMAGED, CONTINUE THE PROCEDURE.	In the absence of damage to the brackets, continue the procedure.
	NO (adj)	IF THERE IS NO CONTAMINATION IN THE FUEL SAMPLE, CONTINUE THE PROCEDURE.	In the absence of contamination in the fuel sample, continue the procedure.
absent (adj)	MISSING (adj)	IF ONE OR MORE BLADES ARE MISSING, MAKE AN ENTRY IN THE ENGINE LOGBOOK.	If one or more blades are absent, make an entry in the engine logbook.
	NO (adj)	IF THERE IS NO CONTAMINATION IN THE FUEL SAMPLE, CONTINUE THE PROCEDURE.	If contamination is absent from the fuel sample, continue the procedure.
absolutely (adv)	FULLY (adv)	MAKE SURE THAT THE LATCH IS FULLY ENGAGED.	Ensure the latch is absolutely engaged.
ABSORB (v), ABSORBS, ABSORBED, ABSORBED	1. To take up or into	ABSORB THE FLUID WITH A CLEAN CLOTH.	
	2. To decrease the effect of	THE SHOCK MOUNT ABSORBS THE VIBRATION.	
absorption (n)	ABSORB (v)	MEASURE THE TIME THAT IS NECESSARY FOR THE SILICA GEL TO ABSORB THE MOISTURE.	Measure the rate of absorption of the moisture by the silica gel.
abundant (adj)	LARGE (adj)	CLEAN YOUR SKIN WITH A LARGE QUANTITY OF CLEAN WATER.	Clean your skin with abundant clean water.

Word (part of speech)	Approved meaning/ ALTERNATIVES	APPROVED EXAMPLE	Not approved example
abut (v)	TOUCH (v)	THE BIN TOUCHES THE FORWARD HINGE SURFACE.	The bin abuts the forward hinge surface.
accelerate (v)	INCREASE (v)	A HIGHER TEMPERATURE INCREASES THE SPEED OF EVAPORATION.	Evaporation is accelerated by a higher temperature.
	FASTER (adj)	TO MAKE THE CURING PROCESS FASTER, APPLY HEAT TO THE COMPOUND.	The curing process of the compound can be accelerated by heating.
ACCEPT (v), ACCEPTS, ACCEPTED, ACCEPTED	To make a decision that something is satisfactory	ACCEPT THE RELAY IF IT IS SERVICEABLE.	
acceptable (adj)	PERMITTED (adj)	A VALUE OF 2 mm IS PERMITTED.	A value of 2 mm is acceptable.
	SATISFACTORY (adj)	REPLACE THE CAP IF ITS CONDITION IS NOT SATISFACTORY.	Replace the cap if its condition is not acceptable.
	SERVICEABLE (adj)	BEFORE INSTALLATION, MAKE SURE THAT THE UNIT IS SERVICEABLE.	Before installation, make sure that the unit is acceptable.
acceptance (n)	ACCEPT (v)	BEFORE YOU ACCEPT THE UNIT, DO THE SPECIFIED TEST PROCEDURE.	Before acceptance of the unit, do the specified test procedure.
ACCESS (n)	The ability to go into or near	GET ACCESS TO THE ACCUMULATOR FOR THE No. 1 HYDRAULIC SYSTEM.	
accessible (adj)	ACCESS (n)	TURN THE COVER UNTIL YOU CAN GET ACCESS TO THE JACKS THAT HAVE + AND - MARKS.	Rotate the cover until the jacks marked by + and - are accessible.
ACCIDENT (n)	An occurrence that causes injury or damage	MAKE SURE THAT THE PINS ARE INSTALLED TO PREVENT ACCIDENTS.	

Word (part of speech)	Approved meaning/ ALTERNATIVES	APPROVED EXAMPLE	Not approved example
ACCIDENTAL (adj)	That does not occur on purpose	TO PREVENT ACCIDENTAL OPERATION OF THE SYSTEM, INSTALL THE SAFETY LOCK.	
ACCIDENTALLY (adv)	That does not occur on purpose	IF YOU ACCIDENTALLY MOVE THE LEVER, SET THE SYSTEM TO THE NEUTRAL POSITION AGAIN.	
accommodate (v)	LET (v)	DIFFERENT LENGTHS OF STUDS LET YOU ATTACH DIFFERENT THICKNESSES OF SKIN.	Studs are available in different lengths to accommodate different thicknesses of skin.
accomplish (v)	DO (v)	DO THIS TASK FIRST.	This task must be accomplished first.
	You can use a more specific command verb.		
	COMPLETE (v)	THE PERSONNEL MUST COMPLETE THE TASK IN THE GIVEN TIME.	The personnel must accomplish the task in the given time.
according to (prep)	REFER (v)	TO CALIBRATE THE TEST SET, REFER TO THE MANUFACTURER'S INSTRUCTIONS.	Calibrate the test set according to the manufacturer's instructions.
account for (v)	MAKE SURE (v)	MAKE SURE THAT YOU REMOVE ALL TOOLS AND EQUIPMENT.	All tools and equipment must be accounted for.
accumulate (v)	COLLECT (v)	IF WATER COLLECTS IN THE FILLER LINE, DRAIN IT.	If water accumulates in the filler line, drain it.
accumulation (n)	QUANTITY (n)	REMOVE LARGE QUANTITIES OF CONTAMINATION.	Large accumulations of contamination must be removed.
	COLLECT (v)	IF FUEL COLLECTS FREQUENTLY, EXAMINE FOR LEAKS.	If accumulation of fuel is frequent, examine for leaks.

Word (part of speech)	Approved meaning/ ALTERNATIVES	APPROVED EXAMPLE	Not approved example
accuracy (n)	PRECISION (n)	THE PRECISION OF THE ADJUSTMENT CAN CHANGE.	The accuracy of the adjustment can vary.
ACCURATE (adj)	Exact	THE ADJUSTMENT MUST BE ACCURATE.	
ACCURATELY (adv)	Exactly	PUT THE REPAIR SHEET ACCURATELY ON THE DAMAGED AREA.	
achieve (v)	GET (v)	SET THE CONTROL TO GET MAXIMUM THRUST.	Set the control to achieve maximum thrust.
acquire (v)	GET (v)	THE COMPUTER GETS THIS DATA FROM FIVE SENSORS.	The computer acquires this data from five sensors.
acrid (adj)	DANGEROUS (adj)	THIS MATERIAL RELEASES DANGEROUS FUMES WHEN IT TOUCHES HOT SURFACES.	On contact with hot surfaces, this material produces acrid fumes.
ACROSS (prep)	From one side to the other side	SAFETY THE CLAMP BLOCK ACROSS THE CONTROL LEVER FORKS WITH SAFETY WIRE.	
act (v)	Use a specific action verb.	THE HYDRAULIC FLUID FLOW OPENS THE VALVE.	The hydraulic fluid flow acts on the inlet valve.
action (n)	STEP (n)	DO THE STEPS THAT FOLLOW:	Do the following actions:
	PROCEDURE (n)	DO NOT DO THIS PROCEDURE IN THE HANGAR.	This action must not be done in the hangar:

Word (part of speech)	Approved meaning/ ALTERNATIVES	APPROVED EXAMPLE	Not approved example
activate (v)	START (v)	START THE MOTOR.	Activate the motor.
	OPERATE (v)	THE LEVERS OPERATE THE MICROSWITCHES.	Microswitches are activated by levers.
	CONNECT (v)	MAKE SURE THAT THE CIRCUIT IS CONNECTED.	Make sure the circuit is activated.
active (adj)	ON (adj)	WHEN THE SYSTEM IS ON, YOU WILL GET A REPORT.	When the system is active, you will get a report.
	OPERATE (v)	THE AUTOTHRUST SYSTEM IS ON BUT IT DOES NOT OPERATE.	The autothrust system is on but not active.
activity (n)	WORK (n)	THIS WORK IS APPLICABLE TO VERTICAL INSTALLATIONS.	This activity is relevant to vertical installations.
actuate (v)	START (v)	START THE MOTOR.	Actuate the motor.
	OPERATE (v)	OPERATE THE HAND PUMP.	Actuate the hand pump.
	PUSH (v)	PUSH THE PUSHBUTTON SWITCH.	Actuate the pushbutton switch.
actuation (n)	OPERATION (n)	MONITOR THE OPERATION OF THE STEERING MOTOR.	Monitor the actuation of the steering motor.
ADAPT (v), ADAPTS, ADAPTED, ADAPTED	To change or adjust to that which is necessary	ADAPT THE PRESSURE CONNECTION TO THE PITOT HEAD.	
		THE SYSTEM INTERFACE CIRCUITS ADAPT TO THE PHYSICAL PROPERTIES OF THE CONNECTED SYSTEMS.	
ADD (v), ADDS, ADDED, ADDED	To increase the number, dimension, or quantity	ADD 5 ml OF HARDENER TO THE COMPOUND.	

Word (part of speech)	Approved meaning/ ALTERNATIVES	APPROVED EXAMPLE	Not approved example
addition (n)	ADD (v)	TO GET THE CORRECT CLEARANCE, ADD SPECIAL SHIMS, AS NECESSARY.	Adjust the clearance by the addition of special shims, as necessary.
additional (adj)	MORE (adj)	THIS CHAPTER GIVES MORE INFORMATION ABOUT SAFETY.	This chapter gives additional information about safety.
adequate (adj)	SUFFICIENT (adj)	MAKE SURE THAT CONTAINERS HAVE SUFFICIENT CAPACITY AND DIAMETER.	Make sure that containers have adequate capacity and diameter.
adhere (v)	BOND (v)	THE SEAL MUST BOND CORRECTLY.	The seal must adhere correctly.
	OBEY (v)	OBEY THE SAFETY INSTRUCTIONS.	Adhere to the safety instructions.
adhesion (n)	BOND (n)	CLEAN THE SURFACE TO MAKE SURE THAT THE BOND IS SATISFACTORY.	Clean the surface to ensure there will be good adhesion.
ADJACENT (adj)	That which is near to an object, with no other object of the same type between the two	MAKE SURE THAT YOU DO NOT OPERATE THE ADJACENT CONTROL.	
ADJACENT TO (prep)	To be near to an object, with no other object of the same type between the two	THE FUEL PUMP IS INSTALLED ADJACENT TO THE BULKHEAD.	
adjoining (adj)	ADJACENT (adj)	ALIGN THE BRACKETS WITH THE ADJACENT COMPONENTS.	Align the brackets with the adjoining components.
ADJUST (v), ADJUSTS, ADJUSTED, ADJUSTED	To put in or come to a specified position or value	ADJUST THE FREQUENCY TO THE VALUE GIVEN IN TABLE 1.	
		THE INTERNAL LOGIC ADJUSTS TO SUDDEN CHANGES IN TEMPERATURE.	
ADJUSTABLE (adj)	That you can adjust	THE TWO STOP BOLTS ARE ADJUSTABLE.	

Word (part of speech)	Approved meaning/ ALTERNATIVES	APPROVED EXAMPLE	Not approved example
ADJUSTMENT (n)	The effect of adjusting	MAKE SURE THAT THE ADJUSTMENT IS IN THE LIMITS GIVEN IN TABLE 1.	
admit (v)	LET (v)	OPEN THE VALVE TO LET NITROGEN GO INTO THE OLEO STRUT.	Open the valve to admit nitrogen into the oleo strut.
adopt (v)	USE (v)	USE THIS PROCEDURE IF THE UNIT IS DAMAGED.	Adopt this procedure if the unit is damaged.
advance (n)	FORWARD (adj)	THE FORWARD MOVEMENT OF THE CONTROL LEVER MUST BE SLOW AND CONTINUOUS.	The advance of the control lever must be slow and continuous.
advance (v)	SET (v)	SET THE THROTTLE TO MAXIMUM POWER.	Advance the throttle to maximum power.
	FORWARD (adv)	MOVE THE LEVER FORWARD.	Advance the lever.
adverse (adj)	BAD (adj) Give accurate and correct conditions if possible.	REFER TO CHAPTER 6 FOR INSTRUCTIONS ABOUT HOW TO PARK THE AIRCRAFT IN BAD WEATHER CONDITIONS.	For parking aircraft in adverse weather conditions, refer to Chapter 6.
advisable (adj)	RECOMMEND (v)	WE RECOMMEND THAT YOU TORQUE THE BOLTS AGAIN AFTER 50 FLIGHT HOURS.	It is advisable to retorque the bolts after 50 flight hours.
advise (v)	TELL (v)	TELL PERSON B THAT THE BRAKES ARE SET.	Advise Person B that the brakes are set.
	RECOMMEND (v)	WE RECOMMEND THAT YOU REFER TO THE REPAIR MANUAL.	We advise you to refer to the repair manual.
affect (v)	EFFECT (n) Be specific if possible.	MAGNETIC TOOLS HAVE AN UNWANTED EFFECT ON THE COMPASS SYSTEM.	Magnetic tools affect the compass system.
AFT (adj)	At or nearer to the rear of an air or sea vehicle	THE PUMP IS IN THE AFT CELL OF THE FUSELAGE TANK.	

Word (part of speech)	Approved meaning/ ALTERNATIVES	APPROVED EXAMPLE	Not approved example
AFT (adv)	In the direction of the rear of an air or sea vehicle	MOVE THE THROTTLE AFT.	
AFTER (conj)	That follows a specified time, sequence, or operation	DO A FUNCTIONAL TEST AFTER YOU INSTALL THE COMPONENT.	
AFTER (prep)	That follows a specified time, sequence, or operation	THE BAR MOVES DOWN AFTER 20 SECONDS.	
AFT OF (prep)	At a position nearer to the rear	THE CONTROL UNIT IS INSTALLED AFT OF THE FLIGHT COMPARTMENT.	
AGAIN (adv)	One more occurrence	MOVE THE CONTROL STICK BACK TO THE CENTER, AND THEN MOVE IT FORWARD AGAIN.	
AGAINST (prep)	In contact with	PUT THE ADAPTER IN POSITION AGAINST ITS SUPPORT.	
AGENT (n)	One of a group of materials made to do a specified task	DO NOT USE THESE CLEANING AGENTS ON HOT SURFACES.	
aggravate (v)	INCREASE (v)	TIRE WEAR INCREASES WITH HIGH SPEED.	Tire wear is aggravated by high speed.
agitate (v)	SHAKE (v)	SHAKE THE SOLUTION.	Agitate the solution.
agitation (n)	SHAKE (v)	IF YOU SHAKE THE SOLUTION, DO NOT USE IT FOR ONE HOUR.	After agitation, do not use the solution for one hour.
AGREE (v), AGREES, AGREED, AGREED	To be consistent with	THE INDICATIONS MUST AGREE WITH THE VALUES IN THE TABLE.	
ahead (adv)	FORWARD (adv)	WHEN YOU INSTALL THE ANTENNA, THE ALIGNMENT ARROW MUST POINT FORWARD.	When you install the antenna, the alignment arrow must point ahead.

Word (part of speech)	Approved meaning/ ALTERNATIVES	APPROVED EXAMPLE	Not approved example
AID (n)	Help that is given	IF YOU GET THE SOLUTION IN YOUR EYES, GET MEDICAL AID IMMEDIATELY.	
aid (v)	HELP (v)	APPLY SAND TO THE PLANKS TO HELP WITH TIRE TRACTION.	Apply sand to the planks to aid tire traction.
aim (v)	POINT (v)	DO NOT POINT THE HOSE AT PERSONS.	Do not aim the hose at persons.
AIRBORNE (adj)	In, or used in, flight	POSITION 2 OF THE SWITCH GIVES A SIGNAL OF THE AIRCRAFT AIRBORNE CONDITION.	
air-dry (v)	DRY (v)	DRY THE FILTER IN THE AIR.	Air-dry the filter.
AIRFLOW (n)	A flow of air	MAKE SURE THAT THE AIRFLOW IS NOT MORE THAN 10 LITERS/MINUTE.	
alert (v)	TELL (v)	TELL ALL PERSONNEL TO GO AWAY FROM THE AREA.	Alert all personnel to go away from the area.
ALIGN (v), ALIGNS, ALIGNED, ALIGNED	To put or come into a specified position in relation to a line	ALIGN THE FLANGE HOLES WITH THE STUDS.	
		ALL THREE BRACKET HOLES MUST ALIGN WITH THE FUEL TUBE.	
alignment (n)	ALIGN (v)	MAKE SURE THAT THE KEYWAY IS AT THE TOP AND ALIGNS WITH THE PEG.	Ensure that the keyway is at the top and in alignment with the peg.
ALL (adj)	Full quantity	DRAIN ALL THE FLUIDS.	
allocate (v)	GIVE (v)	GIVE THE DIFFERENT MAINTENANCE TASKS TO THE TECHNICIANS.	Allocate the different maintenance tasks to technicians.
allow (v)	LET (v)	DO NOT LET THE CABLE TOUCH THE FLOOR.	Do not allow the cable to touch the floor.

Word (part of speech)	Approved meaning/ ALTERNATIVES	APPROVED EXAMPLE	Not approved example
allowable (adj)	PERMITTED (adj)	THE MAXIMUM PERMITTED DIFFERENCE IS 5 mm.	The maximum allowable difference is 5 mm.
	APPROVED (adj)	THE VALUES MUST NOT BE MORE THAN THE APPROVED DIFFERENCES.	The values must not be more than the allowable differences.
ALMOST (adv)	Near the limit or the value	MAKE SURE THAT THE PRESSURE IN THE CABIN IS ALMOST ZERO.	
ALONG (prep)	In a line parallel with the length or direction of	MOVE THE TENSIOMETER ALONG THE CABLE.	
already (adv)	IN PROGRESS (adv)	THE DATABASE SYNCHRONIZATION IS IN PROGRESS.	The database is already synchronizing.
	NO OTHER	YOU CAN DO THIS PROCEDURE ONLY IF THERE IS NO OTHER REPAIR IN THE REPAIR AREA.	Only do this procedure if there is not already a repair in the repair area.
	Frequently, an alternative for this word is not necessary.	MAKE SURE THAT THE SOFTWARE IS INSTALLED IN THE REPOSITORY.	Make sure that the software is already installed in the repository.
ALSO (adv)	In addition	ALSO DO THE APPLICABLE STEPS OF THIS PROCEDURE.	
alter (v)	CHANGE (v)	IT IS POSSIBLE THAT THE INSTALLATION OF THE CONNECTOR PINS CHANGED THE ROUTING OF THE CABLE LOOM.	It is possible that the installation of the connector pins altered the routing of the cable loom.
alteration (n)	CHANGE (v)	IF YOU CHANGE THIS REPAIR PROCEDURE, GET APPROVAL OF THE DESIGN AUTHORITY.	The alteration to this repair procedure must get approval of the Design Authority.

Word (part of speech)	Approved meaning/ ALTERNATIVES	APPROVED EXAMPLE	Not approved example
alternate (adj)	ALTERNATIVE (adj)	AN ALTERNATIVE REPAIR IS AVAILABLE.	An alternate repair is available.
	EACH SECOND	EACH SECOND SQUARE IS RED.	Alternate squares are red.
alternate (v)	THEN (adv)	DO THE PILOT'S INSTRUMENT TEST AND THEN THE COPILOT'S TEST.	Alternate between pilot's and copilot's instrument test.
alternately (adv)	IN ONE (TN) ... AND THEN THE OTHER	TO MAKE PIN REMOVAL EASIER, MANUALLY TURN THE CANOPY IN ONE DIRECTION AND THEN IN THE OTHER.	To make pin removal easier, manually turn the canopy alternately in the two directions.
ALTERNATIVE (adj)	That gives a choice	IT IS PERMITTED TO USE ALTERNATIVE MATERIALS.	
ALTERNATIVE (n)	One of two or more objects or processes that have the same or almost the same properties or functions	IN AN EMERGENCY YOU CAN USE FUEL F-43 AS AN ALTERNATIVE TO FUELS F-34 AND F-40.	
alternatively (adv)	ALTERNATIVE (n)	DO TEST B AS AN ALTERNATIVE TO TEST A.	Alternatively, you can do test B.
ALTHOUGH (conj)	In spite of the fact that	ALTHOUGH THE PRESSURE DECREASES, THE VALVE MUST STAY CLOSED.	
ALWAYS (adv)	At all times	ALWAYS KEEP THE BRIGHTNESS CONTROL ON THE LCD DISPLAY AT MAXIMUM.	

Word (part of speech)	Approved meaning/ ALTERNATIVES	APPROVED EXAMPLE	Not approved example
amend (v)	CHANGE (v)	IF IT IS NECESSARY TO CHANGE THE REPAIR SCHEME, GET THE APPROVAL OF THE DESIGN AUTHORITY.	If it is necessary to amend the Repair Scheme, get the approval of the Design Authority.
amendment (n)	CHANGE (n)	THE DESIGN AUTHORITY MUST GIVE APPROVAL TO CHANGES THAT YOU MAKE TO THE REPAIR SCHEME.	The Design Authority must approve the amendments that you make to the Repair Scheme.
	CHANGE (v)	IF IT IS NECESSARY TO CHANGE THE REPAIR SCHEME, GET THE APPROVAL OF THE DESIGN AUTHORITY.	If an amendment to the Repair Scheme is necessary, get the approval of the Design Authority.
among (prep)	IN (prep)	IF THERE ARE METAL PARTICLES IN THE SAMPLE SOLIDS, REPLACE THE PUMP.	If there are metal particles among the sample solids, replace the pump.
amount (n)	QUANTITY (n)	MAKE SURE THAT THE QUANTITY OF FUEL IN THE TANK IS NOT MORE THAN 500 LITERS.	Make sure that the amount of fuel in the tank is not more than 500 liters.
AN (art)	Function word: indefinite article	MAKE AN ENTRY IN THE LOGBOOK.	
ANALOG (adj)	Of data given by pointers and dials	THE PRE-MOD UNIT HAS ANALOG INDICATORS.	
analyze (v)	ANALYSIS (TN)	MAKE AN ANALYSIS OF THE OIL.	Analyze the oil sample.
anchor (v)	ATTACH (v)	THE STRAPS ARE ATTACHED TO A FULL-LENGTH RAIL.	The straps are anchored to a full-length rail.

Word (part of speech)	Approved meaning/ ALTERNATIVES	APPROVED EXAMPLE	Not approved example
ancillary (adj)	AUXILIARY (adj)	THE ENGINE AND ALL AUXILIARY EQUIPMENT USE THE SAME LUBRICANT.	The engine and all ancillary equipment use the same lubricant.
	SECONDARY (adj)	DO AN OPERATIONAL CHECK OF THE SECONDARY CONTROLS.	Do an operational check of the ancillary controls.
AND (conj)	Function word used to connect words, phrases, or clauses	MAKE SURE THAT THE FLAPS ARE RETRACTED AND THE AILERON TRIM HANDWHEEL IS IN THE ZERO POSITION.	
ANGULAR (adj)	That has one or more angles	MAKE SURE THAT THE ANGULAR POSITION IS CORRECT.	
annotation (n)	WRITE (v)	WRITE THE TEMPERATURE ON THE ENGINE RECORD CARD.	Make an annotation of the temperature on the Engine Record Card.
annunciate (v)	SHOW (v)	THE DISPLAY SHOWS A FAULT MESSAGE.	The display annunciates a fault message.
another (adj)	ONE (TN) MORE	INSTALL ONE MORE WASHER ON THE LINE.	Install another washer on the line.
	DIFFERENT (adj)	REFER TO DETAIL C ON FIG. 4 FOR A DIFFERENT PROCEDURE TO ATTACH THE CABLE.	Refer to detail C on Fig. 4 for another way of attaching the cable.
	ADD (v)	ADD A WASHER IF NECESSARY.	Use another washer if necessary.
	AGAIN (adv)	REMOVE THE FUEL FROM THE GROUND BEFORE YOU TRY TO START THE ENGINE AGAIN.	Remove the fuel from the ground before you try another start.
anticlockwise (adv)	COUNTERCLOCKWISE (adv)	TURN THE CAP COUNTERCLOCKWISE.	Turn the cap anticlockwise.

Word (part of speech)	Approved meaning/ ALTERNATIVES	APPROVED EXAMPLE	Not approved example
any (adj)	Frequently, an alternative for this word is not necessary.	IF YOU HAVE QUESTIONS ABOUT THIS DOCUMENT, SEND AN E-MAIL TO THE ADDRESS THAT FOLLOWS:	If you have any questions about this document, send an e-mail to the address that follows:
any (pron)	ONE (TN)	REMOVE ONE OF THE FOUR BOLTS.	Remove any of the four bolts.
anytime (adv)	WHEN (conj)	WHEN YOU USE THE EMERGENCY PUMP, RECORD THE TIME OF OPERATION.	Anytime you use the emergency pump, record the time of operation.
APART (adv)	At a distance (not specified) from	MOVE THE CABLE STRANDS APART.	
aperture (n)	OPENING (n)	GET ACCESS TO THE CONTROL UNIT THROUGH THE OPENING IN THE SIDE OF THE TANK.	Get access to the control unit through the aperture in the side of the tank.
apparent (adj)	SEE (v)	IF YOU SEE DISCOLORATION, REPLACE THE PART.	If discoloration is apparent, replace the part.
	FIND (v)	IF YOU FIND A CRACK, DO THE APPLICABLE REPAIR PROCEDURE.	If a crack is apparent, do the applicable repair procedure.
appear (v)	SHOW (v)	A WARNING FLAG SHOWS IN THE WINDOW.	A warning flag appears in the window.
	VIEW (n)	THE LOWER SEAL COMES INTO VIEW WHEN YOU PULL THE LOWER BEARING DOWN.	The lower seal appears when you pull the lower bearing down.
	THINK (v)	IF YOU THINK THAT THE ENGINE IS TOO HOT, TURN THE SWITCH TO "OFF".	If the engine appears to be too hot, turn the switch to off.
	POSSIBLE (adj)	IF IT IS POSSIBLE THAT THE TEMPERATURE VALUES ARE INCORRECT, DECREASE THE POWER.	If it appears that the temperature values are incorrect, decrease the power.

Word (part of speech)	Approved meaning/ ALTERNATIVES	APPROVED EXAMPLE	Not approved example
APPLICABLE (adj)	Correct or satisfactory for the task	REFER TO THE TABLE THAT FOLLOWS FOR THE APPLICABLE TORQUE VALUES.	
application (n)	APPLY (v)	APPLY THE SEALANT WITH A WOODEN SPATULA.	The application of sealant must be carried out with a wooden spatula.
APPLY (v), APPLIES, APPLIED, APPLIED	To put or spread something on	APPLY A FORCE OF 100 N ON THE END OF THE LEVER.	
		APPLY THE SEALING COMPOUND.	
appoint (v)	TELL (v)	TELL TWO PERSONS TO BE PREPARED TO OPERATE THE FIRE EXTINGUISHERS.	Appoint two persons to be prepared to operate the fire extinguishers.
approach (v)	GO NEAR	IF IT IS NECESSARY TO GO NEAR THE ENGINE EXHAUST, USE EAR PROTECTION.	If it is necessary to approach the engine exhaust, use ear protection.
appropriate (adj)	APPLICABLE (adj)	LOCK THE WASHER WITH THE APPLICABLE TABS.	Lock the washer with the appropriate tabs.
APPROVAL (n)	That which permits a person or a thing to do something	GET THE APPROVAL OF THE MANUFACTURER BEFORE YOU REPAIR THIS UNIT.	
approve (v)	APPROVAL (n)	GET APPROVAL FROM THE DESIGN AUTHORITY FOR ALL MODIFICATIONS.	All modifications must be approved by the Design Authority.
APPROVED (adj)	Permitted by an authority	DO THE PROCEDURE IN AN APPROVED AREA.	
APPROXIMATE (adj)	Almost correct or accurate	AN APPROXIMATE VALUE IS SATISFACTORY.	
APPROXIMATELY (adv)	Almost correct or accurate	DRAIN APPROXIMATELY 2 LITERS OF FUEL.	

Word (part of speech)	Approved meaning/ ALTERNATIVES	APPROVED EXAMPLE	Not approved example
AREA (n)	A specified surface or location	DO NOT SMOKE IN THE WORK AREA.	
arise (v)	OCCUR (v)	SHOCK LOADING OF THE ENGINE CAN OCCUR DURING A HEAVY LANDING.	Shock loading of the engine can arise from a heavy landing.
	CAUSE (v)	A HEAVY LANDING CAN CAUSE SHOCK LOADING OF THE ENGINE.	Shock loading of the engine can arise from a heavy landing.
ARM (v), ARMS, ARMED, ARMED	1. To install armaments	ARM THE AIRCRAFT.	
	2. To prepare for automatic operation	SET THE SWITCH TO "CLOSE" TO ARM THE CIRCUIT.	
AROUND (prep)	On all sides of	PUT THE STRAP CLAMPS AROUND THE CYLINDER.	
arrange (v)	PUT (v)	PUT THE COMPONENTS IN THIS SEQUENCE:	Arrange the components in this order.
arrangement (n)	CONFIGURATION (TN)	THE CONFIGURATION OF THE INSERTS IS RELATED TO THE DIMENSIONS OF THE CONNECTOR SHELL.	The arrangement of the inserts is related to the dimensions of the connector shell.
	PREPARE (v)	PREPARE FOR THE LEAK TEST.	Make the arrangements for the leak test.
arrest (v)	STOP (v)	STOP THE FLOW FOR 10 SECONDS.	Arrest the flow for 10 seconds.
	HOLD (v)	A COTTER PIN HOLDS THE GIRT BAR.	A cotter pin arrests the girt bar.
arrive (v)	GET (v)	TO GET THE CORRECT VALUE, SET THE SWITCH TO POSITION 2.	To arrive at the correct value, set the switch to position 2.
ARROW (n)	A mark that has a point to show direction	MAKE SURE THAT THE ARROW POINTS IN THE DIRECTION OF THE FUEL FLOW.	

Word (part of speech)	Approved meaning/ ALTERNATIVES	APPROVED EXAMPLE	Not approved example
article (n)	OBJECT (n)	MAKE SURE THAT THERE ARE NO LOOSE OBJECTS IN THE ENGINE AIR INTAKES.	Ensure that there are no loose articles in the engine air intakes.
AS (prep)	In the manner of, to the same quantity, equally	DO NOT USE THE CYLINDER AS A SUPPORT.	
as (conj)	BECAUSE (conj)	THE PUMP DID NOT OPERATE BECAUSE THE SHAFT WAS BROKEN.	The pump did not operate as the shaft was broken.
	WHILE (conj) You can use:	WHILE YOU DO EACH STEP OF THE VALVE TEST, MAKE SURE THAT THERE IS NO LEAKAGE.	As you do each step of the valve test, make sure that there is no leakage.
	AS FOLLOWS	ASSEMBLE THE VALVE AS FOLLOWS:	
	AS NECESSARY	SET THE LENGTH OF THE CONTROL ROD AS NECESSARY.	
	AS SHOWN	MEASURE THE DEPTH OF THE DENT AS SHOWN IN FIGURE 4.	
AS....AS (conj)	Construction used to show comparison or relation For "as long as", refer to "long (as long as)".	MOVE THE CONTROL AS QUICKLY AS POSSIBLE.	
ascertain (v)	MAKE SURE (v)	MAKE SURE THAT THE LIGHT IS OFF.	Ascertain that the light is off.
ask (v)	TELL (v)	TELL PERSON B TO MAKE SURE THAT THE SWITCH IS OFF.	Ask Person B to make sure that the switch is off.
	SPEAK (v)	SPEAK TO THE LOCAL AIR TRAFFIC CONTROLLER.	Ask the local air traffic controller.
ASSEMBLE (v), ASSEMBLES, ASSEMBLED, ASSEMBLED	To attach, connect, or put together the parts of something	ASSEMBLE THE VALVE AS FOLLOWS:	

Word (part of speech)	Approved meaning/ ALTERNATIVES	APPROVED EXAMPLE	Not approved example
ASSEMBLY (n)	Items that are connected for a specified function	REMOVE THE WHEEL BRAKE ASSEMBLY FROM THE AXLE.	
assess (v)	CALCULATE (v)	CALCULATE THE QUANTITY OF FUEL THAT IS NECESSARY.	Assess the quantity of fuel that is necessary.
	ESTIMATE (n)	MAKE AN ESTIMATE OF THE DAMAGE.	Assess the damage.
assessment (n)	ESTIMATE (n)	MAKE AN ESTIMATE OF THE DAMAGE.	Carry out an assessment of the damage.
	CALCULATE (v)	CALCULATE THE NUMBER OF SHIMS THAT ARE NECESSARY.	Do an assessment of the number of shims that are necessary.
assign (v)	GIVE (v)	GIVE A CODE TO EACH CONNECTOR.	Assign a code to each connector.
assist (v)	HELP (v)	KEEP THE DATA TO HELP YOU WITH SUBSEQUENT PROCEDURES.	Keep the data to assist subsequent procedures.
assistance (n)	AID (n)	YOU MUST HAVE THE AID OF ONE MORE PERSON FOR THE STEP THAT FOLLOWS.	You must have the assistance of one more person for the step that follows.
associated (adj)	CORRECT (adj)	INSTALL THE DUMP VALVE AND THE CORRECT GASKET.	Install the dump valve with the associated gasket.
	RELATED (adj)	REMOVE THE PANELS FROM THE RELATED DOORS.	Remove the panels from the associated doors.
	ITS, THEIR (pron)	REMOVE THE BOLT AND ITS WASHER.	Remove the bolt and the associated washer.
		REMOVE THE EIGHT BOLTS AND THEIR WASHERS.	Remove the eight bolts and the associated washers.
assume (v)	THINK (v)	IF YOU THINK THAT YOU CAN REPAIR THE DAMAGE, REFER TO THE REPAIR SECTION.	If damage is assumed to be repairable, refer to the repair section.

Word (part of speech)	Approved meaning/ ALTERNATIVES	APPROVED EXAMPLE	Not approved example
assure (v)	MAKE SURE (v)	USE FERRULES TO MAKE SURE THAT THE LOCATION IS CORRECT.	Ferrules must be used to assure correct location.
as to (prep)	ABOUT (prep)	REFER TO TABLE 7001 FOR INFORMATION ABOUT HOW THE PIVOT ASSEMBLY IS MADE.	Refer to Table 7001 for information as to the construction of the pivot assembly.
asymmetric (adj)	NOT SYMMETRICAL	IF THE RESULT THAT YOU GET IS NOT SYMMETRICAL, DO A RIGGING TEST.	If you get an asymmetric result, do a rigging test.
AT (prep)	Function word that shows the location, position, direction, or time that something occurs	DISCONNECT THE HOSE AT THE PITOT HEAD.	
		AT APPROXIMATELY 300 KNOTS, THE LIGHT COMES ON.	
		WITH THE NOSEWHEEL AT 30 DEG. TO THE CENTER LINE, SET THE LEVER TO "UP".	
at least (adv)	MINIMUM (adj)	THE MINIMUM TIME BETWEEN CHECKS IS TWO MINUTES.	Allow at least two minutes between checks.
	MINIMUM (n)	DO STEPS 13 THRU 16 A MINIMUM OF THREE TIMES.	Do steps 13 thru 16 at least three times.
ATTACH (v), ATTACHES, ATTACHED, ATTACHED	To stay together or to cause different items to stay together	MAKE SURE THAT THE LABEL FULLY ATTACHES TO THE SURFACE AND THAT THERE ARE NO BUBBLES.	
		TEMPORARILY ATTACH THE CABLE TO THE ADJACENT STRUCTURE.	

Word (part of speech)	Approved meaning/ ALTERNATIVES	APPROVED EXAMPLE	Not approved example
attachment (n)	ATTACH (v)	MAKE SURE THAT THE DUCT IS CORRECTLY ATTACHED TO THE FILTER.	Make sure that the attachment of the duct to the filter is correct.
attain (v)	SHOW (v)	WHEN THE VOLTMETER SHOWS 28 VOLTS DC, THE RELAY CLOSES.	When 28 volts DC is attained, the relay closes.
	BE (v)	WHEN THE VOLTAGE IS 28 VOLTS DC, THE RELAY CLOSES.	When 28 volts DC is attained, the relay closes.
	GET (v)	WHEN YOU GET 28 VOLTS DC, THE RELAY CLOSES.	When 28 volts DC is attained, the relay closes.
attempt (n)	TRY (v)	TRY TO LIFT THE COVER MANUALLY.	Make an attempt to lift the cover manually.
attempt (v)	TRY (v)	TRY TO FIND THE DEFECT.	Attempt to find the defect.
attention (n)	AID (n)	IF YOU GET THE SPRAY IN YOUR EYES, GET MEDICAL AID.	If you get the spray in your eyes, get medical attention.
	CAREFUL (adj)	BE CAREFUL WHEN YOU REMOVE THE CONTROL UNIT.	Pay attention when you remove the control unit.
	MONITOR (v)	MONITOR THE RESULTS.	Pay attention to the results.

Word (part of speech)	Approved meaning/ ALTERNATIVES	APPROVED EXAMPLE	Not approved example
audible (adj)	HEAR (v)	IF YOU CANNOT HEAR THE ALARM, ADJUST THE VOLUME CONTROL.	If the alarm is not audible, adjust the volume control.
	LISTEN (v)	LISTEN FOR AIR FROM THE LINES.	Check the lines for audible leaks of air.
augment (v)	INCREASE (v)	A MIXTURE OF WATER AND METHANOL INCREASES THE POWER OF THE ENGINE.	A mixture of water and methanol augments the power of the engine.
aurally (adv)	LISTEN (v)	LISTEN TO MAKE SURE THAT THE FLUID FLOWS.	Check aurally to ensure fluid flow.
	HEAR (v)	YOU CAN HEAR THE WARNING IN THE COCKPIT.	The warning is given aurally in the cockpit.
authentic (adj)	ACCURATE (adj)	IF THE ADJUSTMENT OF THE STOPS IS NOT CORRECT, THE RESULTS OF THE TEST WILL NOT BE ACCURATE.	If the adjustment of the stops is not correct, the results of the test will not be authentic.
AUTHORITY (n)	An official organization that gives approval to something	GET THE APPROVAL OF THE DESIGN AUTHORITY BEFORE YOU CHANGE THE PROCEDURE.	
authorized (adj)	APPROVED (adj)	ONLY APPROVED PERSONS ARE PERMITTED TO STAY IN THIS AREA.	Only authorized persons are permitted to stay in this area.
AUTOMATIC (adj)	That operates without other input	USE THE AUTOMATIC SYSTEM TO OPEN THE DOORS.	
AUTOMATICALLY (adv)	That operates without other input	THE FUEL FLOW STOPS AUTOMATICALLY.	

Word (part of speech)	Approved meaning/ ALTERNATIVES	APPROVED EXAMPLE	Not approved example
AUXILIARY (adj)	1. That operates in a secondary function	LUBRICANTS FOR THE ENGINE AND THE AUXILIARY EQUIPMENT ARE THE SAME.	
	2. That gives help	USE AUXILIARY SCREWS TO ASSEMBLE THE UNIT TEMPORARILY.	
AVAILABLE (adj)	Prepared or supplied for you to use	IF ELECTRICAL POWER IS AVAILABLE, DO A CONTINUITY CHECK.	
avert (v)	PREVENT (v)	IF THE MICROSWITCH BECOMES UNSERVICEABLE, DO THE DEACTIVATION PROCEDURE OF THE ACTUATOR TO PREVENT DAMAGE TO THE VALVE.	If the microswitch becomes unserviceable, do the deactivation procedure of the actuator to avert damage to the valve.
avoid (v)	PREVENT (v)	TURN THE CONTROLS SLOWLY TO PREVENT DAMAGE.	Turn the controls slowly to avoid damage.
	DO NOT	DO NOT PUSH ON THE SEALS.	Avoid pushing on the seals.
aware (adj)	KNOW (v)	ALL PERSONNEL MUST KNOW THE EFFECT.	All personnel must be aware of the effect.
	TELL (v)	TELL ALL PERSONNEL ABOUT THE EFFECT.	All personnel must be aware of the effect.
AWAY FROM (prep)	In or to a different position or direction	DO THE WORK IN AN AREA AWAY FROM EQUIPMENT OR METAL FRAME BUILDINGS.	
AXIAL (adj)	Related to an axis	DO STEPS 1 THRU 4 TO MAKE SURE THAT NO AXIAL LOAD IS APPLIED ON THE ROD.	
AXIALLY (adv)	Related to an axis	APPLY THE LOAD AXIALLY.	

Word (part of speech)	Approved meaning/ ALTERNATIVES	APPROVED EXAMPLE	Not approved example
back (adj)	REAR (adj)	DISCONNECT THE CABLE FROM THE REAR SIDE OF THE INSTRUMENT.	Disconnect the cable from the back side of the instrument.
BACK (adv)	To an initial condition	MOVE THE ENGINE THROTTLE BACK TO 60% RPM.	
back (n)	REAR (n)	THE CONTROL UNIT IS AT THE REAR.	The control unit is at the back.
back and forth (adv)	DIRECTION (n)	MOVE THE SLEEVE IN THE PISTON TO ONE DIRECTION AND THEN BACK AGAIN.	Move the sleeve back and forth in the piston.
back off (v)	LOOSEN (v)	DO NOT LOOSEN THE NUT TO ALIGN IT.	Do not back off the nut to obtain alignment.
backup (n)	EMERGENCY (n)	IN AN EMERGENCY, THE ONE-SHOT BATTERY SUPPLIES DC POWER.	The one-shot battery is a backup for the DC electrical supply.
	AUXILIARY (adj)	USE AUXILIARY SCREWS TO ASSEMBLE THE UNIT TEMPORARILY.	Use some screws as a backup to assemble the unit temporarily.
backward (adv)	REARWARD (adv)	MOVE THE LEVER REARWARD.	Move the lever backward.
BAD (adj) (WORSE, WORST)	Not satisfactory Do not use this word if a more specific word is available.	REFER TO CHAPTER 6 FOR INSTRUCTIONS ABOUT HOW TO PARK IN BAD WEATHER CONDITIONS.	

Word (part of speech)	Approved meaning/ ALTERNATIVES	APPROVED EXAMPLE	Not approved example
badly (adv)	INCORRECTLY (adv)	IF THE SLEEVE IS INCORRECTLY ATTACHED, ADJUST IT.	If the sleeve is badly attached, adjust it.
	UNSATISFACTORILY (adv)	DISCARD ALL SEALS WHICH ARE UNSATISFACTORILY BONDED.	Discard all seals which are badly bonded.
	DECREASE (v)	FUEL LEAKS CAN DECREASE ENGINE PERFORMANCE.	A fuel leak can cause the engine to operate badly.
BALANCE (v), BALANCES, BALANCED, BALANCED	To make equal	BALANCE THE ELECTRICAL LOADS ON EACH GENERATOR.	
bank (v)	BANK (TN)	THE V-BARS GIVE THE INDICATION FOR A BANK.	The V-BARS indicate the command to bank.
BARE (adj)	Without a surface layer or protection	BARE METAL SURFACES ARE USUALLY SATISFACTORY FOR AN ELECTRICAL BOND.	
bared (adj)	BARE (adj)	GIVE TEMPORARY PROTECTION TO AREAS OF BARE METAL.	Give temporary protection to areas of bared metal.
base (n)	BOTTOM (n)	MAKE SURE THAT THE TWO SPIGOTS AT THE BOTTOM OF THE UNIT ENGAGE.	Make sure that the two spigots at the base of the unit engage.
base (v)	USE (v)	USE THE TORQUE VALUES GIVEN IN CHAPTER 10.	Base the torque values on those given in Chapter 10.
BASIC (adj)	Approved minimum	THIS MODIFICATION CHANGES THE BASIC CONFIGURATION OF THE SYSTEM.	

Word (part of speech)	Approved meaning/ ALTERNATIVES	APPROVED EXAMPLE	Not approved example
BE (v), IS, WAS, (also ARE, WERE) No other forms of this verb.	1. To occur, exist	IF THERE IS CORROSION ON THE PUMP VANES, REPLACE THE PUMP.	
	2. To have a property, to be equal to	ACID SOLUTIONS ARE DANGEROUS.	
bear down (v)	APPLY (v)	APPLY WEIGHT ON THE SURFACE OF THE PATCH TO MAKE SURE THAT THE PARTS BOND CORRECTLY.	Bear down on the surface of the patch to make sure that the parts bond correctly.
		APPLY A LOAD TO THE PANEL UNTIL IT MOVES INTO POSITION WITH A CLICK.	Bear down on the panel until it moves into place with a click.
BECAUSE (conj)	As a result of	BECAUSE THE FLUID LEVEL IS INCORRECT, THE SYSTEM WILL NOT OPERATE.	
BECAUSE OF (prep)	As a result of	IF YOU CANNOT REMOVE A BOLT BECAUSE OF CORROSION, APPLY PENETRATING OIL.	
BECOME (v), BECOMES, BECAME	To come to be	THE SCREEN BECOMES AVAILABLE.	
BEFORE (conj)	That precedes a specified time, sequence, or operation	BLEED THE SYSTEM BEFORE YOU DISCONNECT THE COMPONENTS.	
BEFORE (prep)	That precedes a specified time, sequence, or operation	THE BAR MOVES DOWN BEFORE 20 SECONDS.	
begin (v)	START (v)	INCREASE THE PRESSURE UNTIL THE FLOW STARTS AGAIN.	Increase pressure until the flow begins again.

Word (part of speech)	Approved meaning/ ALTERNATIVES	APPROVED EXAMPLE	Not approved example
beginning (n)	START (n)	YOU CAN GET SLOW MOVEMENT AT THE START.	You can get slow movement at the beginning.
	SOURCE (n)	FIND THE SOURCE OF THE FUEL LEAKAGE.	Find the beginning of the fuel leak.
	START (v)	WHEN THE LEVER STARTS TO MOVE, THE MICROSWITCH OPERATES.	At the beginning of the lever movement, the microswitch will operate.
BEHIND (prep)	In a position at the rear of	THE PUMP IS INSTALLED BEHIND THE HYDRAULIC MOTOR.	
BELOW (prep)	In (or to) a position farther down than something	THE DATE IS WRITTEN BELOW THE CYLINDER NECK.	
	For other meanings, use:		
	LESS THAN	MAKE SURE THAT THE DIAMETER OF THE HOLE IS LESS THAN THE SPECIFIED VALUE.	Make sure that the diameter of the hole is below the specified value.
BEND (n)	The area where something is bent	EXAMINE THE BENDS FOR CRACKS.	
BEND (v), BENDS, BENT, BENT	To change or cause to change from straight to curved	BEND THE PIPE CAREFULLY.	
		THESE PARTS CAN EASILY BEND, BREAK OR BECOME INCORRECTLY ALIGNED.	
beneath (prep)	BELOW (prep)	PUT THE JACK BELOW THE AXLE.	Put the jack beneath the axle.
beside (prep)	ADJACENT TO	THE FUEL PUMP IS ADJACENT TO THE SPAR.	The fuel pump is beside the spar.

Word (part of speech)	Approved meaning/ ALTERNATIVES	APPROVED EXAMPLE	Not approved example
BETWEEN (prep)	Related to something before and after in time or position	MAKE SURE THAT THE STOPWATCH INDICATION IS BETWEEN 2 AND 4 MINUTES.	
beware (v)	CAREFUL (adj)	BE CAREFUL OF DANGEROUS VOLTAGES.	Beware of dangerous voltages.
beyond (prep)	MORE THAN	REPLACE ALL COMPONENTS THAT ARE WORN MORE THAN THE MAXIMUM LIMITS.	Replace all components that are worn beyond their maximum limits.
big (adj)	LARGE (adj)	INSTALL THE LARGER BOLTS.	Install the bigger bolts.
bind (v)	FREELY (adv)	MAKE SURE THAT THE VALVES MOVE FREELY.	Check that valves do not bind.
	WIND AROUND	WIND TWO LAYERS OF TAPE AROUND THE PIPE JOINTS.	Bind the pipe joints with two layers of tape.
blank (v)	CAP (TN)	PUT A CAP ON THE CONNECTOR.	Blank the connector.
	COVER (TN)	PUT A COVER ON THE OUTLET PORT	Blank the outlet port.
blank off (v)	SEAL (v)	SEAL THE HOSES.	Blank off the hoses.
blank out (v)	CANCEL (v)	CANCEL THE DISPLAY.	Blank out the display.
BLEED (v), BLEEDS, BLED, BLED	To let a gas out of	BLEED THE SPEEDBRAKE HYDRAULIC SYSTEM.	
blend (v)	MIX (v)	MIX THE ADHESIVE WITH THE HARDENER.	Blend the adhesive with the hardener.
block (v)	CAUSE A BLOCKAGE	MAKE SURE THAT THE SEALANT DOES NOT CAUSE A BLOCKAGE OF THE DRAIN.	Check that the sealant does not block the drain.

Word (part of speech)	Approved meaning/ ALTERNATIVES	APPROVED EXAMPLE	Not approved example
BLOCKAGE (n)	That which prevents correct operation	MAKE SURE THAT THERE IS NO BLOCKAGE OF THE PITOT PIPE.	
BLOCKED (adj)	That prevents correct operation	A BLOCKED OIL JET CAN CAUSE INCORRECT OPERATION.	
blot (v)	DRY (v)	DRY THE SURFACE WITH A TOWEL.	Blot the surface with a towel.
BLOW (v), BLOWS, BLEW, BLOWN	To cause gas to move	BLOW DRY AIR THROUGH THE FEED AND SPRAY TUBES.	
BLUNT (adj)	Not sharp	REMOVE THE SEALANT WITH A BLUNT SCRAPER.	
blunted (adj)	BLUNT (adj)	EXAMINE THE STATIC DISCHARGERS FOR BENT AND BLUNT PINS.	Inspect the static dischargers for bent and blunted pins.
bolt (v)	ATTACH (v) (WITH A BOLT [TN] OR BOLTS [TN])	ATTACH THE TRACK TO THE CHANNELS WITH THE BOLTS.	Bolt the track to the channels.
BOND (n)	1. A connection between structural components for electrical continuity	DO A RESISTANCE CHECK OF THE PRIMARY BONDS.	
	2. A firm connection with another material	MAKE SURE THAT THE BOND BETWEEN THE HONEYCOMB CORE AND THE SKIN IS CORRECT.	

Word (part of speech)	Approved meaning/ ALTERNATIVES	APPROVED EXAMPLE	Not approved example
BOND (v), BONDS, BONDED, BONDED	1. To make an electrical bond	THE STATIC DISCHARGER IS ELECTRICALLY BONDED TO THE FRAME.	
	2. To attach firmly or become firmly attached with another material	IF THE FILLER BONDS TO THE TOOL, MAKE THE TOOL MOIST WITH COLD WATER.	
		WHEN YOU BOND THE MATERIALS, APPLY A THIRD LAYER OF ADHESIVE.	
both (adj)	THE TWO (TN)	ATTACH THE TWO ENDS OF THE HOSE.	Attach both ends of the hose.
BOTTOM (adj)	That is lowermost	REMOVE THE BOTTOM BOLT.	
BOTTOM (n)	The lowermost position	THE DRAIN PLUG IS AT THE BOTTOM OF THE UNIT.	
bottom (v)	TOUCH (v)	OPEN THE DOOR UNTIL IT TOUCHES THE FLOOR.	Open the door until it bottoms on the floor.
bounds (n)	LIMIT (n)	MAKE SURE THAT THE CLEARANCE IS IN THE LIMITS.	Make sure that the clearance is in bounds.
brace (v)	HOLD (v)	SUPPORT STRUTS HOLD THE FLOOR STRUCTURE IN POSITION.	Support struts brace the floor structure.
	STRONG (adj)	REINFORCEMENT PLATES MAKE THE LAP JOINTS STRONGER.	Reinforcement plates brace the lap joints.
branch (v)	DIVIDE (v)	THE PRESSURE LINES DIVIDE IN DIFFERENT DIRECTIONS.	The pressure lines branch in different directions.
breadth (n)	WIDTH (n)	THE WIDTH OF THE MATERIAL MUST BE LARGER THAN 10 mm.	The breadth of the material must be larger than 10 mm.

Word (part of speech)	Approved meaning/ ALTERNATIVES	APPROVED EXAMPLE	Not approved example
break (n)	STOP (v)	IF THE TRANSMISSION STOPS, CANCEL THE TEST.	If there is a break in the transmission, cancel the test.
BREAK (v), BREAKS, BROKE, BROKEN	To cause to separate or become separated into parts by force	BREAK THE BOND OF THE SEALANT.	
		IF THE SHEAR PIN BREAKS, DO AN INSPECTION OF THE TOWBAR.	
breakdown (n)	STOP (v)	IF THE PRESSURE SUPPLY STOPS, CANCEL THE TEST.	If there is a breakdown in the pressure supply, cancel the test.
BREATHE (v), BREATHES, BREATHED, BREATHED	To get gas into or out of the lungs	DO NOT BREATHE THE SOLVENT FUMES.	
		BREATHE INTO THE OXYGEN MASK.	
BRIGHT (adj) (BRIGHTER, BRIGHTEST)	That gives much light	THE NAVIGATION LIGHTS CHANGE FROM DIM TO BRIGHT.	
BRIGHTLY (adv)	In a bright manner	THE LIGHTS COME ON BRIGHTLY.	
brightness (n)	INTENSITY (n)	OPERATE THE DIMMER SWITCH AND MAKE SURE THAT THE INTENSITY OF THE LIGHT CHANGES.	Operate the dimmer switch and check that the brightness of the light changes.
	BRIGHT (adj)	TURN THE ADJUSTER CLOCKWISE TO MAKE THE DISPLAY BRIGHTER.	The display brightness is increased by turning the adjuster clockwise.
brilliance (n)	INTENSITY (n)	SET THE INTENSITY TO THE MINIMUM.	Set the brilliance to the minimum.
	BRIGHT (adj)	TURN THE BRT CONTROL CLOCKWISE TO MAKE THE DISPLAY BRIGHTER.	To increase the display brilliance, turn the BRT CONTROL clockwise.

Word (part of speech)	Approved meaning/ ALTERNATIVES	APPROVED EXAMPLE	Not approved example
bring (v)	MOVE (v)	MOVE THE PANEL INTO POSITION.	Bring the panel into place.
	PUT (v)	PUT THE PANEL INTO POSITION.	Bring the panel into place.
briskly (adv)	QUICKLY (adv)	RUB THE SURFACE QUICKLY WITH A SOFT, DRY CLOTH.	Rub the surface briskly with a soft, dry cloth.
broadcast (v)	TRANSMIT (v)	TRANSMIT THE DATA ON THE TWO CHANNELS.	Broadcast the data on both channels.
brush (v)	BRUSH (TN)	USE A BRUSH TO CLEAN THE FILTER ELEMENT.	Brush the filter element to clean it.
		CLEAN THE FILTER ELEMENT WITH A BRUSH.	Brush the filter element to clean it.
brush on (v)	BRUSH (TN)	APPLY THE PAINT WITH A BRUSH.	Brush the paint on.
BUBBLE (n)	A small quantity of gas in a liquid or a solid	ADJUST THE CLINOMETER UNTIL THE BUBBLE IS IN THE CENTER.	
build (n)	STRUCTURE (n)	THE WING STRUCTURE GIVES MAXIMUM FUEL CAPACITY.	The build of the wing allows maximum fuel storage.
build (v)	ASSEMBLE (v)	ASSEMBLE THE UNIT.	Build the unit.
bump (v)	HIT (v)	THE FUELING HOSE MUST NOT HIT THE EDGE OF THE TANK.	The fueling hose must not bump the edge of the tank.
BURN (v), BURNS, BURNED, BURNED	1. To cause or undergo combustion	THE TURBINE BURNS FUEL.	
		THE SEALANT BURNS AT 109 DEG. F.	
	2. To cause injury or damage through heat	HOT SURFACES CAN BURN YOU.	

Word (part of speech)	Approved meaning/ ALTERNATIVES	APPROVED EXAMPLE	Not approved example
burst (v)	BREAK (v)	IF THE DISK BREAKS, STOP THE SUPPLY.	If the disk bursts, switch off the supply.
BUT (conj)	On the contrary	TIGHTEN THE NUTS BUT DO NOT TORQUE THEM AT THIS TIME.	
butt (v)	TOUCH (v)	THE WIRE MUST TOUCH THE REAR OF THE CONTACT.	The wire must butt against the rear of the contact.
BY (prep)	Mathematical function word	DIVIDE THE VALUE BY TWO.	
		INCREASE THE TEMPERATURE BY 10 DEGREES.	
	Use as a function word to show agency when a passive construction is necessary in descriptive writing.		
by means of (prep)	WITH (prep)	PRESSURIZE THE SYSTEM TO 150 BAR WITH THE HAND PUMP.	By means of the hand pump, pressurize the system to 150 bar.
BYPASS (v), BYPASSES, BYPASSED, BYPASSED	To change a circuit so that it goes past something instead of through it	IN THIS CONFIGURATION, THE HYDRAULIC PRESSURE BYPASSES THE VALVE.	

Word (part of speech)	Approved meaning/ ALTERNATIVES	APPROVED EXAMPLE	Not approved example
CALCULATE (v), CALCULATES, CALCULATED, CALCULATED	To find a result by mathematics	CALCULATE THE PERCENTAGE WITH THE APPLICABLE FORMULA.	
calculation (n)	CALCULATE (v)	IN THIS EXAMPLE, WE CALCULATED THE DATA APPLICABLE ONLY TO A TYPE B UNIT.	The data that we got from the calculations in this example apply only to a Type B unit.
CALIBRATE (v), CALIBRATES, CALIBRATED, CALIBRATED	To measure and adjust the precision of something	CALIBRATE THE THERMOMETER TO THE CELSIUS SCALE.	
CALIBRATION (n)	The procedure that calibrates	WRITE THE RESULTS OF THE CALIBRATION ON THE CORRECTION CHART.	
call (v)	TELL (v)	TELL PERSON A TO REMOVE THE CHOCKS.	Call Person A to remove the chocks.
CAN (v), CAN, COULD	Helping verb that means to be possible, to be able to, or to be permitted to	A MIXTURE OF FUEL AND OXYGEN CAN CAUSE AN EXPLOSION.	
		YOU CAN CLEAN THE DRAIN HOLES WITH THE CLEANING TOOL.	
		YOU CAN OPERATE THE VEHICLE AFTER THE INSPECTION IS COMPLETED.	
CANCEL (v), CANCELS, CANCELED, CANCELED	1. To do the necessary steps to stop a process	IF YOU ACCIDENTALLY PUSH THE SWITCH, CANCEL THE START SEQUENCE.	
	2. To remove the validity of something	WHEN A NEW REVISION IS AVAILABLE, THE STATUS OF THE LAST REVISION CHANGES TO “CANCELED”.	

Word (part of speech)	Approved meaning/ ALTERNATIVES	APPROVED EXAMPLE	Not approved example
CANNOT (v)	Helping verb that means to not be able to	THE WHEEL CANNOT MOVE WHEN THE LOCK IS ENGAGED.	
cap (v)	SEAL (v)	SEAL ALL THE HOSES.	Cap all the hoses.
	CAP (TN)	PUT CAPS ON THE CONNECTORS.	Cap the connectors.
	COVER (TN)	PUT A COVER ON THE INLET PORT.	Cap the inlet port.
capability (n)	FUNCTION (n)	THESE FUNCTIONS OPEN THE APPLICABLE GRAPHIC DEVICE.	These capabilities open the applicable graphic device.
	CAN (v)	THE COMPUTER CAN CALCULATE THE RESULT AUTOMATICALLY.	The computer has the capability to calculate the result automatically.
capable (adj)	APPROVED (adj)	AN APPROVED PERSON MUST DO THE IMPORTANT CHECKS.	Vital checks are to be carried out by a capable person.
	CAN (v)	THE POWER UNIT CAN SUPPLY 28 VDC.	The power unit is capable of producing 28 VDC.
CAPACITY (n)	The maximum quantity that something can hold or make	THE CAPACITY OF THE TANK IS 564 LITERS.	
care (n)	PRECAUTION (n)	OBEY THE SAFETY PRECAUTIONS WHEN YOU DO WORK WITH HIGH VOLTAGES.	You must take care when you work with high voltages.
	CAREFUL (adj)	BE CAREFUL WHEN YOU INSTALL THE NUTS.	Take care when installing the nuts.
	CAREFULLY (adv)	REMOVE THE GEARBOX CAREFULLY.	Remove the gearbox with great care.
CAREFUL (adj)	With precaution	BE CAREFUL WHEN YOU REMOVE THE COVER.	
CAREFULLY (adv)	In a careful manner	CAREFULLY APPLY A BEAD OF SEALING COMPOUND.	

Word (part of speech)	Approved meaning/ ALTERNATIVES	APPROVED EXAMPLE	Not approved example
carry (v)	TRANSMIT (v)	THE CABLE TRANSMITS THE ELECTRICAL ENERGY FROM THE IGNITION EXCITER TO THE IGNITER.	The cable carries the electrical energy from the ignition exciter to the igniter.
	TRANSPORT (TN)	THIS VEHICLE IS FOR THE TRANSPORT OF 5 PASSENGERS.	This vehicle is meant to carry 5 passengers.
carry out (v)	DO (v)	DO THE LEAK TEST.	Carry out the leak test.
	You can use a more specific command verb.		
	MEASURE (v)	MEASURE THE CLEARANCE OF THE IMPELLER.	Carry out the clearance measurement of the impeller.
case (n)	CONDITION (n)	FIGURE 1 SHOWS THE DATA FOR ALL OPERATIONAL CONDITIONS.	The data for all operational cases are in shown in Figure 1.
case (in case of) (conj)	IF (conj)	IF THERE IS A FIRE, OPERATE THE CRASH SWITCHES.	In case of a fire, operate the crash switches.
CATCH (v), CATCHES, CAUGHT, CAUGHT	To stop or prevent the movement of something	THE LUG ON THE PANEL OPENING CATCHES THE BOTTOM OF THE DRAWER.	
		DO NOT LET THE TOOL CATCH ON THE PANEL.	
	For other meanings, use:		
	COLLECT (v)	COLLECT THE LEAKAGE IN A GRADUATED BEAKER.	Catch the leakage in a graduated beaker.
categorize (v)	CATEGORY (TN)	THE TESTING EQUIPMENT DIVIDES THE FAULTS INTO CATEGORIES.	The testing equipment categorizes the faults.

Word (part of speech)	Approved meaning/ ALTERNATIVES	APPROVED EXAMPLE	Not approved example
CAUSE (n)	Something that brings about a result	DO THE TROUBLESHOOTING PROCEDURE TO FIND THE CAUSE OF THE MALFUNCTION.	
CAUSE (v), CAUSES, CAUSED, CAUSED	To be the cause of	METAL OBJECTS CAN CAUSE MAGNETIC INTERFERENCE.	
caution (v)	TELL (v)	TELL ALL PERSONNEL NOT TO SMOKE NEAR FUEL TANKS.	Caution all personnel not to smoke near fuel tanks.
cautiously (adv)	CAREFULLY (adv)	CAREFULLY FEEL THE HEATER WITH YOUR HANDS TO MAKE SURE THAT IT OPERATES.	Cautiously feel the heater with your hands to make sure that it operates.
cavity (n)	HOLE (n)	FILL THE HOLE WITH RESIN.	Fill the cavity with resin.
	SPACE (n)	FILL THE SPACE BETWEEN THE PATCH AND THE ADJACENT STRUCTURE WITH SEALANT.	Fill the cavity between the patch and the adjacent structure with sealant.
cease (v)	STOP (v)	THE WARNING STOPS AT 17 DEG.	The warning ceases at 17 deg.
center (v)	CENTER (TN)	ALIGN THE PISTON WITH THE CENTER OF THE SLEEVE.	Center the piston in the sleeve.
centralize (v)	CENTER (TN)	SET THE CONTROLS TO THE CENTER POSITION.	Centralize the controls.
	NEUTRAL (TN)	SET THE CONTROLS TO THE NEUTRAL POSITION.	Centralize the controls.
certain (adj)	SURE (adj)	IF YOU ARE NOT SURE THAT THE RESULTS ARE CORRECT, DO THE TEST AGAIN.	If you are not certain that the results are correct, do the test again.
	SOME (adj)	SOME FUNCTIONS ARE NOT AVAILABLE.	Certain functions are not available.
	SPECIFIED (adj)	THE SPECIFIED FUNCTIONS ARE NOT AVAILABLE.	Certain functions are not available.

Word (part of speech)	Approved meaning/ ALTERNATIVES	APPROVED EXAMPLE	Not approved example
certify (v)	WRITE (v)	WRITE THE CHANGE IN THE LOGBOOK.	Certify the change in the logbook.
	APPROVAL (n)	GET APPROVAL FROM YOUR SUPERVISOR FOR THE REPAIR.	Get the supervisor to certify the repair.
chafe (v)	RUB (v)	DO NOT LET THE WIRES RUB.	Do not let the wires chafe.
chafed (adj)	WORN (adj)	EXAMINE THE TUBING FOR WORN AREAS.	Examine the tubing for chafed areas.
chance (by chance) (n)	RISK (n)	IF THERE IS A RISK OF LEAKAGE, PUT A CONTAINER BELOW THE UNIT.	If there is a chance of leakage, put a container below the unit.
	ACCIDENTALLY (adv)	IF THE LEVER MOVED ACCIDENTALLY, DO THE TEST AGAIN.	If, by chance, the lever moved, do the test again.
CHANGE (n)	That which occurs when something changes	THE COLOR CHANGE SHOWS THAT THE TEMPERATURE IS TOO HIGH.	
CHANGE (v), CHANGES, CHANGED, CHANGED	To become or to cause to become different	IF THE HUMIDITY CHANGES FREQUENTLY, PUT A COVER ON THE UNIT.	
	For other meanings, use:	CHANGE THE COLOR OF TI IE DISPLAY.	
	REPLACE (v)	REPLACE THE DAMAGED VALVE.	Change the damaged valve.
channel (v)	CHANNEL (TN)	INSTALL THE WIRES IN THE CHANNEL ALONG THE HOUSING.	Channel the wires along the housing.
characteristic (n)	PROPERTY (n)	THE PROPERTIES OF THESE SEALANTS PREVENT CORROSION.	The characteristics of these sealants prevent corrosion.
	QUALITY (n)	DO NOT USE THIS MATERIAL, BECAUSE IT DOES NOT HAVE THE NECESSARY QUALITIES.	Do not use this material, because it doesn't have the necessary characteristics.

Word (part of speech)	Approved meaning/ ALTERNATIVES	APPROVED EXAMPLE	Not approved example
CHARGE (v), CHARGES, CHARGED, CHARGED	To accumulate or add electrical energy	CHARGE THE BATTERY.	
		MAKE SURE THAT THE BATTERY CHARGES.	
	For other meanings, use:		
	FILL (v)	FILL THE TANK WITH 10 LITERS OF METHANOL.	Charge the tank with 10 liters of methanol.
	PRESSURIZE (v)	PRESSURIZE THE ACCUMULATOR WITH NITROGEN.	Charge the accumulator with nitrogen.
CHECK (n)	The procedure you do to make sure that something operates correctly or has no defects	DO A CHECK OF THE HYDRAULIC SYSTEM.	
		DO A CHECK FOR LOOSE FASTENERS.	
check (v)	MAKE SURE (v)	MAKE SURE THAT IT IS SAFE TO SUPPLY ELECTRICAL POWER.	Check that it is safe to supply electrical power.
	MEASURE (v)	MEASURE THE DISTANCE BETWEEN THE FACES.	Check the distance between the faces.
	EXAMINE (v)	EXAMINE THE CASTING FOR CORROSION.	Check the casting for corrosion.
	CHECK (n)	DO A LEAKAGE CHECK OF THE VALVE.	Check the valve for leakage.
CHEMICAL (adj)	Related to a chemical	REMOVE THE CORROSION WITH THE CHEMICAL COMPOUND THAT IS SPECIFIED IN TABLE 6001.	
CHEMICALLY (adv)	Related to a chemical	REMOVE CORROSION CHEMICALLY.	

Word (part of speech)	Approved meaning/ ALTERNATIVES	APPROVED EXAMPLE	Not approved example
chip (n)	PARTICLE (n)	EXAMINE THE FILTER ELEMENT FOR METAL PARTICLES.	Examine the filter element for metal chips.
chip (v)	DAMAGED (adj)	IF THE ENAMEL IS DAMAGED, REPLACE THE UNIT.	If the enamel is chipped, replace the unit.
chock (v)	CHOCK (TN)	PUT THE CHOCKS AGAINST THE WHEELS.	Chock the wheels.
choice (n)	SELECTION (n)	MAKE A SELECTION FROM THE AVAILABLE SHIMS TO GET THE CORRECT THICKNESS.	Make a choice from the available shims to get the correct thickness.
	ALTERNATIVE (adj)	TABLE 1 GIVES ALTERNATIVE ADHESIVES WHICH YOU CAN USE.	In Table 1, there is a choice of adhesives which you are allowed to use.
choose (v)	SELECT (v)	SELECT THE HYDRAULIC SYSTEM THAT YOU WILL PRESSURIZE.	Choose the hydraulic system that you will pressurize.
circa (prep)	APPROXIMATELY (adv)	PUT A CONTAINER (APPROXIMATELY 5 LITERS) BELOW THE WASTE WATER OUTLET.	Put a container (circa 5 liters) under the waste water outlet.
circle (v)	AROUND (prep)	THE POINTER MOVES AROUND THE DIAL.	The pointer circles the dial.
CIRCULAR (adj)	That has the shape of a circle	POLISH THE WINDOW WITH A CIRCULAR MOVEMENT.	
circulate (v)	SUPPLY (v)	SUPPLY AIR AT 24 °C TO THE AREA.	Circulate air at 24 °C in the area.
clamp (v)	ATTACH (v) (WITH A CLAMP [TN] OR CLAMPS [TN])	ATTACH THE ADJACENT CABLES WITH CLAMPS.	Clamp the adjoining cables together.
	PUT (v)	PUT CLAMPS ON THE CABLE TO HOLD IT IN POSITION.	Clamp the cable in position.

Word (part of speech)	Approved meaning/ ALTERNATIVES	APPROVED EXAMPLE	Not approved example
classification (n)	CLASS (TN)	THERE ARE FOUR GENERAL CLASSES OF FASTENERS.	Fasteners can be separated into four general classifications.
	CATEGORY (TN)	YOU CAN DIVIDE THE FASTENERS INTO FOUR GENERAL CATEGORIES.	Fasteners can be separated into four general classifications.
CLEAN (adj)	That does not include dirt or unwanted material	MAKE SURE THAT THE AREA IS CLEAN.	
CLEAN (v), CLEANS, CLEANED, CLEANED	To remove dirt or unwanted materials	CLEAN THE GASKET CONTACT SURFACE.	
cleanliness (n)	CLEAN (adj)	CONDITIONS FOR A SATISFACTORY BOND ARE: - A CLEAN SURFACE - A CORRECT TEMPERATURE - A CORRECT PRESSURE.	A satisfactory bond depends on the cleanliness of the surface, the temperature, and the pressure.
CLEAR (adj)	Without blockage or interference	MAKE SURE THAT THE PITOT AIR INLET IS CLEAR.	
	For other meanings, use:		
	TRANSPARENT (adj)	PUT THE UNIT IN A TRANSPARENT PLASTIC BAG.	Put the unit in a clear plastic bag.
clear (v)	CLEAN (v)	CLEAN THE DRAIN HOLE.	Clear the drain hole.
	You can also use a different construction.	LIFT THE SEAT UNTIL IT IS AWAY FROM THE TRACK LOCKS.	Lift the seat so that it clears the track locks.
CLEARANCE (n)	The space between two objects that must be clear to make sure that they do not touch	MAKE SURE THAT THERE IS CLEARANCE BETWEEN THE LEVER AND THE ROLLER.	
CLEARLY (adv)	In a clear manner	MAKE SURE THAT YOU HEAR THE SIGNALS CLEARLY AT ALL STATIONS.	

Word (part of speech)	Approved meaning/ ALTERNATIVES	APPROVED EXAMPLE	Not approved example
CLICK (n)	A short, sharp sound	TURN THE ADJUSTER FIVE CLICKS CLOCKWISE.	
click (v)	CLICK (n)	WHEN THE SOLENOID MAKES A CLICK, THE VALVE IS OPEN.	When the solenoid clicks, the valve is open.
	You can use this word as a technical verb for computer processes and applications.	CLICK "NEXT" TO CONTINUE.	
clip (v)	CUT (v)	CUT THE WIRE TO THE CORRECT LENGTH.	Clip the wire to the correct length.
	ATTACH (v) (WITH A CLIP [TN] OR CLIPS [TN])	ATTACH THE CONDUIT TO THE STRUCTURE WITH CLIPS.	The conduit should be clipped to the structure.
CLOCKWISE (adv)	In the direction in which the hands of a clock turn when seen from the front	TURN THE NOZZLE 40 DEG. CLOCKWISE.	
CLOGGED (adj)	Blocked with unwanted material.	IF THE FILTER IS CLOGGED, REPLACE IT.	
close (adj)	NEAR (prep)	DO NOT USE METHYL ETHYL KETONE NEAR FLAMES OR SPARKS.	Do not use methyl ethyl ketone in close contact with flames or sparks.
CLOSE (v), CLOSES, CLOSED, CLOSED	1. To move together, or to move to a position that stops or prevents materials from going in or out	CLOSE THE INSTRUMENT PANEL.	
		MAKE SURE THAT THE DOORS CLOSE FULLY.	
	2. To operate a circuit breaker to make an electrical circuit	MAKE SURE THAT THE CIRCUIT BREAKER IS CLOSED.	
		MAKE SURE THAT THE CIRCUIT BREAKER CLOSES.	

Word (part of speech)	Approved meaning/ ALTERNATIVES	APPROVED EXAMPLE	Not approved example
closely (adv)	CAREFULLY (adv)	CAREFULLY MONITOR THE OPERATION OF THE SYSTEM.	Closely monitor the operation of the system.
close to (adv)	NEAR (prep)	PUT THE CLINOMETER NEAR THE FUSELAGE CENTERLINE.	Put the clinometer close to the fuselage centerline.
coat (n)	LAYER (n)	APPLY A LAYER OF SEALANT PR 1440 B2 ON THE SURFACE.	Spread a coat of sealant PR 1440 B2 on the surface.
coat (v)	LAYER (n)	APPLY A LAYER OF GREASE TO THE BATTERY CABLES.	Coat the battery cables with grease.
coating (n)	LAYER (n)	THE METAL HAS A LAYER OF PLASTIC ON IT TO PREVENT CORROSION.	The metal has a coating of plastic on it to prevent corrosion.
cock (v)	SET (v)	SET THE RACK HOOKS.	Cock the rack hooks.
CODE (n)	A sequence of symbols, letters, and/or numbers used for identification	EACH ELECTRICAL SYSTEM HAS A CODE TO IDENTIFY IT.	
code (v)	IDENTIFY (v)	IDENTIFY THE CABLES WITH A CODE.	Code the cables.
	CODE (n)	PUT A CODE ON THE CABLES.	Code the cables.
COIL (n)	A sequence of loops	MAKE SURE THAT THE COILS OF THE OXYGEN HOSE DO NOT RUB TOGETHER.	
coil (v)	WIND (v)	WIND THE CABLES CAREFULLY.	Coil the cables carefully.
coincide (v)	ALIGN (v)	THE FASTENERS MUST ALIGN WITH THE HOLES IN THE TRACK.	The fasteners must coincide with the holes in the track.
coincident (adj)	SYNCHRONIZED (adj)	MAKE SURE THAT THE MOVEMENT OF THE TWO SURFACES IS SYNCHRONIZED.	Make sure that the movement of the two surfaces is coincident.
COLD (adj) (COLDER, COLDEST)	At low temperature	MAKE SURE THAT COLD AIR COMES OUT OF THE OUTLETS.	

Word (part of speech)	Approved meaning/ ALTERNATIVES	APPROVED EXAMPLE	Not approved example
collapse (v)	CLOSE (v)	CLOSE THE BOX.	Collapse the box.
	FALL (v)	THE BOARDING BRIDGE FELL.	The boarding bridge collapsed.
COLLECT (v), COLLECTS, COLLECTED, COLLECTED	To come, or cause to come, together in one location	FLUID COLLECTS IN THE BOTTOM OF THE COWLING.	
		COLLECT ALL LOOSE ITEMS AND PUT THEM IN THE CONTAINER.	
COLOR (n)	A property of light	THE COLOR CHANGES FROM BLUE TO RED.	
color code (v)	IDENTIFY (v)	IDENTIFY THE ASSEMBLY WITH THE CORRECT COLOR.	The assembly must be color coded.
	CODE (n)	PUT A COLOR CODE ON EACH OF THE ASSEMBLIES.	Color code the assemblies.
colored (adj)	COLOR (n)	THE SCALE HAS TWO AREAS OF DIFFERENT COLOR.	The scale is divided into two colored areas.
combine (v)	MIX (v)	MIX THE WATER AND THE GLYCOL.	Combine water and glycol.
	PUT TOGETHER	DO NOT PUT DIFFERENT HPT SEGMENTS TOGETHER.	Do not combine different HPT segments.
combustible (adj)	FLAMMABLE (adj)	KEEP FLAMMABLE MATERIALS AT A MINIMUM DISTANCE OF 50 FEET FROM THE WORK AREA.	Keep combustible materials at a minimum distance of 50 feet from the work area.
COME (v), COMES, CAME	To move to your location	FUMES MUST NOT COME INTO THE CABIN.	
COME ON (v), COMES ON, CAME ON	To become bright with light when an internal power source is energized	THE LIGHTS MUST COME ON.	

Word (part of speech)	Approved meaning/ ALTERNATIVES	APPROVED EXAMPLE	Not approved example
commence (v)	START (v)	THE WARNING STARTS AND THE INDICATORS FLASH.	The warning commences and the indicators flash.
commencement (n)	START (v)	AFTER THE TEST STARTS, MONITOR THE RESULTS ON THE SCREEN.	After the test commencement, monitor the results on the screen.
common (adj)	SAME (adv)	THE REMOVAL PROCEDURE IS THE SAME FOR THE TWO ITEMS.	The removal procedure is common to both items.
communicate (v)	TELL (v)	TELL PERSON B TO SET THE SWITCH TO "OFF".	Communicate with Person B to set the switch to OFF.
	SPEAK (v)	USE A TELEPHONE TO SPEAK WITH PERSON B.	Use a telephone to communicate with Person B.
communication (n)	SPEAK (v)	MAKE SURE THAT PERSON A CAN SPEAK TO PERSON C.	Make sure that there is communication between Person A and Person C.
	HEAR (v)	MAKE SURE THAT AIR TRAFFIC CONTROL CAN HEAR YOU.	Make sure that you are in communication with ATC.
COMPARE (v), COMPARES, COMPARED, COMPARED	To examine for differences	COMPARE THE TWO VALUES.	
comparison (n)	COMPARE (v)	COMPARE THE TWO INDICATIONS.	Make a comparison of the two readings.
COMPATIBLE (adj)	That can operate together satisfactorily or be used together safely	THIS SOFTWARE IS COMPATIBLE ONLY WITH THE SPECIFIED OPERATING SYSTEM.	
		THE COOLANT IS COMPATIBLE WITH ALL AEROSPACE ALLOYS.	

Word (part of speech)	Approved meaning/ ALTERNATIVES	APPROVED EXAMPLE	Not approved example
compile (v)	MAKE A LIST	MAKE A LIST OF THE NECESSARY TOOLS.	Compile a list of the required tools.
	RECORD (v)	RECORD THE AILERON MOVEMENTS ON FORM B.	Compile the aileron movements on Form B.
	COLLECT (v)	COLLECT ALL LOG RECORDS FOR THE COMPONENT.	Compile all log records for the component.
complete (adj)	FULL (adj)	EXTEND AND RETRACT THE ACTUATOR THROUGH THREE FULL CYCLES.	Extend and retract the actuator through three complete cycles.
	ALL (adj)	SEND ALL THE TEST RESULTS TO THE MANUFACTURER.	The complete test results must be sent to the manufacturer.
	COMPLETED (adj)	THE GREEN INDICATION SHOWS THAT SERVICING IS COMPLETED.	The green indication shows that servicing is complete.
COMPLETE (v), COMPLETES, COMPLETED, COMPLETED	To bring to an end	COMPLETE THE FUNCTIONAL TEST.	
COMPLETED (adj)	Successfully brought to an end	SEND THE COMPLETED REPORT TO THE SUPERVISOR FOR APPROVAL.	
completely (adv)	FULLY (adv)	LET THE FIRST LAYER OF PAINT DRY FULLY BEFORE YOU APPLY THE SECOND LAYER.	Allow the first layer of paint to dry completely before applying the second layer.
completion (n)	END (n)	AT THE END OF THIS PROCEDURE, REFER TO 72-00-00, TEST 001.	See 72-00-00, Testing 001 upon completion of this procedure.
	COMPLETE (v)	TWO HOURS ARE NECESSARY TO COMPLETE THIS TASK.	Completion of the task will require two hours.
complicated (adj)	NOT EASY	THIS TEST PROCEDURE IS NOT EASY WITHOUT THE CORRECT EQUIPMENT.	This test procedure is complicated without the right equipment.

Word (part of speech)	Approved meaning/ ALTERNATIVES	APPROVED EXAMPLE	Not approved example
comply (v)	OBEY (v)	OBEY THESE INSTRUCTIONS.	You must comply with these instructions.
	Used in regulatory language. Refer to Rule 1.12.	THE SHIP CERTIFICATE MUST COMPLY WITH ALL THE REQUIREMENTS OF THE AUTHORITIES.	
COMPONENT (n)	A part, subassembly, or unit that has a specified function	DO NOT CLEAN THESE COMPONENTS WITH TRICHLORO-ETHYLENE.	
comprehensive (adj)	FULL (adj)	DO A FULL INSPECTION OF THE FAIRING.	Do a comprehensive inspection of the fairing.
COMPRESS (v), COMPRESSES, COMPRESSED, COMPRESSED	To decrease or cause to decrease in dimension or volume	COMPRESS THE SPRING AND CAREFULLY INSTALL THE COVER.	
		WHEN THE SPRING COMPRESSES, THE VALVE OPENS.	
comprise (v)	HAVE (v)	THE VALVE ASSEMBLY HAS THESE PARTS:	The valve assembly comprises these parts:
	CONTAIN (v)	EACH SURVIVAL KIT CONTAINS THESE ITEMS.	Each survival kit comprises these items.
compulsory (adj)	MANDATORY (adj)	BEFORE YOU PUT NEW OIL IN THE SYSTEM, IT IS MANDATORY TO REPLACE THE FILTER.	Before you put new oil in the system, it is compulsory to replace the filter.
compute (v)	CALCULATE (v)	CALCULATE THE VOLTAGE IN THE CIRCUIT.	Compute the voltage in the circuit.
concentrate (v)	ALL (adj)	ALL THE WEIGHT MUST BE ON THE SPAR TIP.	The weight must be concentrated on the spar tip.
CONCENTRATION (n)	The strength of something contained in a mixture	IN A HIGH CONCENTRATION, THIS MATERIAL IS POISONOUS.	

Word (part of speech)	Approved meaning/ ALTERNATIVES	APPROVED EXAMPLE	Not approved example
CONCENTRIC (adj)	That has a common center	THE EXTERNAL DIAMETER MUST BE CONCENTRIC TO THE INTERNAL DIAMETER.	
concern (v)	APPLICABLE (adj)	THIS PROCEDURE IS APPLICABLE ONLY TO A TYPE B UNIT.	This procedure only concerns a Type B unit.
conclusion (n)	END (n)	AT THE END OF THE TEST, SET THE SWITCH TO "NORMAL".	At the conclusion of the test, you must switch to NORMAL.
	RESULT (n)	IF THE RESULTS OF THE TEST ARE UNSATISFACTORY, REPLACE THE UNIT.	If the conclusions of the test are unsatisfactory, replace the unit.
concurrently (adj)	AT THE SAME TIME	DO THESE TWO PROCEDURES AT THE SAME TIME.	These two procedures must be applied concurrently.
CONDENSATION (n)	The result when a gas changes into a liquid	SET THE SWITCH TO THE "ON" POSITION TO REMOVE THE CONDENSATION.	
CONDITION (n)	1. Something that is necessary for what occurs	THESE INITIAL CONDITIONS ARE NECESSARY TO GET CORRECT RESULTS.	
	2. The state of an item	DO A CHECK OF THE UNIT CONDITION.	
	Use IF (conj) for "on the condition that."	IF THERE IS NO WIND, DO THE TEST OUTDOORS.	On the condition that there is no wind, do the test outdoors.
conduct (v)	DO (v)	DO THE LEAK TEST.	Conduct the leak test.
confine to (v)	ONLY (adv)	IF THE DAMAGE IS ONLY TO THE MIDDLE TREADS, THE TIRE IS SERVICEABLE.	If the damage is confined to the middle treads, the tire is serviceable.
confirm (v)	MAKE SURE (v)	IF POSSIBLE, DO AN ENGINE GROUND RUN TO MAKE SURE THAT YOU GET THE SAME INDICATIONS.	If possible, ground run the engine to confirm the indications.

Word (part of speech)	Approved meaning/ ALTERNATIVES	APPROVED EXAMPLE	Not approved example
conflict (n)	DIFFERENT (adj)	IF THE TWO INDICATIONS ARE DIFFERENT, DO THE TEST AGAIN	If there is a conflict between the two indicators, repeat the test.
	AGREE (v)	IF THE TWO INDICATIONS DO NOT AGREE, DO THE TEST AGAIN.	If there is a conflict between the two indicators, repeat the test.
conform (v)	AGREE (v)	IF THE FLAP EXTENSION DOES NOT AGREE WITH THE POSITION OF THE FLAP CONTROL LEVER, DO THE RIGGING PROCEDURE AGAIN.	If the flap extension does not conform to the setting of the flap control lever, do the rigging procedure again.
conformance (n)	AGREE (v)	THE INDICATIONS MUST AGREE WITH THOSE IN TABLE 2.	The indications must be in conformance with those in Table 2.
conformity (n)	AGREE (v)	THE PRECISION OF THE GAUGES MUST AGREE WITH THE APPLICABLE QUALITY STANDARD.	The accuracy of the gauges must be in conformity with the applicable quality standard.
CONNECT (v), CONNECTS, CONNECTED, CONNECTED	To come together or cause to come together to make one unit or system	CONNECT THE ELECTRICAL CONNECTORS TO THE INSTRUMENT.	
		PLUG "A" CONNECTS TO SOCKET "B".	
CONNECTION (n)	That which connects or is connected	MAKE SURE THAT ALL THE FUEL AND AIR CONNECTIONS ARE INSTALLED.	
consecutive (adj)	ONE (TN) AFTER THE OTHER	DO NOT DO MORE THAN THREE STARTS, ONE AFTER THE OTHER.	Do not do more than three consecutive starts.
consecutively (adv)	ONE (TN) AFTER THE OTHER	DO THESE STEPS ONE AFTER THE OTHER.	These actions must be done consecutively.
consequence (n)	BECAUSE OF (prep)	BECAUSE OF THIS PROBLEM, THE UNIT WILL NOT OPERATE.	As a consequence of this problem, the unit will not operate.

Word (part of speech)	Approved meaning/ ALTERNATIVES	APPROVED EXAMPLE	Not approved example
consider (v)	THINK (v)	IF YOU THINK THAT THE INDICATION IS NOT CORRECT, DO THE TEST AGAIN.	If you consider that the indication is not correct, do the test again.
considerable (adj)	LARGE (adj)	CLEAN YOUR SKIN WITH A LARGE QUANTITY OF WATER.	Clean your skin with a considerable quantity of water.
	IMPORTANT (adj)	CONTAMINATION OF CABIN AIR CAN BE AN IMPORTANT PROBLEM.	Contamination of cabin air can be a considerable problem.
	DANGEROUS (adj)	HYDRAULIC FLUID IS DANGEROUS FOR YOUR EYES.	Hydraulic fluid can cause considerable eye injury.
consist of (v)	HAVE (v)	EACH ANTENNA HAS THREE PARTS.	Each antenna consists of three parts.
consistent (adj)	AGREE (v)	MAKE SURE THAT THE INDICATIONS AGREE WITH THE POSITION OF THE FLAPS.	Check that the indications are consistent with the position of the flaps.
CONSTANT (adj)	That continues to be the same, without change	MAKE SURE THAT THE CABIN PRESSURE STAYS CONSTANT.	
CONSTANTLY (adv)	Incessantly	THE INSTRUMENT WILL CONSTANTLY SHOW THIS VALUE.	
construct (v)	ASSEMBLE (v)	ASSEMBLE THE UNIT.	Construct the unit.
consult (v)	REFER (v)	FOR MORE INFORMATION, REFER TO THE USER'S MANUAL.	For more information, consult the User's Manual.
contact (v)	TOUCH (v)	THE TOOL MUST NOT TOUCH THE CABLE.	The tool must not contact the cable.
	SPEAK (v)	SPEAK TO THE CREW CHIEF ON THE INTERPHONE.	Contact the crew chief on the interphone.
	WRITE (v)	FOR MORE INFORMATION ABOUT SPARE PARTS, WRITE TO THE SALES DEPARTMENT.	For more information about spare parts, contact the sales department.

Word (part of speech)	Approved meaning/ ALTERNATIVES	APPROVED EXAMPLE	Not approved example
CONTAIN (v), CONTAINS, CONTAINED, CONTAINED	To have in something or hold in something	EACH SURVIVAL KIT CONTAINS THESE ITEMS:	
CONTAINER (n)	Something that holds fluids, materials, or objects	PUT THE CONTAINER BELOW THE DRAIN PLUG.	
contaminant (n)	CONTAMINATION (n)	THIS FILTER REMOVES ALL CONTAMINATION FROM THE AIR SUPPLY.	This filter removes all contaminants from the air supply.
contaminated (adj)	DIRTY (adj)	REPLACE THE DIRTY OIL.	Contaminated oil must be replaced.
	CONTAMINATION (n)	REMOVE ALL CONTAMINATION FROM THE HYDRAULIC FLUID WITH A CLASS 5 FILTER.	Filter the contaminated hydraulic fluid through a Class 5 filter.
CONTAMINATION (n)	Unwanted change of the original condition of a material because of chemical or physical agents or a material that causes such a change	DUST CAN CAUSE CONTAMINATION ON ALL FILTERS.	
CONTENTS (n)	Something that is in a container	EXAMINE THE CONTENTS OF THE BOX FOR MISSING ITEMS.	
contiguous (adj)	ADJACENT (adj)	REMOVE THE SENSOR TO GET ACCESS TO THE ADJACENT STRUCTURE.	Remove the sensor to get access to the contiguous structure.
CONTINUE (v), CONTINUES, CONTINUED, CONTINUED	To stay or keep in current condition or operation	CONTINUE THE TEST.	
		IF THE LOW PRESSURE CONDITION CONTINUES, REPLACE THE UNIT.	

Word (part of speech)	Approved meaning/ ALTERNATIVES	APPROVED EXAMPLE	Not approved example
CONTINUOUS (adj)	That continues	MAKE SURE THAT THERE IS CONTINUOUS MOVEMENT OF THE PROBE.	
CONTINUOUSLY (adv)	In a continuous manner	APPLY PRESSURE CONTINUOUSLY.	
CONTOUR (n)	The outer line of something	CUT ALONG THE CONTOUR TO REMOVE THE DAMAGED SKIN.	
CONTROL (n)	Something that controls	IN AN EMERGENCY, USE THE MANUAL CONTROL.	
CONTROL (v), CONTROLS, CONTROLLED, CONTROLLED	To give or send signals that adjust, operate, or keep something to a limit, or that cause something to operate	RADIO MASTER SWITCH 1 CONTROLS THE BUS BAR.	
conventional (adj)	STANDARD (adj)	PULL THE VEHICLE TO A SAFE AREA WHERE YOU CAN USE THE STANDARD PROCEDURES.	Pull the vehicle to a safe location where more conventional procedures can be applied.
converse (adj)	OPPOSITE (adj)	THE RESET DEVICE OPERATES IN THE OPPOSITE DIRECTION.	The reset device operates in the converse direction.
convert (v)	CHANGE (v)	CHANGE THE UNITS FOR THE FUEL LOAD TO U.S. GALLONS.	Convert the fuel load to U.S. gallons.
convey (v)	MOVE (v)	MOVE THE EQUIPMENT TO A SAFE AREA.	Convey the equipment to a safe area.
	TELL (v)	TELL THE INFORMATION TO YOUR MANAGER.	Convey the information to your manager.
COOL (adj)	Moderately cold	WHEN THE AREA IS SUFFICIENTLY COOL, POLISH THE SURFACE.	

Word (part of speech)	Approved meaning/ ALTERNATIVES	APPROVED EXAMPLE	Not approved example
cool (v)	DECREASE (v)	LET THE TEMPERATURE OF THE COMPONENT DECREASE UNTIL IT IS THE SAME AS THE AMBIENT TEMPERATURE.	Let the temperature of the component cool to ambient.
	COOL (adj)	OPERATE THE PARKING BRAKE WHEN THE BRAKES ARE COOL.	Apply the parking brake when the brakes have cooled.
coordinate (v)	SYNCHRONIZED (adj)	MAKE SURE THAT COMPASS SYSTEM 1 AND COMPASS SYSTEM 2 ARE SYNCHRONIZED.	Coordinate compass system 1 and compass system 2.
COPY (n)	An object made the same as another object	SEND FIVE COPIES OF THE DEFECT REPORT TO THE ENGINEERING AUTHORITY.	
copy (v)	WRITE (v)	WRITE THE TAIL NUMBER ON THE RECORD FORM.	Copy the tail number on the record form.
	COPY (n)	MAKE A COPY OF THE DATA THAT IS ON THE IDENTIFICATION PLATE.	Copy the data on the identification plate.
	RECORD (v)	RECORD THE DATA THAT IS ON THE IDENTIFICATION PLATE.	Copy the data on the identification plate.
CORNER (n)	The point or the angle made when lines, sides, or edges touch	EXAMINE THE OPENING FOR CRACKS THAT START AT ITS CORNERS.	
CORRECT (adj)	That agrees with all that is necessary to the maximum	MAKE SURE THAT THE POLARITIES ARE CORRECT.	
CORRECT (v), CORRECTS, CORRECTED, CORRECTED	To make correct	CORRECT THE TENSION.	

Word (part of speech)	Approved meaning/ ALTERNATIVES	APPROVED EXAMPLE	Not approved example
CORRECTION (n)	A change to make something correct	DO NOT MAKE THE CORRECTION AT THIS HEADING.	
CORRECTLY (adv)	In a correct manner	MAKE SURE THAT THE CIRCLIP IS CORRECTLY INSTALLED.	
correctness (n)	CORRECT (adj)	THIS ADJUSTMENT MUST BE CORRECT.	The correctness of this adjustment is important.
	CORRECTLY (adv)	IT IS IMPORTANT TO DO THIS ADJUSTMENT CORRECTLY.	The correctness of this adjustment is important.
correspond (v)	AGREE (v)	MAKE SURE THAT THE POSITION OF THE CLAMP AGREES WITH THE POSITION THAT YOU RECORDED DURING THE REMOVAL PROCEDURE.	Make sure that the position of the clamp corresponds to the position recorded during the removal procedure.
	SAME (adj)	CLOSE THE VALVE WHEN THE TWO GAUGES SHOW THE SAME INDICATION.	Close the valve when the indications on the two gauges correspond.
	DIFFERENT (adj)	IF THE TWO INDICATIONS ARE DIFFERENT, DO THE TEST AGAIN.	If the two indications do not correspond, repeat the test.
corresponding (adj)	RELATED (adj)	THE RELATED CAUTION LIGHT MUST GO OFF AFTER 30 SECONDS.	The corresponding caution light shall go off after 30 seconds.
corrode (v)	CORROSION (TN)	THIS MATERIAL CAUSES CORROSION OF SURFACES WITH NO PROTECTION.	This material corrodes unprotected surfaces.
corrosive (adj)	CORROSION (TN)	HYDRAULIC FLUID CAN CAUSE CORROSION.	Hydraulic fluid is corrosive.

Word (part of speech)	Approved meaning/ ALTERNATIVES	APPROVED EXAMPLE	Not approved example
COUNT (v), COUNTS, COUNTED, COUNTED	To add the number of objects or occurrences to get a total	COUNT THE NUMBER OF TREADS ON THE TIRE.	
COUNTERCLOCK-WISE (adv)	In the opposite direction to clockwise	MAKE SURE THAT YOU TURN THE CONTROL KNOB COUNTERCLOCKWISE.	
couple (v)	CONNECT (v)	CONNECT THE FUELING NOZZLE TO THE AIRCRAFT.	Couple the fueling nozzle to the aircraft.
	ATTACH (v)	ATTACH THE FUELING NOZZLE TO THE AIRCRAFT.	Couple the fueling nozzle to the aircraft.
cover (v)	INCLUDE (v)	THIS PROCEDURE INCLUDES THE INSTRUCTIONS TO REMOVE THE CONSTANT SPEED DRIVE (CSD).	This procedure covers the removal of the Constant Speed Drive (CSD).
	HAVE (v)	THE BOTTOM EDGE OF THE PANEL HAS CARPET ON IT.	The bottom edge of the panel is covered with carpet.
	COVER (TN)	PUT THE COVER ON THE CONTAINER.	Cover the container.
crack (v)	OPEN (v)	WHEN THE RELIEF VALVE OPENS, THE PRESSURE DECREASES.	When the relief valve cracks open, the pressure decreases.
	CRACK (TN)	IF THERE ARE CRACKS IN THE CENTER PLY, REPLACE THE WINDSHIELD PANEL.	If a center ply is cracked, replace the windshield panel.
create (v)	MAKE (v)	CUT THE BRACKET TO MAKE A CLEARANCE.	Cut the bracket to create a clearance.
	CAUSE (v)	VIBRATION CAN CAUSE CRACKS.	Vibration can create cracks.

Word (part of speech)	Approved meaning/ ALTERNATIVES	APPROVED EXAMPLE	Not approved example
critical (adj)	VERY IMPORTANT	THE CONDITION OF THE RADOME IS VERY IMPORTANT FOR ITS PERFORMANCE.	The condition of the radome is critical to its performance.
	CAREFUL (adj)	MAKE A CAREFUL ESTIMATE OF THE DAMAGED AREA.	Make a critical estimate of the damaged area.
cross (v)	ACROSS (adv)	DO NOT GO ACROSS THE SAFETY LINE.	Do not cross the safety line.
	CORRECTLY (adv)	CONNECT THE WIRES CORRECTLY	Do not cross the wires.
curvature (n)	BEND (n)	THE BEND IN THE PIPE MUST HAVE A 30 mm RADIUS.	The curvature of the pipe must have a 30 mm radius.
	CURVE (n)	THE RADIUS OF THE CURVE MUST BE A MINIMUM OF 8 mm.	The radius of the curvature must be a minimum of 8 mm.
	RADIUS (TN)	BEND THE PIPE TO THE CORRECT RADIUS.	Bend the pipe to the correct curvature.
CURVE (n)	The shape of something which is bent but which has no angles	THE RADIUS OF THE CURVE MUST BE A MINIMUM OF 8 mm.	
curve (v)	MAKE A CURVE	BEND THE STRIP OF METAL TO MAKE A CURVE.	Curve the strip of metal.
CUT (v), CUTS, CUT, CUT	1. To divide into parts	CUT THE CABLE INTO THREE EQUAL LENGTHS.	
	2. To remove with a sharp tool	CUT AND DISCARD THE LOCKWIRE.	
CYCLE (n)	One complete sequence of operation	SET THE SWITCH TO THREE CYCLES A MINUTE.	
cycle (v)	OPERATE (v)	OPERATE THE ACTUATOR THROUGH THREE CYCLES.	Cycle the actuator three times.

Blank Page

Word (part of speech)	Approved meaning/ ALTERNATIVES	APPROVED EXAMPLE	Not approved example
DAMAGE (n)	The result of an occurrence that causes deterioration of the condition of something	LOOK FOR DAMAGE.	
damage (v)	DAMAGE (n)	DISCONNECT THE SPRING TO PREVENT DAMAGE TO THE ROD.	Disconnect the spring so as not to damage the rod.
DAMAGED (adj)	That has or can be affected by damage	REPLACE THE DAMAGED PARTS.	
damp (adj)	MOIST (adj)	REMOVE ALL LOOSE DIRT FROM THE SEAT WITH A MOIST CLOTH.	Remove all loose dirt from the seat with a damp cloth.
dampen (v)	DECREASE (v)	THE MOUNTING PADS DECREASE VIBRATION.	The mounting pads dampen vibration.
	MAKE MOIST	MAKE THE CLOTH MOIST WITH SOLVENT.	Dampen the cloth with solvent.
danger (n)	RISK (n)	THERE IS A RISK OF FIRE.	There is a danger of fire.
	DANGEROUS (adj) This word is a technical name when it identifies a safety instruction	HYDRAULIC FLUID IS DANGEROUS FOR YOUR HEALTH.	Hydraulic fluid is a danger to your health.
DANGEROUS (adj)	That can cause injury, damage to health or can kill	IT IS DANGEROUS TO USE AIR IN THIS SYSTEM.	
dangerously (adv)	DANGEROUS (adj)	DO NOT OPEN THE CARGO DOOR IN STRONG WINDS. STRONG WINDS ARE DANGEROUS.	Do not open the cargo door if the wind-speed is dangerously high.

Word (part of speech)	Approved meaning/ ALTERNATIVES	APPROVED EXAMPLE	Not approved example
DATA (n)	1. Known facts	MAKE SURE THAT YOU GET ALL THE DATA ABOUT THE PROBLEM.	
	2. Numbers and symbols used by computers	DO A TEST TO FIND IF THE MANAGEMENT UNIT TRANSMITS THESE DATA.	
	You can also use the singular.	DO A TEST TO FIND IF THE MANAGEMENT UNIT TRANSMITS THIS DATA	
DATE (n)	The time at which an event occurs	ON THE LABEL, READ THE DATE OF THE LAST INSPECTION.	
deactivate (v)	STOP (v)	THE WARNING BELL AND THE HORN WILL OPERATE AND THEN STOP.	The warning bell and the horn will be activated and then deactivated.
	DISCONNECT (v)	MAKE SURE THAT THE MICROSWITCHES ARE DISCONNECTED.	Make sure that the microswitches are deactivated.
	ISOLATE (v)	OPEN THE CIRCUIT BREAKER TO ISOLATE THE LANDING GEAR SYSTEM.	Open the circuit breaker to deactivate the landing gear system.
deadly (adj)	KILL (v)	HIGH VOLTAGES CAN KILL YOU.	High voltages are deadly.
deaerate (v)	BLEED (v)	BLEED THE No. 1 HYDRAULIC SYSTEM.	Deaerate the No. 1 hydraulic system.
	AIR (TN)	THE VACUUM EQUIPMENT REMOVES THE AIR FROM THE COMPOUND.	The vacuum equipment deaerates the compound.
dearm (v)	DISARM (v)	DISARM THE ESCAPE SLIDE.	Dearm the escape slide.
	REMOVE (v)	REMOVE THE CARTRIDGES FROM THE FIRE BOTTLES.	Dearm the fire bottles.

Word (part of speech)	Approved meaning/ ALTERNATIVES	APPROVED EXAMPLE	Not approved example
decay (v)	DECREASE (v)	WHEN THE ENGINE SPEED DECREASES TO 500 RPM, STOP THE ENGINE.	When the engine speed decays to 500 rpm, stop the engine.
decelerate (v)	DECREASE (v)	MAKE SURE THAT THE ENGINE SPEED DECREASES TO 500 RPM.	Make sure that the engine decelerates to 500 rpm.
	SLOWER (adj)	TO MAKE THE CURING PROCESS SLOWER, DECREASE THE HEAT THAT YOU APPLY TO THE COMPOUND.	The curing process of the compound can be decelerated by decreasing the heat applied.
decide (v)	SELECT (v)	SELECT THE HYDRAULIC SYSTEM THAT YOU WILL PRESSURIZE.	Decide which hydraulic system you will pressurize.
	DECISION (n)	IF YOUR DECISION IS NOT TO REPLACE THE UNIT, MONITOR ITS TEMPERATURE FOR A MINIMUM OF 15 CYCLES.	If you decide not to replace the existing unit, watch its temperature for at least 15 cycles.
DECISION (n)	The result after you think about a problem	MAKE A DECISION ABOUT WHICH PROCEDURE YOU WILL USE.	
decontamination (n)	REMOVE CONTAMINATION	REMOVE THE CONTAMINATION FROM THE INSTRUMENTS AND MEDICAL EQUIPMENT.	Do the decontamination of the instruments and medical equipment.
	CLEAN (v)	CLEAN THE AIRCRAFT IN AN APPROVED AREA.	The decontamination of the aircraft shall be done in an authorized area.
decrease (n)	DECREASE (v)	THE PRESSURE MUST DECREASE BY MORE THAN 2 BAR.	There must be a decrease in pressure of more than 2 bar.

Word (part of speech)	Approved meaning/ ALTERNATIVES	APPROVED EXAMPLE	Not approved example
DECREASE (v), DECREASES, DECREASED, DECREASED	To make or become smaller or lower	DECREASE THE HEADING INDICATIONS.	
decrement (n)	INCREMENT (n)	TURN THE ADJUSTER COUNTERCLOCKWISE IN INCREMENTS OF 60 DEG.	Turn the adjuster anticlockwise in decrements of 60 deg.
DEDICATED (adj)	That is used only for one purpose	A DEDICATED TOOL IS NECESSARY FOR THE INSTALLATION OF THE UNIT.	
DE-ENERGIZE (v), DE-ENERGIZES, DE-ENERGIZED, DE-ENERGIZED	To remove electrical power from	DE-ENERGIZE THE AC AND DC BUSBARS.	
deep (adj)	DEPTH (n)	MEASURE THE DEPTH OF THE DENT WITH A STRAIGHTEDGE AND FEELER GAUGES.	Measure how deep the dent is with a straightedge and feeler gauges.
DEFECTIVE (adj)	Unserviceable for operation	REPLACE THE DEFECTIVE BOOSTER PUMP.	
define (v)	CALCULATE (v)	CALCULATE THE QUANTITY OF FUEL THAT IS NECESSARY FOR THE FLIGHT.	Define the amount of fuel required for the flight.
	GIVE (v)	THE BRAKE INDICATOR GIVES THE WEAR ON THE BRAKE SHOE.	The brake indicator defines the wear on the brake shoe.
	SPECIFIED (adj)	THE FLAP ANGLES ARE SPECIFIED IN TABLE 1.	Table 1 defines the flap angles.

Word (part of speech)	Approved meaning/ ALTERNATIVES	APPROVED EXAMPLE	Not approved example
DEFLATE (v), DEFLATES, DEFLATED, DEFLATED	To make or become smaller as a result of depressurization	DEFLATE THE TIRE TO A PRESSURE OF 2 BAR (29 PSI).	
		IF THE WHEEL TEMPERATURE INCREASES TO MORE THAN 177 DEG., THE CORE OF THE FUSIBLE PLUG MELTS AND THE TIRE DEFLATES.	
deflect (v)	MOVE (v)	THE BARS MOVE TO THE RIGHT SIDE.	Bars deflect to the right.
deflection (n)	MOVEMENT (n)	REMOVE ALL THE RIGGING PINS AND EQUIPMENT THAT PREVENT FREE MOVEMENT OF THE CONTROL SURFACES.	Remove all rigging pins and equipment preventing free deflection of control surfaces.
	POSITION (n)	MAKE SURE THAT THE POSITION IS 23 DEG.	Ensure that deflection is 23 deg.
	TRAVEL (n)	MAKE SURE THAT THE TRAVEL IS 23 DEG.	Ensure that deflection is 23 deg.
deformed (adj)	DAMAGED (adj)	THE COVER ON THE GUN CAMERA IS DAMAGED.	The cover on the gun camera is deformed.
	DEFORMATION (TN)	MAKE SURE THAT THE TUBES HAVE NO DEFORMATION.	Make sure that the tubes are not deformed.
DEFUEL (v), DEFUELS, DEFUELED, DEFUELED	To remove fuel	DEFUEL THE TANK.	
defueling (n)	DEFUEL (v)	DEFUEL ONLY IN AN OPEN AREA THAT HAS A GOOD AIRFLOW.	Only do the defueling in an open area that is well-ventilated.
degrease (v)	GREASE (TN)	REMOVE ALL GREASE FROM THE FAYING SURFACE WITH SOLVENT.	Degrease the faying surface with solvent.
de-ice (v)	ICE (TN)	REMOVE THE ICE FROM THE AIRCRAFT.	De-ice the aircraft.

Word (part of speech)	Approved meaning/ ALTERNATIVES	APPROVED EXAMPLE	Not approved example
delay (n)	INTERVAL (n)	AN INTERVAL OF 5 MINUTES IS NECESSARY BEFORE YOU CONTINUE THE TEST.	A delay of 5 minutes is necessary before continuing with the test.
	IMMEDIATELY (adv)	THE JAWS OF THE UPLOCK UNIT MUST OPEN IMMEDIATELY.	The jaws of the uplock unit must open with no delay.
delay (v)	AFTER (prep)	ONLY USE THE STARTER AFTER A PERIOD OF 20 MINUTES.	Use of the starter must be delayed for 20 minutes.
delete (v)	ERASE (v)	ERASE THE FLIGHT DATA FROM THE TAPE.	Delete the flight data from the tape.
	REMOVE (v)	REMOVE THE PENCIL MARKS WITH AN ERASER.	Delete the pencil marks with an eraser.
	You can use this word as a technical verb for computer processes and applications.	DELETE THE FILE.	
delicate (adj)	EASILY DAMAGED	BE CAREFUL WITH EASILY DAMAGED PARTS.	Be careful with delicate parts.
deliver (v)	SUPPLY (v)	SET THE RIG TO SUPPLY A PRESSURE OF 800 kPa.	Set the rig so that it delivers a pressure of 800 kPa.
delivery (n)	SUPPLY (v)	A CENTRIFUGAL PUMP SUPPLIES HYDRAULIC FLUID TO THE UNIT.	Hydraulic fluid delivery to the unit is ensured by a centrifugal pump.
demand (v)	MANDATORY (adj)	IN THIS CONFIGURATION, A DIFFERENT TEST PROCEDURE IS MANDATORY.	This configuration demands a different test procedure.
denote (v)	SHOW (v)	FIGURE 2 SHOWS THAT YOU CAN INSTALL AN ALTERNATIVE RELAY.	Figure 2 denotes that an alternative relay may be installed.

Word (part of speech)	Approved meaning/ ALTERNATIVES	APPROVED EXAMPLE	Not approved example
dent (v)	DENT (TN)	DO NOT MAKE DENTS IN THE BELLOWS.	Do not dent the bellows.
depend (v)	IF (conj)	THE TENSION IN THE CABLES DECREASES IF THE TEMPERATURE INCREASES.	Cable tension changes depending on the temperature.
deplete (v)	DECREASE (v)	DO NOT OPERATE THE BRAKES TO DECREASE THE PRESSURE IN THE ACCUMULATORS.	Do not deplete accumulators by operating the brakes.
DEPLOY (v), DEPLOYS, DEPLOYED, DEPLOYED	To move or cause to move from a specified position of storage and into operation	KEEP PERSONNEL AWAY FROM THE AIRCRAFT WHEN THE THRUST REVERSER DEPLOYS.	
		DEPLOY THE SLIDE RAFT.	
deposit (n)	PARTICLE (n)	IF THERE ARE METAL PARTICLES IN THE OIL FILTER, EXAMINE THE FILTER ELEMENT FOR WEAR.	If there are metal deposits in the oil filter, check the filter element for wear.
	CONTAMINATION (n)	FLUSH THE PIPES TO REMOVE CONTAMINATION.	Flush the pipes to remove deposits.
depress (v)	PUSH (v)	PUSH THE BUTTON AND THEN RELEASE IT.	Depress and release the button.
depressurize (v)	PRESSURE (TN)	RELEASE THE PRESSURE FROM THE HYDRAULIC SYSTEM.	Depressurize the hydraulic system.
DEPTH (n)	The distance from the top down, from the surface down, from the front surface to the rear surface	MEASURE THE DEPTH OF THE DENT AS SHOWN IN FIGURE 4.	
describe (v)	GIVE (v)	THIS SECTION GIVES THE PROCEDURES FOR THE LEAK TEST OF THE SYSTEM.	This section describes the procedures for testing the system for leaks.
deselect (v)	CANCEL (v)	PUSH THE BUTTON TO CANCEL THE FREEZE MODE.	Press the button to deselect the freeze mode.

Word (part of speech)	Approved meaning/ ALTERNATIVES	APPROVED EXAMPLE	Not approved example
design (v)	HAVE (v)	THE MAIN LANDING GEAR DOORS HAVE STEPS FOR ACCESS TO THE LANDING GEAR BAY.	The main landing gear doors are designed with steps for access to landing gear bay.
desired (adj)	NECESSARY (adj)	ADD THE NECESSARY QUANTITY OF OIL.	Add the desired quantity of oil.
	CORRECT (adj)	TIGHTEN THE TURNBUCKLE UNTIL THE TENSION IN THE CABLE IS CORRECT.	Tighten the turnbuckle until you have the desired tension in the cable.
destroy (v)	UNSERVICEABLE (adj)	MAKE THE CONTAINER UNSERVICEABLE TO MAKE SURE THAT YOU CANNOT USE IT AGAIN.	To prevent further use, destroy the container.
detach (v)	DISCONNECT (v)	DISCONNECT THE PIPELINE FROM THE UNION.	Detach the pipeline from the union.
	DISENGAGE (v)	BEFORE YOU MOVE THE CONTROLS, DISENGAGE THE LOCKS.	Before moving the controls, detach the locks.
	REMOVE (v)	DO NOT REMOVE THE PROTECTIVE LAYER.	Do not detach the protective film.
detail (n)	INSTRUCTION (n)	THIS PROCEDURE GIVES THE INSTRUCTIONS TO ATTACH THE PYLON.	This procedure gives details on how to attach the pylon.
detail (v)	GIVE (v)	THIS SECTION GIVES THE REMOVAL / INSTALLATION INSTRUCTIONS.	This section details the Removal / Installation instructions.
	REFER (v)	DO THE LEAK TEST (REFER TO PARAGRAPH 4.0).	Do the leak test as detailed in paragraph 4.0.
	SPECIFIED (adj)	DO THE LEAK TEST AS SPECIFIED IN PARAGRAPH 4.0.	Do the leak test as detailed in paragraph 4.0.

Word (part of speech)	Approved meaning/ ALTERNATIVES	APPROVED EXAMPLE	Not approved example
detect (v)	FIND (v)	TO FIND DAMAGE ON WIRES, RUB A CLOTH ALONG THEM.	To detect damage to wires, rub a cloth along them.
	SENSE (v)	THE COMPUTER SENSES THE DIFFERENCES BETWEEN THE SIGNALS.	The computer detects discrepancies between the signals.
DETERIORATION (n)	A worse condition	THIS CONDITION WILL CAUSE DETERIORATION OF THE SURFACE.	
determine (v)	FIND (v)	FIND THE DEFECTS IN THE SYSTEM.	Determine the defects in the system.
	GIVE (v)	DIMENSIONS 1 AND 2 GIVE THE CORRECT POSITION.	The correct position is determined by dimensions 1 and 2.
	SELECT (v)	SELECT THE HYDRAULIC SYSTEM THAT YOU WILL PRESSURIZE.	Determine which hydraulic system you will pressurize.
	CALCULATE (v)	CALCULATE THE DIMENSION OF THE SHIM WITH THIS FORMULA:	Determine the dimension of the shim with this formula:
detrimental (adj)	DANGEROUS (adj)	HYDRAULIC FLUID IS DANGEROUS FOR YOUR HEALTH.	Hydraulic fluid is detrimental to health.
	DAMAGE (n)	SMALL QUANTITIES OF SURFACE BLOOMING, WHICH CAN OCCUR ON ITEMS IN STORAGE, DO NOT CAUSE DAMAGE TO THE COMPONENT.	Light surface blooming, which can occur on items in storage, is not detrimental to the component.

Word (part of speech)	Approved meaning/ ALTERNATIVES	APPROVED EXAMPLE	Not approved example
develop (v)	START (v)	IF A FIRE STARTS, OPERATE THE FIRE SHUTOFF VALVE.	If a fire develops, operate the fire shutoff valve.
	CAUSE (v)	DO NOT LET HYDRAULIC FLUID STAY ON YOUR SKIN. HYDRAULIC FLUID CAN CAUSE DERMATITIS.	Do not leave hydraulic fluid on your skin, as dermatitis could develop.
deviate (v)	OBEY (v)	PERSONNEL MUST OBEY THIS PROCEDURE.	Personnel must not deviate from this procedure.
DEVICE (n)	Something used to do a task	INSTALL THE SAFETY DEVICES.	
devise (v)	MAKE (v)	IF THE HOLDING FIXTURE IS NOT AVAILABLE, MAKE A SUPPORT FROM THE SHIPPING CONTAINER.	If the holding fixture is not available, devise one from the shipping container.
	FIND (v)	FIND NEW TROUBLESHOOTING PROCEDURES.	Devise new troubleshooting procedures.
diagnose (v)	FIND (v)	FIND THE CAUSE OF THE MALFUNCTION.	Diagnose the cause of the malfunction.
DIAGONALLY (adv)	In a diagonal direction	APPLY THE FIRST LAYER OF TAPE ACROSS THE CUTOUT AND THE SECOND AND THIRD LAYERS DIAGONALLY.	
diametrically (adv)	OPPOSITE (adj)	MAKE SURE THAT THE TWO OPPOSITE V-GROOVES ALIGN WITH THE RAILS.	Ensure the two diametrically opposed V-grooves are in line with the rails.
differ (v)	DIFFERENT (adj)	IF THE VALUES ARE DIFFERENT, DO THIS TEST.	If the values differ, do the following test.
DIFFERENCE (n)	That which is different between two quantities, numbers, or functions	THE DIFFERENCE BETWEEN THE TWO DIMENSIONS MUST NOT BE MORE THAN 0.10 mm.	

Word (part of speech)	Approved meaning/ ALTERNATIVES	APPROVED EXAMPLE	Not approved example
DIFFERENT (adj)	Not the same	IF THE INDICATIONS ARE DIFFERENT, DO THESE STEPS.	
DIFFERENTLY (adv)	In a different manner	THE TWO EXTRACTORS OPERATE DIFFERENTLY.	
differentiate (v)	IDENTIFY (v)	LETTERS IDENTIFY THE WIRE SEGMENTS.	Letters are used to differentiate between the wire segments.
difficult (adj)	NOT EASY	IF IT IS NOT EASY TO INSTALL THE RIGGING PIN, ADJUST THE LENGTH OF THE ROD.	If the rigging pin is difficult to install, adjust the length of the rod.
	NOT EASILY	IF YOU CANNOT INSTALL THE RIGGING PIN EASILY, ADJUST THE LENGTH OF THE ROD.	If the rigging pin is difficult to install, adjust the length of the rod.
difficulty (n)	NOT EASY	IF IT IS NOT EASY TO REMOVE THE BOLT, APPLY PENETRANT OIL.	If there is difficulty in removing the bolt, apply penetrant oil.
	NOT EASILY	IF YOU CANNOT REMOVE THE BOLT EASILY, APPLY PENETRANT OIL.	If there is difficulty in removing the bolt, apply penetrant oil.
DIGITAL (adj)	Related to digits	CONNECT THE PRODS OF THE DIGITAL VOLTMETER.	
DIGITALLY (adv)	Given by digits	THE VALUE IS SHOWN DIGITALLY ON THE DISPLAY.	
dilute (v)	MIX (v)	MIX THE PRIMER WITH SOLVENT.	Dilute the primer with solvent.
	ADD (v)	DO NOT ADD MORE THAN AN EQUAL PART OF WATER TO THE ANTIFREEZE.	Do not dilute the antifreeze with more than an equal part of water.

Word (part of speech)	Approved meaning/ ALTERNATIVES	APPROVED EXAMPLE	Not approved example
dim (v)	DECREASE (v)	MAKE SURE THAT THE INTENSITY OF THE RETICLE LIGHT GRADUALLY DECREASES TO ZERO.	Check that the reticle brightness dims gradually until it disappears.
DIM (adj) (DIMMER, DIMMEST)	Not bright	DURING NIGHT OPERATION, MAKE SURE THAT THE PANEL LIGHTS ARE DIM.	
DIMENSION (n)	The result when something is measured in one direction	DIMENSION "A" MUST NOT BE MORE THAN 50 PER CENT OF DIMENSION "D".	
diminish (v)	DECREASE (v)	IF THE HYDRAULIC PRESSURE DECREASES, CLOSE THE VALVE.	If the hydraulic pressure diminishes, close the valve.
DIMLY (adv)	Not bright	MAKE SURE THAT THE LIGHT COMES ON DIMLY.	
dip (v)	MOMENTARILY (adv)	PUT THE ELEMENT INTO THE SOLVENT MOMENTARILY.	Dip the element in solvent.
direct (v)	POINT (v)	POINT THE OPEN END OF THE DRAIN HOSE AWAY FROM THE STRUCTURE.	Direct the open end of the drain hose away from the structure.
DIRECTION (n)	The line on which something moves or is pointed	THE ARROW ON THE VALVE BODY SHOWS THE DIRECTION OF THE FLOW.	
DIRECTLY (adv)	In a direct procedure or by direct effect	DO NOT FILL THE SYSTEM DIRECTLY FROM A HIGH-PRESSURE STORAGE-CYLINDER.	
DIRTY (adj) (DIRTIER, DIRTIEST)	Not clean	A DIRTY CONTACT IN THE CONNECTION UNITS WILL CAUSE THEM TO OPERATE INCORRECTLY.	

Word (part of speech)	Approved meaning/ ALTERNATIVES	APPROVED EXAMPLE	Not approved example
disappear (v)	VIEW (n)	THE HORIZONTAL BAR GOES OUT OF VIEW.	The horizontal bar disappears from the window.
DISARM (v), DISARMS, DISARMED, DISARMED	1. To remove armaments	DISARM THE AIRCRAFT.	
	2. To prevent automatic operation	DISARM THE LIFT DUMPER.	
DISASSEMBLE (v), DISASSEMBLES, DISASSEMBLED, DISASSEMBLED	To take an assembly apart	DISASSEMBLE THE VALVE ASSEMBLY.	
DISCARD (v), DISCARDS, DISCARDED, DISCARDED	To not use again	REMOVE THE O-RINGS AND DISCARD THEM.	
discharge (v)	RELEASE (v)	RELEASE PRESSURE FROM THE TANKS.	Discharge pressure from the tanks.
	GO (v)	THE UNWANTED AIR GOES OVERBOARD.	The unwanted air discharges overboard.
DISCONNECT (v), DISCONNECTS, DISCONNECTED, DISCONNECTED	To separate something from the thing that it is connected to, or to become separated	DISCONNECT THE POWER SUPPLY.	
		WHEN YOU TOUCH THIS ICON, THE COMPUTER DISCONNECTS FROM THE INTERNET.	
discontinue (v)	STOP (v)	STOP THE TEST PROCEDURE.	Discontinue the test procedure.
discover (v)	FIND (v)	IF YOU FIND CORROSION DURING THE INSPECTION, REMOVE IT.	If you discover corrosion during the inspection, remove it.
discrepancy (n)	DIFFERENCE (n)	IF THERE IS A DIFFERENCE OF MORE THAN 5 mm, MAKE SURE THAT THE ADJUSTMENT IS CORRECT.	If the discrepancy is greater than 5 mm, check the adjustment.

Word (part of speech)	Approved meaning/ ALTERNATIVES	APPROVED EXAMPLE	Not approved example
DISENGAGE (v), DISENGAGES, DISENGAGED, DISENGAGED	To release or become released from something that engages	DISENGAGE THE CLUTCH BEFORE YOU START THE ENGINE.	
		IF THERE IS A MALFUNCTION, THE GEARS WILL AUTOMATICALLY DISENGAGE.	
disinfect (v)	DISINFECTANT (TN)	USE DISINFECTANT TO CLEAN THE MASK ASSEMBLIES.	Disinfect the mask assemblies.
dismantle (v)	DISASSEMBLE (v)	DISASSEMBLE THE BRAKE UNIT IN THE WORKSHOP.	Dismantle the brake unit in the workshop.
dispatch (v)	SEND (v)	SEND A REPORT TO THE INSPECTION DEPARTMENT.	Dispatch a report to the Inspection Department.
disperse (v)	REMOVE (v)	BLOW THROUGH THE TUBES TO REMOVE THE REMAINING FLUID.	Blow through the tubes to disperse the residual fluid.
displace (v)	MOVE (v)	THE PISTON MOVES THE FLUID.	Fluid is displaced by the piston.
DISPLAY (n)	A visual indication	A NUMBER IN THE TOP LEFT CORNER OF THE DISPLAY REFERS TO THE RELATED PAGE.	
display (v)	SHOW (v)	THE COMPASS INDICATORS SHOW THE HEADINGS.	The headings are displayed on the compass indicators.
dispose of (v)	DISCARD (v)	DISCARD THE USED SOLVENT AS SPECIFIED IN THE APPLICABLE PROCEDURE.	Dispose of the used solvent according to the applicable procedure.

Word (part of speech)	Approved meaning/ ALTERNATIVES	APPROVED EXAMPLE	Not approved example
disposition (n)	LOCATION (n)	MAKE SURE THAT THE FIRE EXTINGUISHERS ARE IN THEIR CORRECT LOCATIONS BEFORE YOU START THE ENGINE.	Check the disposition of the fire extinguishers before you start the engine.
	POSITION (n)	MAKE SURE THAT THE POSITION OF THE SWITCHES IS CORRECT.	Ensure that the switch disposition is correct.
disregard (v)	IGNORE (v)	DO THE GENERATOR DRIVE TEST, BUT IGNORE STEPS 16 THRU 20.	Perform the generator drive test, but disregard steps 16 thru 20.
DISTANCE (n)	The dimension between two points or objects	MEASURE THE DISTANCE BETWEEN POINTS "A" AND "B".	
distinct (adj)	CLEAR (adj)	MAKE SURE THAT THE SOUND IS CLEAR.	Ensure the sound is distinct.
distinctly (adv)	CLEARLY (adv)	YOU MUST HEAR THE SOUND CLEARLY.	You must hear the sound distinctly.
distort (v)	TWIST (v)	DO NOT TWIST THE PIPE WHEN YOU TURN THE CONNECTORS.	Do not distort the pipe when turning the connectors.
distribute (v)	APPLY (v)	APPLY THE LOAD EQUALLY ON THE AREA.	Distribute the load evenly over the area.
	SUPPLY (v)	THE SYSTEM SUPPLIES HYDRAULIC POWER.	The system distributes hydraulic power.
distribution (n)	SUPPLY (n)	THE FUEL SUPPLY IS FROM THE COLLECTOR TANKS.	The fuel distribution is from the collector tanks.
disturb (v)	MOVE (v)	DO NOT MOVE THE TIMING WHEN YOU INSTALL THE DISTRIBUTOR.	Do not disturb the timing during the distributor installation.
	CHANGE (v)	DO NOT CHANGE THE SEQUENCE OF THE WORK STEPS IN THE TEST.	Do not disturb the sequence of the work steps in the test.

Word (part of speech)	Approved meaning/ ALTERNATIVES	APPROVED EXAMPLE	Not approved example
DIVIDE (v), DIVIDES, DIVIDED, DIVIDED	1. To separate into parts or groups	YOU CAN DIVIDE THE DRAINS INTO THREE GROUPS.	
		THE PRESSURE LINES DIVIDE AND GO IN DIFFERENT DIRECTIONS.	
	2. To do mathematical division	TO FIND THE MIDDLE POINT, DIVIDE THE MEASURED CLEARANCE BY TWO.	
DO (v), DOES, DID, DONE	1. To complete a procedure, task, or step	DO A FUNCTIONAL TEST.	
	2. As a helping verb		
	a. As part of a negative command or statement	DO NOT BREATHE THE SOLVENT FUMES.	
	b. As part of a question	DOES THE LIGHT COME ON?	
don (v)	USE (v)	USE PROTECTIVE GOGGLES WHEN YOU DO WORK ON FIRE EXTINGUISHERS.	You must don protective goggles when working on fire extinguishers.
	PUT ON (v)	PUT THE MASK ON AND BREATHE THROUGH IT TO MAKE SURE THAT THE OXYGEN SUPPLY IS CORRECT.	Don the mask and breathe through it to check the oxygen supply.
double (adj)	TWO (TN)	YOU MUST SEE TWO MARKS ON THE STAND.	Double marks must appear on the stand.
doubt (n)	NOT SURE	IF YOU ARE NOT SURE THAT THE UNIT IS SERVICEABLE, DISCARD IT.	If there are doubts about the serviceability of the unit, discard it.
	THINK (v)	IF YOU THINK THAT THE RESULTS ARE INCORRECT, DO THE TEST AGAIN.	If you are in doubt about the results, repeat the test.

Word (part of speech)	Approved meaning/ ALTERNATIVES	APPROVED EXAMPLE	Not approved example
DOWN (adj)	In a position below	MAKE SURE THAT THE ANCHOR IS DOWN.	
DOWN (adv)	To a position below	THE POINTER MOVES DOWN.	
DOWN (prep)	To a position below	LET THE SLEEVE MOVE DOWN THE GUIDE TUBE.	
DOWNSTREAM (adj)	In the direction of the flow	DISCONNECT THE DOWNSTREAM CONNECTION.	
DOWNSTREAM (adv)	In the direction of the flow	MOVE THE PROBE DOWNSTREAM AND MONITOR THE TEMPERATURE AGAIN.	
DOWNSTREAM OF (prep)	In the direction of the flow	DISCONNECT THE CONNECTION WHICH IS DOWNSTREAM OF THE VALVE.	
downward (adv)	DOWN (adv)	MAKE SURE THAT THE AIR OUTLET POINTS DOWN.	Make sure that the air outlet is positioned downward.
drag (v)	PULL (v)	PULL THE WIRES THROUGH THE CONDUIT.	Drag the wires through the conduit.
	You can use this word as a technical verb for computer processes and applications.	DRAG THE ICON TO A NEW POSITION.	
DRAIN (v), DRAINS, DRAINED, DRAINED	To remove liquid	DRAIN THE SYSTEM FULLY.	
draw (v)	MAKE (v)	MAKE A LINE ON THE BLADE.	Draw a line on the blade.
	PULL (v)	PULL THE WIRE OUT OF THE CONDUIT.	Draw the wire out of the conduit.

Word (part of speech)	Approved meaning/ ALTERNATIVES	APPROVED EXAMPLE	Not approved example
dress (v)	USE (v)	USE PROTECTIVE CLOTHING BEFORE YOU GO INTO THE FUEL TANK.	Dress suitably before entering the fuel tank.
	PUT ON (v)	PUT ON PROTECTIVE CLOTHING BEFORE YOU GO INTO THE FUEL TANK.	Dress suitably before entering the fuel tank.
drift (v)	MOVE (v)	THE HORIZON BAR CAN MOVE A SMALL DISTANCE.	The horizon bar may drift slightly.
	REMOVE (v) (WITH A DRIFT [TN])	REMOVE THE PIN FROM ITS HOLE. USE A DRIFT.	Drift the pin from its hole.
DRINK (v), DRINKS, DRANK, DRUNK	To consume liquid For safety instructions, use:	DO NOT DRINK IN THE WORK AREA.	
	SWALLOW (v)	DO NOT SWALLOW THE SOLVENT.	Do not drink the solvent.
drive (v)	MOVE (v)	THE ACTUATOR MOVES TO ZERO.	The actuator is driven to zero.
	REMOVE (v)	REMOVE THE LOCKBOLT FROM THE HOLE.	Drive the lockbolt out of the hole.
	OPERATE (v)	A HYDRAULIC MOTOR OPERATES THE PUMP.	A hydraulic motor drives the pump.
droop (v)	MOVE (v)	IF YOU DO NOT INSTALL THE RIGGING PIN, THE ELEVATORS WILL MOVE DOWN WHEN THERE IS NO PRESSURE IN THE SYSTEM.	If you do not install the rigging pin, the elevators will droop when there is no pressure in the system.
DROP (n)	A small quantity of liquid in a spherical shape	MAKE SURE THAT NO DROPS OF ADHESIVE FALL ONTO THE MATERIAL.	

Word (part of speech)	Approved meaning/ ALTERNATIVES	APPROVED EXAMPLE	Not approved example
drop (v)	FALL (v)	PARTICLES THAT FALL INTO THE INTAKE CAN CAUSE DAMAGE.	Particles that drop into the intake may cause damage.
	DECREASE (v)	THE PRESSURE DECREASES WHILE THE TIRES BECOME COOL.	The pressure drops as the tires cool.
DRY (adj) (DRIER, DRIEST)	Without liquid or moisture	MAKE SURE THAT THE SURFACE IS DRY.	
DRY (v), DRIES, DRIED, DRIED	To remove liquid or moisture, or to become dry	DRY THE FILTER IN AN OVEN.	
		LET THE PAINT DRY AT AMBIENT TEMPERATURE.	
due to (prep)	BECAUSE OF (prep)	BECAUSE OF THE SMALL DIAMETER OF THE HOLE, THE FLOW IS LOW.	Due to the small diameter of the hole, the flow is low.
	BECAUSE (conj)	THE SHAFT MOVED BECAUSE THE ACTUATOR OPERATED.	The shaft moved due to the actuator action.
dull (adj)	DIM (adj)	THE LAMP COMES ON WITH A DIM LIGHT.	The lamp comes on with a dull light.
	MATT (adj)	THE LETTERS WILL BE MATT BLACK ON WHITE.	The letters will be dull black on a white background.
duration (n)	DURING (prep)	DO NOT MAKE ADJUSTMENTS DURING THIS TEST.	Do not make adjustments for the duration of this test.
DURING (prep)	In or for a specified time	MONITOR THE INDICATIONS DURING THE RETRACTION TEST.	
dust (v)	APPLY (v)	APPLY TALCUM POWDER TO THE AREA.	Dust the area with talcum powder.

Blank Page

Word (part of speech)	Approved meaning/ ALTERNATIVES	APPROVED EXAMPLE	Not approved example
EACH (adj)	Every one of two or more objects or persons, seen together as a group but identified separately	MAKE SURE THAT THE SWITCH IN EACH PANEL IS SET TO "OFF".	
EACH (pron)	Every one of two or more objects or persons, seen together as a group but identified separately	INSTALL TWO SPACERS, EACH WITH A THICKNESS OF 3.5 mm.	
EACH OTHER (pron)	Each of two or more in reciprocal action or relation	WHEN YOU REMOVE THE PROBES, DO NOT LET THEM TOUCH EACH OTHER.	
early (adj)	SHORT (adj)	IF THE FILTERS BECOME CLOGGED AFTER AN UNUSUALLY SHORT TIME, SEND THEM TO THE LABORATORY FOR ANALYSIS OF THE CONTAMINATION.	If the filters clog at an unusually early stage, send them to the laboratory for analysis of the contamination.
earth (n)	GROUND (n)	MAKE SURE THAT THE ELECTRICAL SUPPLY GOES TO GROUND.	Make sure the electric supply goes to earth.
earth (v)	GROUND (v)	MAKE SURE THAT THE FUEL TANKS ARE CORRECTLY GROUNDED.	Make sure the fuel tanks are correctly earthed.
ease (n)	EASILY (adv)	MAKE SURE THAT YOU CAN MOVE THE HANDLE EASILY.	The handle must move with ease.
	EASY (adj)	MAKE SURE THAT IT IS EASY TO MOVE THE HANDLE.	The handle must move with ease.
ease (v)	CAREFULLY REMOVE	CAREFULLY REMOVE THE TRIM COVER FROM THE ADHESIVE TAPE.	Ease the trim cover from the adhesive tape.
	CAREFULLY MOVE	CAREFULLY MOVE THE PIPE INTO THE CORRECT POSITION.	Ease the pipe into the right position.

Word (part of speech)	Approved meaning/ ALTERNATIVES	APPROVED EXAMPLE	Not approved example
EASILY (adv)	Without difficulty	OXIDATION OCCURS VERY EASILY.	
EASY (adj) (EASIER, EASIEST)	Without difficulty	USE THE LEFT OPENING FOR EASY ACCESS.	
EAT (v), EATS, ATE, EATEN	To consume food	DO NOT EAT IN THE WORK AREA.	
	For safety instructions, use:		
	SWALLOW (v)	DO NOT SWALLOW THE POTTING COMPOUND.	Do not eat the potting compound.
EDGE (n)	A line that is the intersection of two surfaces of a solid object	THE DISTANCE BETWEEN THE EDGE OF THE PANEL AND THE PARTITION MUST NOT BE MORE THAN 0.05 mm.	
EFFECT (n)	The result of a cause	WHEN DUST MIXES WITH OIL, IT HAS AN ABRASIVE EFFECT.	
effect (v)	DO (v)	DO THE TIGHTENING PROCEDURE.	Effect the tightening procedure.
effective (adj)	GOOD (adj)	THIS MATERIAL GIVES GOOD PROTECTION FROM CORROSION.	This material provides effective protection from corrosion.
efficacious (adj)	GOOD (adj)	THIS IS A GOOD PROCEDURE TO REMOVE PAINT.	This is an efficacious way to remove paint.
efficient (adj)	SATISFACTORY (adj)	THE TRANSMISSION CONTROL PROTOCOL IS NOT SATISFACTORY FOR THE TRANSMISSION OF INTERACTIVE TRAFFIC.	The Transmission Control Protocol is not efficient for the transmission of interactive traffic.

Word (part of speech)	Approved meaning/ ALTERNATIVES	APPROVED EXAMPLE	Not approved example
efficiently (adv)	SATISFACTORILY (adv)	MAKE SURE THAT THE UNIT OPERATES SATISFACTORILY.	Make sure the unit operates efficiently.
efflux (n)	EXHAUST (n)	MAKE SURE THAT THERE ARE SAFETY BARRIERS AROUND THE ENGINE EXHAUST AREA.	Ensure that there are safety barriers around the engine efflux area.
effort (n)	FORCE (TN)	DO NOT TIGHTEN THE BOLTS WITH TOO MUCH FORCE.	Do not use too much effort to tighten the bolts.
	TRY (v)	FIRST, TRY TO TURN THE SHAFT WITH YOUR HANDS.	First, make an effort to turn the shaft with your hands.
either (adj)	ONE (TN) OF THE TWO (TN)	IF THERE IS MERCURY CONTAMINATION IN ONE OF THE TWO COMPARTMENTS, CLEAN THE AREA IMMEDIATELY.	If there is mercury spillage in either compartment, clean the area immediately.
either (conj)	ONE (TN) OF THE TWO (TN)	YOU CAN USE A RED PEN OR A GREEN PEN.	You can use either a red pen or a green pen.
either (pron)	ONE (TN) OF THE TWO (TN)	APPLY ELECTRICAL POWER TO ONE OF THE TWO SOLENOIDS.	Apply electrical power to either of the solenoids.
EJECT (v), EJECTS, EJECTED, EJECTED	To move or to cause a person or item to move from an aircraft or equipment with force	IF YOU PULL THE EJECTION SEAT HANDLE, THE SEAT WILL EJECT.	
ejection (n)	EJECT (v)	PUSH THE APPLICABLE BUTTON TO EJECT THE DISK DRIVE.	Push the applicable button for disk drive ejection.
elapse (v)	TIME (n)	MAKE SURE THAT THE TIME BETWEEN STARTS IS A MINIMUM OF 30 SECONDS.	Check that 30 seconds have elapsed between starts.
	AFTER (prep)	AFTER 30 SECONDS, DO THE TEST AGAIN.	When 30 seconds have elapsed, do the test again.
	INTERVAL (n)	AT INTERVALS OF SIX MONTHS, DO THE CHECK OF THE OIL LEVEL.	When six months have elapsed, do the check of the oil level.

Word (part of speech)	Approved meaning/ ALTERNATIVES	APPROVED EXAMPLE	Not approved example
ELECTRIC (adj)	Operated by electricity	IF THE HYDRAULIC PUMP DOES NOT SUPPLY A PRESSURE OF 3000 PSI, USE THE ELECTRIC PUMP.	
ELECTRICAL (adj)	Related to electricity	SUPPLY ELECTRICAL POWER TO THE SYSTEM.	
ELECTRICALLY (adv)	Related to or operated by electricity	THE SYSTEM IS HYDRAULICALLY OPERATED AND ELECTRICALLY CONTROLLED.	
ELECTRO-MAGNETIC (adj)	Related to electromagnetism	THE ELECTRO-MAGNETIC TEST PROCEDURE USES ELECTROMAGNETIC FORCE.	
ELECTRO-MAGNETICALLY (adv)	Related to electromagnetism	THE SYSTEM HOLDS THE POSITION ELECTRO-MAGNETICALLY.	
ELECTRONIC (adj)	Related to or operated by electronics	THIS BAY CONTAINS SOME OF THE ELECTRONIC EQUIPMENT.	
ELECTRONICALLY (adv)	Related to or operated by electronics	THIS UNIT IS ELECTRONICALLY CONTROLLED.	
eliminate (v)	REMOVE (v)	REMOVE ALL SHARP EDGES.	Eliminate all sharp edges.
	STOP (v)	PUT A BAG OF DESICCANT IN THE PACKAGE TO STOP CONDENSATION.	Put a bag of desiccant in the package to eliminate condensation.
	PREVENT (v)	PUT A BAG OF DESICCANT IN THE PACKAGE TO PREVENT CONDENSATION.	Put a bag of desiccant in the package to eliminate condensation.

Word (part of speech)	Approved meaning/ ALTERNATIVES	APPROVED EXAMPLE	Not approved example
EMERGENCY (n)	A condition that occurs suddenly and is dangerous	IN AN EMERGENCY, BREAK THE GLASS.	
emit (v)	FROM (prep)	THE VAPORS FROM THIS MATERIAL ARE DANGEROUS TO THE SKIN.	The vapors that this material emits are dangerous to the skin.
employ (v)	USE (v)	TYPE B USES A LOOSE COLLAR.	Type B employs a loose collar.
EMPTY (adj)	Without contents	MAKE SURE THAT THE RESERVOIR IS EMPTY.	
empty (v)	REMOVE (v)	REMOVE ALL FUEL FROM THE FUEL TANKS.	Empty the fuel tanks.
enable (v)	LET (v)	THE DISASSEMBLY FIXTURE LETS YOU DISASSEMBLE THE MODULE.	The disassembly fixture enables you to disassemble the module.
encircle (v)	AROUND (prep)	PUT SEALANT AROUND EACH SCREW HOLE.	Encircle each screw hole with sealant.
enclosed (adj)	CLOSED (adj)	IF YOU USE SOLVENT IN A CLOSED SPACE, MAKE SURE THAT THERE IS GOOD AIRFLOW IN THE WORK AREA.	The work area must be well-ventilated if solvent is used in an enclosed space.
encounter (v)	BE (v)	IF THERE ARE ROUGH AREAS, REMOVE THEM WITH EMERY CLOTH.	If rough spots are encountered, remove them with emery cloth.
	FIND (v)	IF YOU FIND ROUGH AREAS, REMOVE THEM WITH EMERY CLOTH.	If rough spots are encountered, remove them with emery cloth.
END (n)	Where time, distance, or sequence are completed	EXAMINE THE SHACKLE AT THE END OF THE SLING.	

Word (part of speech)	Approved meaning/ ALTERNATIVES	APPROVED EXAMPLE	Not approved example
end (v)	STOP (v)	STOP THE TEST.	End the test.
	COMPLETE (v)	WHEN YOU COMPLETED THE TEST, REMOVE THE SENSOR.	When the test has ended, remove the sensor.
		WHEN THE TEST IS COMPLETED, REMOVE THE SENSOR.	When the test has ended, remove the sensor.
ENERGIZE (v), ENERGIZES, ENERGIZED, ENERGIZED	To supply electrical power to	ENERGIZE THE AC AND DC BUSBARS.	
enforce (v)	OBEY (v)	ALL PERSONNEL MUST OBEY THE SAFETY PRECAUTIONS.	Safety precautions must be enforced.
ENGAGE (v), ENGAGES, ENGAGED, ENGAGED	To correctly align and come together	ENGAGE THE CLUTCH.	
		MAKE SURE THAT THE CLUTCH ENGAGES.	
engagement (n)	ENGAGE (v)	MAKE SURE THAT THE SPLINES ENGAGE FULLY.	Make sure there is full engagement of the splines.
enlarge (v)	INCREASE (v)	INCREASE THE DIAMETER OF THE RIVET HOLE.	Enlarge the rivet hole diameter.
	LARGE (adj)	MAKE THE HOLES LARGER.	Enlarge the holes.
enough (adj)	SUFFICIENT (adj)	MAKE SURE THAT THERE IS SUFFICIENT PRESSURE IN THE ACCUMULATOR.	Make sure that there is enough pressure in the accumulator.
ensue (v)	CAUSE (v)	INCORRECT INSTALLATION WILL CAUSE DAMAGE.	Damage will ensue from incorrect installation.
ensure (v)	MAKE SURE (v)	MAKE SURE THAT THE CORRECT SEALS ARE INSTALLED.	Ensure that the correct seals are installed.

Word (part of speech)	Approved meaning/ ALTERNATIVES	APPROVED EXAMPLE	Not approved example
entail (v)	MUST (v)	FOR THIS PROCEDURE, YOU MUST USE SPECIAL TOOLS.	This procedure entails the use of special tools.
enter (v)	GO INTO	DO NOT GO INTO THE ENGINE TEST AREA WITHOUT APPROVAL.	Do not enter the engine test area without approval.
	RECORD (v)	RECORD THE DATA IN THE LOGBOOK.	Enter the data in the logbook.
	ENTRY (n)	MAKE AN ENTRY IN THE LOGBOOK TO SHOW THAT YOU DID THE DEACTIVATION PROCEDURE.	Enter in the logbook that you did the deactivation procedure.
	You can use this word as a technical verb for computer processes and applications.	ENTER YOUR PASSWORD.	
entire (adj)	FULL (adj)	MAKE A SCAN OF THE FULL RANGE OF FREQUENCIES AVAILABLE.	Scan the entire range of frequencies available.
	ALL (adj)	EXAMINE ALL OF THE SYSTEM TO FIND THE CAUSE OF THE INTERFERENCE.	Check the entire system to determine the cause of interference.
entirely (adv)	FULLY (adv)	THE SYSTEM IS FULLY AUTOMATIC.	The system is entirely automatic.
ENTRANCE (n)	An opening to go into a space	PUT SAFETY BARRIERS AT THE ENTRANCE TO THE WORK AREA.	
ENTRY (n)	A record of an occurrence	MAKE AN ENTRY IN THE LOG CARD.	
EQUAL (adj)	The same dimension, quality, quantity, or number	MAKE SURE THAT THERE IS AN EQUAL DISTANCE BETWEEN THE HOLES.	

Word (part of speech)	Approved meaning/ ALTERNATIVES	APPROVED EXAMPLE	Not approved example
equal (v)	SAME (adj)	DIMENSION "X" MUST BE THE SAME AS DIMENSION "Z".	Dimension X must equal dimension Z.
	EQUAL (adj)	DIMENSION "X" MUST BE EQUAL TO DIMENSION "Y".	Dimension X must equal dimension Y.
EQUALLY (adv)	The same dimension, quality, quantity, or number	ADJUST THE TURNBUCKLES EQUALLY.	
equip (v)	INSTALL (v)	INSTALL THE SHUTOFF VALVES ON EACH HP MANIFOLD.	Equip each HP manifold with shutoff valves.
EQUIPMENT (n)	The objects or tools used for operation and maintenance	REMOVE ALL THE EQUIPMENT FROM THE WORK AREA.	
equipped (adj)	HAVE (v)	BE CAREFUL WHEN YOU MOVE JET PIPES OR EXHAUST CONES THAT HAVE INSULATING BLANKETS.	Be careful when handling jet pipes or exhaust cones equipped with insulating blankets.
EQUIVALENT (adj)	Has the same properties, functions, or values	MATERIALS (REF. No. 2) AND (REF. No. 18) ARE EQUIVALENT.	
EQUIVALENT (n)	Something that has the same properties, functions, or values	USE CLEANING COMPOUND ARDROX 6025, OR AN EQUIVALENT.	
ERASE (v), ERASES, ERASED, ERASED	To remove data from a medium	ERASE THE TAPE.	
erect (v)	ASSEMBLE (v)	ASSEMBLE THE MOVABLE HOIST BEFORE YOU LIFT THE GEARBOX.	Erect the movable hoist before lifting the gearbox.
	BECOME STABLE	LET THE VERTICAL GYRO BECOME STABLE.	Give the vertical gyro time to erect.
erratic (adj)	IRREGULAR (adj)	THE OPERATION OF THE COMPASS IS IRREGULAR.	The compass is erratic.

Word (part of speech)	Approved meaning/ ALTERNATIVES	APPROVED EXAMPLE	Not approved example
ERROR (n)	The difference from that which is correct or accurate	IF THE ERROR IS MORE THAN 3 DEG., CALIBRATE THE COMPASS.	
escape (v)	DEFLATE (v)	DEFLATE THE TIRE.	Let the air escape from the tire.
	RELEASE (v)	RELEASE ALL THE NITROGEN.	Allow all the nitrogen to escape.
essential (adj)	NECESSARY (adj)	HYDRAULIC POWER IS NOT NECESSARY.	Use of hydraulic power is not essential.
	MUST (v)	YOU MUST COMPLETE THE TEST.	It is essential to complete the test.
establish (v)	MAKE SURE (v)	IN THIS STEP, YOU WILL MAKE SURE THAT THE FUEL PUMP OPERATES CORRECTLY.	The following step establishes the correct functioning of the fuel pump.
ESTIMATE (n)	A calculated, approximate result	MAKE AN ESTIMATE OF THE NECESSARY QUANTITY.	
estimate (v)	ESTIMATE (n)	MAKE AN ESTIMATE OF THE QUANTITY OF CLEANING FLUID THAT IS NECESSARY.	Estimate the amount of cleaning fluid that is necessary.
evaluate (v)	EXAMINE (v)	EXAMINE THE DAMAGE AFTER AN ACCIDENT.	Evaluate the damage after an accident.
	ANALYSIS (TN)	MAKE AN ANALYSIS OF THE TEST RESULTS.	Evaluate the test results.
evaluation (n)	EXAMINE (v)	EXAMINE THE DAMAGE.	Do an evaluation of the damage.
	ANALYSIS (TN)	MAKE AN ANALYSIS OF THE TEST RESULTS.	Do an evaluation of the test results.
evaporate (v)	DRY (v)	LET THE COMPOUND DRY.	Allow the solvent to evaporate from the compound.

Word (part of speech)	Approved meaning/ ALTERNATIVES	APPROVED EXAMPLE	Not approved example
even (adj)	SMOOTH (adj)	A GUIDE ROD IS ATTACHED TO THE METAL BELLOWS DISC FOR A SMOOTH MOVEMENT OF THE BELLOWS.	A guide rod is attached to the metal bellows disc for an even movement of the bellows.
	SAME (adj)	APPLY THE SAME LOAD ON THE TWO BRAKES.	Braking must be even on the two brakes.
	FLAT (adj)	MAKE SURE THAT THE SURFACE IS FLAT.	Ensure the surface is even.
	FLUSH (adj)	MAKE SURE THAT THE FASTENER HEADS ARE FLUSH WITH THE SKIN.	Make sure that the fastener heads are even with the skin.
	LEVEL (adj)	MAKE SURE THAT THE MARK ON THE BOLT IS LEVEL WITH THE MARK ON THE HANDLE.	Make sure that the mark on the bolt is even with the mark on the handle.
evenly (adv)	GRADUALLY (adv)	INCREASE THE TEMPERATURE GRADUALLY.	Increase the temperature evenly.
	EQUALLY (adv)	APPLY THE LOAD EQUALLY ON THE AREA.	Apply the load evenly on the area.
event (n)	IF (conj)	IF THE PRESSURE DECREASES SUDDENLY, CLOSE THE VALVE.	In the event of a sudden dropping of pressure, close the valve.
eventually (adv)	SOME TIME	IF THE CABLE RUBS, IT WILL BREAK AFTER SOME TIME.	If the cable chafes, it will eventually break.
every (adj)	ALL (adj)	REMOVE ALL SAFETY PINS BEFORE FLIGHT.	Every safety pin must be removed before flight.
	EACH (adj)	SET THE BTRY SWITCH TO “ON” EACH TIME.	It is necessary to set the BTRY switch to ON every time.
	INTERVAL (n)	DO AN INSPECTION AT INTERVALS OF 20 000 FLIGHT HOURS.	Do an inspection every 20 000 flight hours.

Word (part of speech)	Approved meaning/ ALTERNATIVES	APPROVED EXAMPLE	Not approved example
evidence (n)	INDICATION (n)	IF THERE IS AN INDICATION OF A LEAK, REPLACE THE VALVE.	If there is evidence of a leak, replace the valve.
	SIGN (n)	ONE SIGN OF A LEAK OF HOT AIR FROM A JOINT IS THAT THE TEMPERATURE INDICATION DECREASES.	Evidence of a leak of hot air from a joint is that the temperature reading drops.
	SHOW (v)	REPLACE THE INSULATION BLANKETS THAT SHOW CHAFING.	Replace the insulation blankets with evidence of chafing.
evidence (v)	SHOW (v)	IF THE TEST SHOWS NO FAULT, CONTINUE THE PROCEDURE.	If no fault is evidenced by the test, continue the procedure.
	FIND (v)	IF YOU DO NOT FIND DAMAGE DURING THE INSPECTION PROCEDURE, ASSEMBLE THE UNIT.	If the inspection does not evidence any damage, assemble the unit.
evident (adj)	FIND (v)	IF YOU DO NOT FIND DAMAGE, CONTINUE THE INSPECTION.	If no damage is evident, continue the inspection.
exact (adj)	ACCURATE (adj)	IF THE INDICATION IS NOT ACCURATE, DO THE TEST AGAIN.	If the indication is not exact, do the test again.
	CORRECT (adj)	IF THE INDICATION IS NOT CORRECT, DO THE TEST AGAIN.	If the indication is not exact, do the test again.
exactly (adv)	ACCURATELY (adv)	PUT THE REPAIR SHEET ACCURATELY ON THE DAMAGED AREA.	Put the repair sheet exactly on the damaged area.
	FULLY (adv)	FIND THE BOLT HOLE THAT ALIGNS FULLY WITH ONE OF THE OUTER SLOTS.	Find the bolt hole that aligns exactly with one of the outer slots.
	CORRECT (adj)	THE SEAL MUST BE OF THE CORRECT DIMENSION FOR THE GROOVE.	The seal must fit the groove exactly.

Word (part of speech)	Approved meaning/ ALTERNATIVES	APPROVED EXAMPLE	Not approved example
examination (n)	EXAMINE (v)	EXAMINE THE BLADE FOR CRACKS.	Do an examination of the blades for cracks.
	FIND (v)	IF YOU FIND CORROSION, REMOVE IT WITH THE SPECIFIED MATERIAL.	If the examination reveals corrosion, it must be removed with the specified material.
EXAMINE (v), EXAMINES, EXAMINED, EXAMINED	To look carefully at	EXAMINE THE COMPUTER FOR DAMAGE.	
EXAMPLE (n)	Something from a group which has the same qualities as the group	FIGURE 2 IS AN EXAMPLE OF A REPAIR PROCEDURE FOR A DAMAGED WEB.	
exceed (v)	MORE THAN	MAKE SURE THAT THE INDICATION IS NOT MORE THAN 400 KNOTS.	Ensure that the indication does not exceed 400 knots.
except (prep)	To replace this word, use a different construction. Be specific.	KEEP ALL THE PARTS THAT ARE SERVICEABLE AND DISCARD THOSE THAT ARE BROKEN.	Keep all parts except those that are broken.
		REMOVE ALL SCREWS, BUT NOT THE ONE ON THE LEFT SIDE OF THE FLANGE.	Remove all screws except the one on the left side of the flange.
exception (n)	To replace this word, use a different construction. Be specific.	KEEP ALL THE PARTS THAT ARE SERVICEABLE AND DISCARD THOSE THAT ARE BROKEN.	Keep all parts with the exception of those that are broken.
		REMOVE ALL SCREWS, BUT NOT THE ONE ON THE LEFT SIDE OF THE FLANGE.	Remove all screws with the exception of the one on the left side of the flange.
exceptional (adj)	UNUSUAL (adj)	IF YOU MUST APPLY UNUSUAL FORCE TO OPEN THE UNIT, EXAMINE THE CONTROL LINKAGE.	If exceptional force is needed to open the unit, examine the control linkage.

Word (part of speech)	Approved meaning/ ALTERNATIVES	APPROVED EXAMPLE	Not approved example
exceptionally (adv)	UNUSUALLY (adv)	IF THE TIRE PRESSURE IS UNUSUALLY HIGH, OPEN THE VALVE.	If the tire pressure is exceptionally high, open the valve.
excess (adj)	TOO MUCH	DO NOT APPLY TOO MUCH ADHESIVE ON THE GASKET.	Do not apply excess adhesive on the gasket.
	MORE THAN	DO NOT PUT MORE THAN THE PERMITTED WEIGHT ON THE TROLLEY.	Do not put excess weight on the trolley.
	UNWANTED (adj)	REMOVE UNWANTED COMPOUND FROM THE GASKET.	Wipe excess compound from the gasket.
	NOT NECESSARY	REMOVE ALL REMAINING SEALANT THAT IS NOT NECESSARY.	Remove all the excess sealant.
excess (n)	TOO MUCH	IF YOU USE TOO MUCH PAINT, THE FINISH WILL BE UNSATISFACTORY.	An excess of paint will give a poor finish.
	MORE THAN	THE HYDRAULIC PRESSURE MUST NOT BE MORE THAN 200 BARS.	The hydraulic pressure must not be in excess of 200 bars.
excessive (adj)	TOO MUCH	TOO MUCH FORCE CAN CAUSE DAMAGE TO THE SEAL.	Excessive force can damage the seal.
	MORE THAN	IF THE PRESSURE IS MORE THAN 1000 kPa, STOP THE TEST.	If the pressure is excessive, stop the test.
excessively (adv)	TOO MUCH	BE CAREFUL THAT YOU DO NOT TURN THE INNER RACES TOO MUCH.	Be careful not to excessively rotate the inner races.
	MORE THAN	IF THE OIL LEAKAGE IS MORE THAN FIVE DROPS A MINUTE, EXAMINE THE SEALANT.	If oil leaks excessively, check the sealant.

Word (part of speech)	Approved meaning/ ALTERNATIVES	APPROVED EXAMPLE	Not approved example
exchange (v)	INTERCHANGE (v)	INTERCHANGE THE No. 1 AND No. 2 COMPUTERS. THEN DO THE TEST AGAIN.	Exchange the No. 1 and No. 2 computers. Then do the test again.
	REPLACE (v)	REPLACE THE O-RINGS.	Exchange the O-rings.
exclude (v)	NOT INCLUDE	THIS PROCEDURE DOES NOT INCLUDE THE REMOVAL OF THE PISTON.	This procedure excludes the removal of the piston.
	NOT USE	DO NOT USE OXYGEN.	Exclude the use of oxygen.
excluding (prep)	WITHOUT (prep)	THE WEIGHT OF THE UNIT, WITHOUT THE PUMP, IS 10 kg.	The weight of the unit, excluding the pump, is 10 kg.
execute (v)	DO (v)	DO THESE STEPS.	Execute the following operations.
exempt (adj)	NO (adj)	MAKE SURE THAT THERE ARE NO METAL PARTICLES IN THE OIL.	Ensure the oil is exempt from metal particles.
	NOT SHOW	MAKE SURE THAT THE SEALS DO NOT SHOW SIGNS OF DETERIORATION.	Ensure the seals are exempt from signs of deterioration.
exercise (v)	MAKE SURE (v)	MAKE SURE THAT YOU DO NOT GET ACID ON YOUR SKIN.	Exercise caution not to allow acid to contact skin.
	CAREFUL (adj)	BE CAREFUL THAT YOU DO NOT GET ACID ON YOUR SKIN.	Exercise caution not to allow acid to contact skin.
exert (v)	APPLY (v)	DO NOT APPLY PRESSURE TO THE AFT FLANGE.	Do not exert pressure on the aft flange.
exhale (v)	BREATHE (v)	BREATHE INTO THE OXYGEN MASK.	Exhale into the oxygen mask.
EXHAUST (n)	The gas from an engine	THE EXHAUST FROM THE APU IS DANGEROUS.	

Word (part of speech)	Approved meaning/ ALTERNATIVES	APPROVED EXAMPLE	Not approved example
exhaust (v)	REMOVE (v)	USE A FAN TO REMOVE THE GASES.	Exhaust the gases using a fan.
	DECREASE (v)	DECREASE THE NITROGEN PRESSURE TO 120 PSI.	Exhaust the nitrogen pressure to 120 psi.
	RELEASE (v)	RELEASE THE HYDRAULIC PRESSURE IN THE ACCUMULATOR.	Exhaust all the hydraulic pressure in the accumulator.
exhaustive (adj)	ALL (adj)	ALL THE TEST CONDITIONS ARE GIVEN IN PARAGRAPH 2.0.	An exhaustive list of test conditions is given at paragraph 2.0.
	FULL (adj)	THIS IS NOT A FULL LIST.	The list is not exhaustive.
exist (v)	BE (v)	THERE ARE TWO DIFFERENT PROCEDURES FOR THIS REPAIR.	Two different procedures exist for this repair.
EXIT (n)	An opening to go out of a space	MAKE SURE THAT ALL THE EMERGENCY EXITS ARE CLEAR OF UNWANTED OBJECTS.	
EXPAND (v), EXPANDS, EXPANDED, EXPANDED	To increase in dimension, volume, or time	THE BELLOWS MUST EXPAND.	
		HEAT EXPANDS THE GAS IN THE CONTAINER.	
expect (v)	POSSIBLE (adj)	IF IT IS POSSIBLE THAT THE TEMPERATURE WILL DECREASE TO -20 DEGREES C, REMOVE THE BATTERIES.	If temperature is expected to drop to -20 degrees C, remove the batteries.

Word (part of speech)	Approved meaning/ ALTERNATIVES	APPROVED EXAMPLE	Not approved example
expel (v)	REMOVE (v)	REMOVE THE AIR BUBBLES FROM THE COMPOUND WITH A SPATULA.	Expel the air bubbles from the compound with a spatula.
	BLEED (v)	BLEED ALL NITROGEN FROM THE SYSTEM.	Expel all nitrogen from the system.
	RELEASE (v)	RELEASE ALL NITROGEN.	Expel all nitrogen.
EXPIRED (adj)	More than the permitted life	MAKE SURE THAT THE SHELF LIFE IS NOT EXPIRED.	
explain (v)	TELL (v)	TELL THE GROUND CREW THAT YOU WILL OPERATE THE SYSTEM.	Explain to the ground crew that you will operate the system.
EXPLOSION (n)	The effect when an explosive material quickly releases its energy	THIS CONDITION CAN CAUSE AN EXPLOSION.	
EXPLOSIVE (adj)	That can cause an explosion	THE SAFETY PRECAUTIONS THAT FOLLOW ARE APPLICABLE TO EXPLOSIVE ITEMS.	
expose (v)	GET (v)	REMOVE THE COVER ASSEMBLY TO GET ACCESS TO THE LAMP ASSEMBLY.	Remove cover assembly to expose lamp assembly.
EXTEND (v), EXTENDS, EXTENDED, EXTENDED	To increase, or cause something to increase, in dimension or range	FULLY EXTEND THE JACK RAM.	
		MAKE SURE THAT THE RODS EXTEND AND RETRACT.	
EXTENSION (n)	1. The action when something extends	MAKE SURE THAT THE EXTENSION OF THE ACTUATOR IS SMOOTH.	
	2. The result when something extends	THE TOTAL EXTENSION OF THE PISTON IS 125 mm.	

Word (part of speech)	Approved meaning/ ALTERNATIVES	APPROVED EXAMPLE	Not approved example
extent (n)	DEPTH (n)	MEASURE THE DEPTH OF THE CRACK.	Measure the extent of the crack.
	LENGTH (n)	MEASURE THE LENGTH OF THE CRACK.	Measure the extent of the crack.
	WIDTH (n)	MEASURE THE WIDTH OF THE CRACK.	Measure the extent of the crack.
exterior (n)	EXTERNAL (adj)	EXAMINE THE EXTERNAL SURFACE OF THE CONTAINER.	Examine the exterior of the container.
EXTERNAL (adj)	Of, or on, the outer side	CONNECT THE EXTERNAL POWER SUPPLY.	
EXTERNALLY (adv)	Of, or on, the outer side	THE BOX IS INSTALLED EXTERNALLY.	
EXTINGUISH (v), EXTINGUISHES, EXTINGUISHED, EXTINGUISHED	To stop burning	TO EXTINGUISH A FIRE IN THE ENGINE, PULL THE "FIRE" HANDLE.	
		WHEN THE FLAME EXTINGUISHES, STOP THE ENGINE TEST.	
	For lights or lamps, use:		
	GO OFF	THE AUTO BRK PUSHBUTTON LIGHT GOES OFF.	The AUTO BRK pushbutton light extinguishes.
	BE OFF	MAKE SURE THAT THE GREEN LIGHT ON THE PANEL IS OFF.	Make sure that the green light on the panel extinguishes.
extra (adj)	MORE (adj)	THIS MATERIAL GIVES MORE PROTECTION.	This material provides extra protection.
extract (v)	REMOVE (v)	REMOVE THE BOLT.	Extract the bolt.

Word (part of speech)	Approved meaning/ ALTERNATIVES	APPROVED EXAMPLE	Not approved example
extreme (adj)	VERY HIGH (adj)	DO NOT APPLY VERY HIGH LOADS.	Do not apply extreme loads.
	VERY BAD (adj)	WHEN THE WEATHER IS VERY BAD, KEEP THESE PARTS IN A DRY AREA.	In extreme weather conditions, keep these parts in a dry area.
	VERY COLD (adj)	MAKE SURE THAT YOU KNOW THE RISKS RELATED TO VERY COLD TEMPERATURES.	Make sure that you know the risks related to extreme cold temperatures.
	VERY HOT (adj)	MAKE SURE THAT YOU KNOW THE RISKS RELATED TO VERY HOT TEMPERATURES.	Make sure that you know the risks related to extreme hot temperatures.
extremely (adv)	VERY (adv)	IT IS VERY IMPORTANT TO OBEY THE FUEL SAFETY PRECAUTIONS.	It is extremely important to observe the fuel safety precautions.
exude (v)	OUT OF (prep)	MAKE SURE THAT NEW GREASE COMES OUT OF THE JOINT.	Make sure fresh grease exudes from the joint.

Word (part of speech)	Approved meaning/ ALTERNATIVES	APPROVED EXAMPLE	Not approved example
fabricate (v)	MAKE (v)	USE COPPER TO MAKE THE CHILL BLOCK.	Use copper to fabricate the chill block.
FACE (n)	The front surface of an object	CLEAN THE FACE OF THE INDICATOR.	
	This word is a technical name when it is referred to a part of the body.	WHEN YOU DO MAINTENANCE ON THE WASTE WATER SYSTEM, DO NOT TOUCH YOUR FACE WITH YOUR GLOVES.	
face (v)	POINT (v)	WHEN YOU INSTALL THE PITOT HEAD, MAKE SURE THAT THE OPENING POINTS FORWARD.	When installing the pitot head, make sure that the opening is facing forward.
	TURN (v)	TURN YOUR FACE AWAY FROM THE DRAIN VALVE BEFORE YOU PUSH THE VALVE STEM DOWN.	Face away from the drain valve when pushing the valve stem down.
facilitate (v)	HELP (v)	IDENTIFY THE COMPONENT WITH A CODE TO HELP YOU TO CORRECTLY INSTALL IT AGAIN.	Mark the component with a code that will facilitate its correct reinstallation.
	MAKE ... EASIER	THIS MAKES FAULT ISOLATION EASIER.	This facilitates fault isolation.
facility (n)	Be specific. Use the name of the facility as a technical name.	WEIGH THE AIRCRAFT ON A WEIGHBRIDGE.	You need a special facility to weigh the aircraft.
FACT (n)	Something that occurred or is correct	GET ALL THE FACTS ABOUT THE INCIDENT FROM THE PILOT.	
factor (n)	CAUSE (n)	THERE CAN BE MANY CAUSES FOR CORROSION.	Corrosion can be caused by several factors.

Word (part of speech)	Approved meaning/ ALTERNATIVES	APPROVED EXAMPLE	Not approved example
fail (v)	IF ... NOT	IF YOU DO NOT OBEY THESE INSTRUCTIONS, THERE IS A RISK OF INJURY.	If you fail to comply with these instructions, there is a risk of injury.
	FAILURE (TN)	IF YOU DO NOT OBEY THE PROCEDURE CORRECTLY, FAILURE OF THE EMERGENCY FLOTATION GEAR CAN OCCUR.	If you do not follow the procedure correctly, the emergency flotation gear can fail.
	UNSATISFACTORY (adj)	IF THE TEST RESULTS ARE UNSATISFACTORY, DO THE FAULT ISOLATION PROCEDURE.	If the test fails, do the fault isolation procedure.
failed (adj)	DEFECTIVE (adj)	REPLACE THE DEFECTIVE PUMP.	Replace the failed pump.
	UNSERVICEABLE (adj)	REPLACE THE UNSERVICEABLE VALVE.	Replace the failed valve.
failure (n)	IF ... NOT	IF YOU DO NOT OBEY THESE INSTRUCTIONS, THERE IS A RISK OF INJURY.	Failure to comply with these instructions, will result in a risk of injury.
	This word when it means a performance error or loss of serviceability is a technical name.	CONTAMINATION CAN CAUSE A FAILURE OF THE PUMP.	
faint (adj)	DIM (adj)	IF THE LIGHT IS TOO DIM, REPLACE THE BATTERY.	If the light is faint, replace the battery.
	INTENSITY (n) Refer also to BRIGHT (adj).	TURN THE KNOB COUNTERCLOCKWISE TO DECREASE THE INTENSITY OF THE LIGHT.	Turn the knob anticlockwise to make the light fainter.

Word (part of speech)	Approved meaning/ ALTERNATIVES	APPROVED EXAMPLE	Not approved example
fall (n)	DECREASE (v)	WHEN THE PRESSURE DECREASES, THE INDICATOR SHOWS THAT THERE IS A LEAK.	When there is a fall in pressure, the indicator shows a leak.
FALL (v), FALLS, FELL, FALLEN	To move down by the force of gravity	MAKE SURE THAT THE TOOLS DO NOT FALL INTO THE ENGINE.	
false (adj)	INCORRECT (adj)	MAKE SURE THAT THERE ARE NO INCORRECT INDICATIONS ON THE INSTRUMENTS.	Make sure there are no false indications on the instruments.
falter (v)	SMOOTHLY (adv)	MAKE SURE THAT THE GENERATOR OPERATES SMOOTHLY.	Make sure that the generator does not falter.
	CONSTANT (adj)	MAKE SURE THAT THE PRESSURE STAYS CONSTANT.	Make sure that the pressure does not falter.
familiar (adj)	KNOW (v)	ALL PERSONNEL MUST FULLY KNOW THE SAFETY PRECAUTIONS.	All personnel must be thoroughly familiar with the safety precautions.
FAR (adj) (FARTHER, FARTHEST)	At or to a relatively large distance	REMOVE THE BOLT THAT IS FARTHEST FROM THE CENTER.	
fashion (n)	PROCEDURE (n)	USE THIS PROCEDURE TO DO THE TASK.	Carry out the task in this fashion.
FAST (adj) (FASTER, FASTEST)	At high speed	A FAST MOVEMENT OF THE THROTTLE WILL CAUSE A SURGE.	
fast (adv)	QUICKLY (adv)	DO NOT LET THE CABIN ALTITUDE CHANGE TOO QUICKLY.	Do not let the cabin altitude change too fast.

Word (part of speech)	Approved meaning/ ALTERNATIVES	APPROVED EXAMPLE	Not approved example
fasten (v)	ATTACH (v)	ATTACH THE TWO STRINGERS TOGETHER WITH RIVETS.	Fasten the two stringers together with rivets.
	CONNECT (v)	CONNECT THE HOSE TO THE TEST EQUIPMENT.	Fasten the hose to the test equipment.
fatal (adj)	KILL (v)	HIGH VOLTAGE IN THE ELECTRONIC SYSTEM CAN KILL YOU.	High voltage in the electronic system can be fatal.
faulty (adj)	DEFECTIVE (adj)	REPLACE ALL DEFECTIVE ELECTRONIC COMPONENTS.	All faulty electronic components must be replaced.
	INCORRECT (adj)	AN INCORRECT ADJUSTMENT OF THE PRESSURE REGULATOR CAN BE DANGEROUS.	Faulty adjustment of the pressure regulator can be dangerous.
	UNSERVICEABLE (adj)	REPLACE THE UNSERVICEABLE CONTROL UNIT.	Replace the faulty control unit.
	UNSATISFACTORY (adj)	UNSATISFACTORY OPERATION OF THE SYSTEM CAN CAUSE DAMAGE.	Faulty operation of the system can cause damage.
fear (v)	THINK (v)	DO A TEST OF THE UNIT IF YOU THINK THAT THERE IS A PROBLEM WITH ITS OPERATION.	Test the unit if you fear that there is a problem with its operation.
	MAKE SURE (v)	DO A TEST OF THE UNIT TO MAKE SURE THAT THERE IS NO PROBLEM WITH ITS OPERATION.	Test the unit if you fear that there is a problem with its operation.
feasible (adj)	POSSIBLE (adj)	DO THE PROCEDURE MANUALLY IF IT IS NOT POSSIBLE TO USE THE AUTOMATIC CONTROL.	Do the manual procedure if it is not feasible to use the automatic control.
	CAN (CANNOT) (v)	IF YOU CANNOT REMOVE THE BOLT, YOU MUST CUT IT.	If it is not feasible to remove the bolt, you must cut it.

Word (part of speech)	Approved meaning/ ALTERNATIVES	APPROVED EXAMPLE	Not approved example
FEATHER (v), FEATHERS, FEATHERED, FEATHERED	To put a propeller to a position of minimum drag	PUT THE CONDITION LEVER IN THE FTR POSITION TO FEATHER THE PROPELLER.	
feature (v)	HAVE (v)	THESE NUTS HAVE A FLAT FLANGE.	These nuts feature a flat flange.
feed (v)	PUT (v)	PUT THE WIRE THROUGH THE CONDUIT.	Feed the wire through the conduit.
	SUPPLY (v)	SUPPLY COOL AIR TO THE CABIN.	Feed cool air to the cabin.
FEEL (v), FEELS, FELT, FELT	To touch to find	CAREFULLY LIFT THE MAGNETIC LEVEL INDICATOR UNTIL YOU FEEL THE MAGNETS.	
few (adj)	SMALL NUMBER If possible, give a specific number.	THIS PROCEDURE IS ONLY FOR REPAIRS THAT INCLUDE A VERY SMALL NUMBER OF RIVETS.	This procedure is only for repairs that include very few rivets.
few (a few) (adj)	SOME (adj)	LET THE ENGINE OPERATE FOR SOME MINUTES.	Let the engine run for a few minutes.
	If this alternative is not sufficient, give the range, number, or quantity.	AFTER APPROXIMATELY 5 SECONDS, THE LIGHT COMES ON AGAIN.	After a few seconds, the light comes on again.
file (v)	REMOVE (v) (WITH A FILE (TN)	DO NOT REMOVE TOO MUCH MATERIAL WITH THE FILE.	Take care not to file too deeply.
filing (n)	PARTICLE (n)	REMOVE ALL THE METAL PARTICLES.	Remove all the metal filings.
FILL (v), FILLS, FILLED, FILLED	To put into a container or a space to the maximum level or, if specified, to a given level, pressure, or quantity	FILL THE TANK WITH METHANOL.	
		MAKE SURE THAT THE RESERVOIR FILLS WITH WATER.	

Word (part of speech)	Approved meaning/ ALTERNATIVES	APPROVED EXAMPLE	Not approved example
film (n)	LAYER (n)	APPLY A LAYER OF COMPOUND ON THE SURFACE OF THE DISC.	Spread a film of compound on the surface of the disc.
filter (v)	FILTER (TN)	PUT THE OIL THROUGH A FILTER.	Filter the oil.
FILTERED (adj)	That has gone through a filter	DRY THE UNIT WITH LOW-PRESSURE FILTERED AIR.	
final (adj)	LAST (adj)	THE LAST STEP WILL OPEN THE VALVE.	The final action will open the valve.
FIND (v), FINDS, FOUND, FOUND	To discover, to examine something so that you know	YOU CAN FIND THE NUMBERS ON THE SWITCH.	
finding (n)	RESULT (n)	TELL THE QUALITY DEPARTMENT ABOUT THE RESULTS OF THE ANALYSIS.	Report the findings of the analysis to the quality department.
fine (adj)	SMALL (adj)	ONLY PARTICLES SMALLER THAN 5 MICRONS CAN GO THROUGH THE FILTER.	Only particles finer than 5 microns can go through the filter.
finger-tighten (v)	TIGHTEN ... WITH YOUR FINGERS	TIGHTEN THE NUT WITH YOUR FINGERS.	Finger-tighten the nut.
FINISH (n)	The result of a finishing procedure	DO NOT CAUSE DAMAGE TO THE SURFACE FINISH OF THE HOUSING.	
finish (v)	COMPLETE (v)	COMPLETE THE TEST.	Finish the test.
FIRE (n)	The light and/or heat from a material when it burns	IF THERE IS A FIRE IN THE ENGINE COMPARTMENT, YOU WILL HEAR THE GONG.	

Word (part of speech)	Approved meaning/ ALTERNATIVES	APPROVED EXAMPLE	Not approved example
FIRE (v), FIRES, FIRED, FIRED	To ignite, or to operate items that contain an explosive material	FIRE THE CARTRIDGES.	
		WHEN YOU PUSH THE AGENT PUSHBUTTON, THE FIRE EXTINGUISHER CARTRIDGE FIRES.	
firmly (adv)	TIGHTLY (adv)	HOLD THE CYLINDER TIGHTLY.	Hold the cylinder firmly.
FIRST (adv)	Before other persons or things	FILL THE INNER TANKS FIRST.	
FIT (n)	The relation between two related parts, a limit of tolerance	YOU WILL FIND DATA ABOUT FITS AND CLEARANCES IN SECTION 9001.	
fit (v)	INSTALL (v)	INSTALL THE DUCT.	Fit the duct.
	ATTACH (v)	ATTACH THE PANEL TO THE STRUCTURE WITH THE BOLTS.	Fit the panel to the structure with the bolts.
	CORRECT (adj)	MAKE SURE THAT THE REPAIR PLUG HAS THE CORRECT DIMENSIONS.	Ensure the repair plug fits.
fix (v)	ATTACH (v)	ATTACH THE TRIM WITH DOUBLE-SIDED ADHESIVE TAPE.	Fix the trim with double-sided adhesive tape.
	SET (v)	SET THE LENGTH OF THE CONTROL ROD AS NECESSARY.	Fix the length of the control rod as required.
	REPAIR (v)	REPAIR ALL LOOSE OR DAMAGED UPHOLSTERY.	Fix any loose or torn upholstery.
	INSTALL (v)	INSTALL THE RETAINER BOLTS ON THE RETAINER.	Fix the retainer bolts to the retainer.
FLAME (n)	Burning gas	IF YOU SEE FLAMES FROM THE EXHAUST PIPE, STOP THE ENGINE.	

Word (part of speech)	Approved meaning/ ALTERNATIVES	APPROVED EXAMPLE	Not approved example
FLAMMABLE (adj)	That burns easily	THE SOLVENT IS FLAMMABLE.	
FLANGE (n)	An end surface at an angle	MAKE SURE THAT THE FLANGE IS NOT DAMAGED.	
FLASH (v), FLASHES, FLASHED, FLASHED	To come on and go off frequently	THE RED LIGHTS FLASH IN HALF-SECOND INTERVALS.	
FLAT (adj) (FLATTER, FLATTEST)	That has a continuous surface in the same plane	LUBRICATE ONE SIDE OF THE FLAT WASHERS.	
flatness (n)	FLAT (adj)	THE SURFACE MUST BE FLAT TO GET A CORRECT BOND.	A correct bond depends on the surface flatness.
flatten (v)	FLAT (adj)	MAKE SURE THAT YOU DO NOT CAUSE THE TUBES TO BECOME FLAT.	Make sure that you do not flatten the tubes.
flaw (n)	DAMAGE (n)	MAKE SURE THAT THERE IS NO DAMAGE TO THE SURFACE FINISH.	Make sure that there are no flaws in the surface finish.
	DEFECTIVE (adj) Be specific if possible.	MAKE SURE THAT THE SURFACE FINISH IS NOT DEFECTIVE.	Make sure that there are no flaws in the surface finish.
flex (v)	BEND (v)	THE WING CAN BEND UP AND DOWN DURING FLIGHT.	The wing can flex up and down during flight.
FLEXIBLE (adj)	That can easily bend again and again	CONNECT THE FLEXIBLE HOSES TO THE PUMP.	
FLIGHT (n)	The time an aircraft is in the air	AFTER EACH FLIGHT, READ THE COUNTER UNIT INDICATIONS.	
flood (v)	USE TOO MUCH	DO NOT USE TOO MUCH SOLVENT ON THE AREA.	Do not flood the area with solvent.

Word (part of speech)	Approved meaning/ ALTERNATIVES	APPROVED EXAMPLE	Not approved example
FLOW (n)	A continuous movement of fluid	AN ARROW ON THE VALVE SHOWS THE DIRECTION OF FUEL FLOW.	
FLOW (v), FLOWS, FLOWED, FLOWED	To move as a fluid moves	THE OIL MUST FLOW OUT OF THE VALVE.	
fluctuate (v)	CONSTANT (adj)	MAKE SURE THAT THE PRESSURE STAYS CONSTANT.	Make sure that the pressure does not fluctuate.
FLUID (n)	Liquid or gas	THESE FLUIDS CAN CAUSE CORROSION.	
FLUSH (adj)	Of a surface that is level or continuous with another surface	MAKE SURE THAT THE DOOR IS FLUSH WITH THE FUSELAGE SKIN.	
FLUSH (v), FLUSHES, FLUSHED, FLUSHED	To remove something or to operate with a flow of liquid	FLUSH THE TUBES WITH PRESSURIZED FUEL.	
		MAKE SURE THAT THE TOILET ASSEMBLY FLUSHES CORRECTLY.	
focus (v)	FOCUS (TN)	PUT THE OPTICAL FIXTURE INTO FOCUS.	Focus the optical fixture.
fold (n)	FOLD (v)	FOLD THE MATERIAL.	Make a fold in the material.
FOLD (v), FOLDS, FOLDED, FOLDED	To double over or to cause to double over on itself	DISENGAGE THE UPPER AND LOWER GUIDE ARMS AND FOLD THEM BACK.	
		FOLD THE ROTOR BLADES.	

Word (part of speech)	Approved meaning/ ALTERNATIVES	APPROVED EXAMPLE	Not approved example
FOLLOW (v), FOLLOWS, FOLLOWED, FOLLOWED	To come after, to go after	THIS TEST FOLLOWS THE INSTALLATION OF THE HP PUMP.	
		DO THE PROCEDURE THAT FOLLOWS.	
	For other meanings, use:	FOLLOW THE GREEN LIGHTS TO THE NEAREST STAIRCASE.	
	OBEY (v)	WHEN YOU USE THIS MATERIAL, OBEY THE MANUFACTURER'S INSTRUCTIONS.	When you use this material, follow the manufacturer's instructions.
following (adj)	THESE (adj)	MAKE SURE THAT THESE CIRCUIT BREAKERS ARE CLOSED.	Make sure that the following circuit breakers are closed.
	FOLLOW (v)	OBEY THE PRECAUTIONS THAT FOLLOW.	Obey the following precautions.
FOR (prep)	Function word that shows purpose, intent, time, result, or the object of an action	LET THE POTTING COMPOUND DRY FOR ONE HOUR.	
		OBEY THE REMOVAL PROCEDURES FOR THE COMPONENTS.	
FOR EXAMPLE	Used to introduce something chosen as a typical case	WHEN YOU REMOVE THE STUD, USE APPLICABLE TOOLS. FOR EXAMPLE, USE A PLASTIC MALLET AND A PUNCH.	
forbidden (adj)	NOT PERMITTED	IT IS NOT PERMITTED TO INSTALL MORE THAN TWO WASHERS FOR EACH BOLT.	It is forbidden to install more than two washers for each bolt.

Word (part of speech)	Approved meaning/ ALTERNATIVES	APPROVED EXAMPLE	Not approved example
force (v)	PUSH (v)	PUSH THE CLIP OUT TO REMOVE IT.	Clip may be removed by forcing it out.
	FORCE (TN)	DO NOT USE FORCE TO PUT THE COVER INTO POSITION.	Do not attempt to force the cover into position.
forecast (v)	POSSIBLE (adj)	IF BAD WEATHER IS POSSIBLE, MOOR THE AIRCRAFT SAFELY.	If unfavorable weather conditions are forecast, moor the aircraft securely.
foreign (adj)	UNWANTED (adj)	THIS COVER WILL KEEP DIRT, PARTICLES, OR OTHER UNWANTED MATERIAL OUT OF THE SYSTEM.	This cover will prevent dirt, particles, or other foreign material from entering the system.
form (n)	SHAPE (n)	THE SEAL HAS THE SHAPE OF AN ELLIPSE.	The seal is in the form of an ellipse.
form (v)	BE (v)	IF THERE IS CONTAMINATION IN THE FUEL TANKS, REMOVE IT IMMEDIATELY.	If contamination has formed in the fuel tanks, remove it immediately.
fortify (v)	STRONG (adj)	CARBON-FIBER STRUTS MAKE THE FLOOR STRUCTURE STRONGER.	Carbon-fiber struts fortify the floor structure.
FORWARD (adj)	Nearer to the front	THE CONTROL UNIT IS INSTALLED IN THE FORWARD PART OF THE CARGO COMPARTMENT.	
FORWARD (adv)	In the direction of the front	MOVE THE LEVERS FORWARD.	
FORWARD OF (prep)	At a position nearer to the front	THE TENSION REGULATOR IS INSTALLED FORWARD OF THE PRESSURE BULKHEAD.	

Word (part of speech)	Approved meaning/ ALTERNATIVES	APPROVED EXAMPLE	Not approved example
foul (v)	CATCH (v)	MAKE SURE THAT THE RODS CANNOT CATCH ON THE FLAPS.	Make sure that the rods cannot foul the flaps.
	HIT (v)	IF YOU INSTALL THE UPPER DRAG STRUT INCORRECTLY, IT WILL HIT THE LOWER DRAG STRUT DURING RETRACTION.	Do not install the upper drag strut so that it fouls the gear retraction.
	TOUCH (v)	IF THE BRACKET TOUCHES THE CAM, ADD A SHIM WASHER.	Should the bracket foul the cam, add a shim washer.
	CLEARANCE (n)	MAKE SURE THAT THE RODS HAVE SUFFICIENT CLEARANCE.	Check that the rods do not foul each other.
fouled (adj)	DIRTY (adj)	REMOVE ALL DIRTY IGNITER PLUGS.	Remove all fouled igniter plugs.
	CLOGGED (adj)	REMOVE ALL CLOGGED FILTER ELEMENTS.	Remove all fouled filter elements.
	CATCH (v)	MAKE SURE THAT THE CABLES DO NOT CATCH ON THE LEVERS.	Make sure that the levers are not fouled by the cables.
fragile (adj)	EASILY DAMAGED	THE WATER QUANTITY TRANSMITTER IS EASILY DAMAGED.	The water quantity transmitter is fragile.
FREE (adj)	That can move easily	MAKE SURE THAT THE NUTS ARE FREE TO TURN ON THE SLEEVES.	
	Do not use compound adjectives with free, for example, lintfree, dustfree, unless they are technical names.	MAKE SURE THAT THERE IS NO DUST ON THE SURFACE.	Make sure that the surface is dustfree.
free from (adj)	NO (adj)	MAKE SURE THAT THERE IS NO DUST ON THE SURFACE.	Make sure that the surface is free from dust.

Word (part of speech)	Approved meaning/ ALTERNATIVES	APPROVED EXAMPLE	Not approved example
free (v)	RELEASE (v)	REMOVE THE SLEEVE TO RELEASE THE SLIDING TUBE.	Remove the sleeve to free the sliding tube.
FREELY (adv)	That can move easily	OXYGEN MUST FLOW FREELY FROM THE END OF THE HOSE.	
FREEZE (v), FREEZES, FROZE, FROZEN	To go or cause to go to a temperature below the freezing point of a liquid (usually of water)	FREEZE THE BUSHINGS AS FOLLOWS:	
		IF YOU THINK THAT THE WATER WILL FREEZE, ADD ANTIFREEZE.	
FREQUENT (adj)	That occurs or is found often at short intervals	FREQUENT INSPECTIONS OF THESE AREAS ARE NECESSARY.	
FREQUENTLY (adv)	At frequent or short intervals	MONITOR THE INDICATORS FREQUENTLY.	
fresh (adj)	CLEAN (adj)	FLUSH THE TUBES WITH CLEAN WATER.	Flush the tubes with fresh water.
	NEW (adj)	FILL WITH NEW DESICCANT.	Recharge with fresh desiccant.
FROM (prep)	Function word that shows a point of departure for movement, time, distance, action, or separation	REMOVE THE PROTECTIVE COVER FROM THE NEW ASSEMBLY.	
FRONT (adj)	Nearer to the person that looks or a reference point Refer also to IN FRONT OF.	THERE IS A “FAULT” LIGHT ON THE FRONT FACE OF THE COMPUTER.	
FRONT (n)	The front part or the front surface	CLEAN THE FRONT OF THE MOUNTING FLANGE WITH SOLVENT.	
fuel (v)	REFUEL (v)	REFUEL THE AIRCRAFT.	Fuel the aircraft.

Word (part of speech)	Approved meaning/ ALTERNATIVES	APPROVED EXAMPLE	Not approved example
fulfill (v)	OBEY (v)	OBEY THE INSTRUCTIONS IN THE TEST PROCEDURE.	Fulfill the instructions in the test procedure.
		THIS PROCEDURE OBEYS ALL THE REQUIREMENTS OF SERVICE BULLETIN No. 97.	This procedure fulfills all the requirements of Service Bulletin No. 97.
FULL (adj) (FULLER, FULLEST)	At or related to maximum travel, length, quantity, or detail	MAKE SURE THAT THE RESERVOIR IS FULL.	
FULLY (adv)	At or related to maximum travel, length, quantity, or detail	TURN THE CONTROL TO THE FULLY CLOSED POSITION.	
FUNCTION (n)	The purpose of an operation or an item	THE PRIMARY FUNCTION OF THIS UNIT IS TO CONTROL THE FUEL FLOW.	
function (v)	OPERATE (v)	THE ELEVATOR CONTROLS MUST OPERATE SMOOTHLY.	The elevator controls must function smoothly.
	MOVE (v)	THE SYSTEM CONTROLS MUST MOVE FREELY.	The system controls must function freely.
fundamental (adj)	IMPORTANT (adj)	IT IS IMPORTANT TO KEEP ALL PARTS CLEAN DURING THIS PROCEDURE.	Cleanliness of all parts is fundamental in this procedure.
furnish (v)	GIVE (v)	GIVE THE TECHNICIAN SUFFICIENT PROTECTIVE CLOTHING.	Furnish the technician with sufficient protective clothing.
	SUPPLY (v)	A JACK IS SUPPLIED WITH THE TOOL ASSEMBLY.	A jack is furnished with the tool assembly.

Word (part of speech)	Approved meaning/ ALTERNATIVES	APPROVED EXAMPLE	Not approved example
further (adj)	MORE (adj)	IF NECESSARY, ADD MORE GASKETS.	If required, add further gaskets.
further (adv)	MORE (adv)	IF IT IS NECESSARY TO DECREASE THE CABLE TENSION MORE, USE THE TURNBUCKLE.	Cable tension can be further reduced if necessary by means of the turnbuckle.

Blank Page

Word (part of speech)	Approved meaning/ ALTERNATIVES	APPROVED EXAMPLE	Not approved example
gain (n)	INCREASE (v)	THE RESULT IS THAT THE TEMPERATURE INCREASES.	The result is a gain in temperature.
gain (v)	GET (v)	GET ACCESS TO THE CONNECTORS AT THE REAR OF THE INSTRUMENT.	Gain access to the connectors at the back of the instrument.
gall (v)	GALLING (TN)	THERE MUST BE NO SCRATCHES OR GALLING ON THE SURFACE.	The surface must not be scratched or galled.
gap (n)	CLEARANCE (n)	MEASURE THE CLEARANCE BETWEEN THE IMPELLER AND THE BODY.	Measure the gap between the impeller and the body.
	DISTANCE (n)	MEASURE THE DISTANCE BETWEEN THE ROTARY ACTUATOR AND THE STOP BOLT.	Measure the gap between the rotary actuator and the stop bolt.
	SPACE (n)	PUT THE BLADE OF A SMALL SCREWDRIVER INTO THE SPACE BETWEEN THE BALL MAT AND THE BALL UNIT.	Put the blade of a small screwdriver into the gap between the ball mat and the ball unit.
garble (adj)	CLEAR (adj)	IF THE RADIO SIGNAL IS NOT CLEAR, ADJUST THE FREQUENCY.	If the radio signal is garbled, adjust the frequency.
	CLEARLY (adv)	YOU MUST HEAR THE ANNOUNCEMENT CLEARLY.	The announcement must be audible and not garbled.

Word (part of speech)	Approved meaning/ ALTERNATIVES	APPROVED EXAMPLE	Not approved example
garner (v)	COLLECT (v)	THESE COMPUTERS COLLECT DATA FROM THE VHF SYSTEM.	These computers garner data from the VHF system.
		THE BITE FUNCTION COLLECTS THE FAILURES AND SENDS THE FAILURE MESSAGE TO THE CENTRAL MAINTENANCE COMPUTER.	The BITE function garners the failures and sends the failure message to the Central Maintenance Computer.
GAS (n)	A fluid that has no shape or volume and that can expand	THE TEMPERATURE SENSING UNIT IS A CAPILLARY TUBE FILLED WITH GAS.	
gaseous (adj)	GAS (n)	THE TOOL CYLINDER CONTAINS A MIXTURE OF GASES (10% HELIUM AND 90% NITROGEN).	The tool cylinder contains a gaseous mixture of 10% helium and 90% nitrogen.
gash (v)	DAMAGED (adj)	IF THE THERMAL BLANKET IS DAMAGED, DO REPAIR No. 9.	If the thermal blanket is gashed, do repair No. 9.
gather (v)	COLLECT (v)	ALL FLUIDS COLLECT AT THE LOWEST POINT IN THE COMPARTMENT.	All fluids gather at the lowest point in the compartment.
		THREE COMPUTERS COLLECT DATA AND PARAMETERS FROM THE SYSTEMS AND PUT THEM TOGETHER ON THE DATA LINK TO THE DISPLAY UNIT.	Three computers gather data and parameters from the systems and put them together on the data link to the display unit.
gauge (v)	MEASURE (v)	MEASURE THE DISTANCE BETWEEN THE FUSELAGE AND THE KRUEGER FLAP.	Gauge the width of the gap between the fuselage and the Krueger flap.
GENERAL (adj)	Of, or related to something as a whole, not detailed	THIS SECTION GIVES GENERAL MAINTENANCE PROCEDURES.	

Word (part of speech)	Approved meaning/ ALTERNATIVES	APPROVED EXAMPLE	Not approved example
generally (adv)	USUALLY (adv)	YOU USUALLY REPLACE MEMORY MODULES ON AIRCRAFT.	Memory modules are generally replaced on aircraft.
generate (v)	BE (v)	THERE MUST BE NO ERRORS WHEN YOU OPEN THE FILE.	The file opening must not generate errors.
	GIVE (v)	IF THE HYDRAULIC PUMP DOES NOT GIVE A PRESSURE OF 3000 PSI, USE THE ELECTRIC PUMP.	If the hydraulic pump fails to generate a pressure of 3000 psi, use the electric pump.
	SUPPLY (v)	THE BATTERIES SUPPLY 28 VDC.	The batteries generate 28 VDC.
generous (adj)	LARGE (adj)	A LARGE QUANTITY OF COMPOUND IS NECESSARY.	A generous amount of compound will be necessary.
	THICK (adj)	APPLY A THICK LAYER OF GREASE.	Apply a generous amount of grease.
	Be specific if possible.		
generously (adv)	LARGE (adj)	APPLY A LARGE QUANTITY OF SEALANT TO THE REPAIR PATCH.	Apply sealant generously to the repair patch.
	THICK (adj)	APPLY A THICK LAYER OF GREASE.	Apply grease generously.
	Be specific if possible.		
gentle (adj)	CAREFULLY (adv)	TURN THE POTENTIOMETER CAREFULLY.	Rotate the potentiometer with a gentle movement.
	LIGHTLY (adv)	TAP THE SURFACE LIGHTLY.	Give the surface a gentle tap.
gently (adv)	CAREFULLY (adv)	CAREFULLY PULL THE ELECTRICAL WIRE OUT FROM THE RUBBER SEAL.	Gently pull out the electrical wire from the rubber seal.
	LIGHTLY (adv)	LIGHTLY TAP THE FLAP SURFACE.	Gently tap the flap surface.

Word (part of speech)	Approved meaning/ ALTERNATIVES	APPROVED EXAMPLE	Not approved example
genuine (adj)	CORRECT (adj)	MAKE SURE THAT THE INDICATION ON THE GAUGE IS CORRECT.	Make sure that the reading on the gauge is genuine.
GET (v), GETS, GOT No other forms of this verb.	To obtain, to come into the state of having Do not use this word to mean BECOME, GO, DECREASE, INCREASE. Refer to rule 9.2 and the entries from "get" until "get to" that follow.	IF YOU GET THE FLUID ON YOUR SKIN, FLUSH YOUR SKIN IMMEDIATELY WITH CLEAN WATER.	
get (v)	BECOME (v)	DO NOT LET THE HEATER BECOME TOO HOT.	Do not let the heater get too hot.
get away (v)	GO AWAY (v)	BEFORE THE ENGINE STARTS, GO AWAY FROM THE EXHAUST AREA.	Get away from the exhaust area before the engine starts.
get down (v)	DECREASE (v)	IF THE TEMPERATURE DECREASES TO LESS THAN ZERO, DRAIN THE WATER SYSTEM.	If the temperature gets down to below zero, drain the water system.
	GO DOWN (v)	BE CAREFUL WHEN YOU GO DOWN THE LADDER FROM THE COCKPIT TO THE AVIONICS COMPARTMENT.	Be careful when you get down the ladder from the cockpit into the avionics compartment.
get into (v)	GO INTO (v)	INSTALL CAPS ON THE RECEPTACLES TO MAKE SURE THAT UNWANTED MATERIAL DOES NOT GO INTO THE RECEPTACLES.	Install caps to make sure that foreign objects do not get into the receptacles.
get off (v)	MOVE OFF (v)	BEFORE YOU MOVE THE AILERONS, TELL ALL PERSONS TO MOVE OFF THE WINGS.	Before you move the ailerons, tell all persons to get off the wings.

Word (part of speech)	Approved meaning/ ALTERNATIVES	APPROVED EXAMPLE	Not approved example
get to (v)	BE (v)	THE MOTOR STOPS WHEN THE FLAPS ARE AT THE END OF THEIR TRAVEL.	The motor stops when the flaps get to the end of their travel.
	GO (v)	GO TO THE AFT CARGO COMPARTMENT TO DO THIS TEST.	Get to the aft cargo compartment to do this test.
	INCREASE (v)	WHEN THE TEMPERATURE OF THE VENTILATION AIR INCREASES TO 27 °C, THE DISCHARGE VALVE OPENS.	When the temperature of the ventilation air gets to 27 °C, the discharge valve opens.
GIVE (v), GIVES, GAVE, GIVEN	To provide	THIS SECTION GIVES YOU THE PROCEDURES FOR THE DISASSEMBLY.	
gleam (v)	SHINY (adj)	POLISH THE SURFACE WITH A SOFT CLOTH UNTIL IT BECOMES SHINY.	Polish the surface with a soft cloth until it gleams.
glitch (n)	ERROR (n)	IF THERE IS AN ERROR IN THE SYSTEM, THE SCREEN SHOWS: “NO GO”.	If there is a glitch in the system, the screen shows: NO GO.
	FAILURE (TN)	IF THERE IS A FAILURE IN THE SYSTEM, THE SCREEN SHOWS: “NO GO”.	If there is a glitch in the system, the screen shows: NO GO.
	UNSERVICEABLE (adj)	IF THE SYSTEM IS UNSERVICEABLE, THE SCREEN SHOWS: “NO GO”.	If there is a glitch in the system, the screen shows: NO GO.
gloss (n)	SHINY (adj)	POLISH THE SURFACE UNTIL IT IS VERY SHINY.	Polish the surface to a high gloss.
GLOSSY (adj) (GLOSSIER, GLOSSIEST)	Smooth and shiny	APPLY THE PATCH WITH THE MATT SIDE AGAINST THE GLOSSY SURFACE.	

Word (part of speech)	Approved meaning/ ALTERNATIVES	APPROVED EXAMPLE	Not approved example
glow (v)	BE (v)	MAKE SURE THAT THE INDICATOR LIGHT IS ON.	Make sure that the indicator light glows.
		ADJUST THE POTENTIOMETER UNTIL THE LIGHT IS DIM.	Adjust the potentiometer until the light glows dimly.
glue (v)	BOND (v)	BOND THE PATCH TO THE SURFACE WITH THE APPLICABLE GLUE.	Glue the patch to the surface.
	ATTACH (v)	ATTACH THE PROTECTIVE PLATE TO THE SURFACE WITH ADHESIVE MATERIAL.	Glue the protective plate to the surface.
GO (v), GOES, WENT	To move to or from something	MAKE SURE THAT THE POINTER GOES OUT OF VIEW.	
GO OFF (v), GOES OFF, WENT OFF	To become dark when an internal power source is de-energized	THE ANNUNCIATOR LIGHT GOES OFF.	
GOOD (adj) (BETTER, BEST)	That is satisfactory	THIS MATERIAL GIVES GOOD PROTECTION FROM CORROSION.	
gouge (v)	GOUGE (TN)	IF A GOUGE OCCURS IN THE BLADE DURING REMOVAL, REPLACE THE BLADE.	If the blade was gouged during removal, replace it.
govern (v)	CONTROL (v)	THIS COMPONENT CONTROLS THE ENGINE SPEED.	This component governs the engine speed.
grab (v)	USE (v)	USE A SCREWDRIVER TO RELEASE THE CLIPS.	Grab a screwdriver to release the clips.
	HOLD (v)	HOLD THE HANDLE.	Grab the handle.
gradient (n)	SLOPE (n)	IF YOU MUST PARK THE AIRCRAFT ON A SLOPE, MAKE SURE THAT WHEEL CHOCKS ARE IN FRONT OF THE WHEELS.	If you have to park the aircraft on a gradient, make sure that wheel chocks are in front of the wheels.

Word (part of speech)	Approved meaning/ ALTERNATIVES	APPROVED EXAMPLE	Not approved example
gradual (adj)	GRADUALLY (adv)	GRADUALLY ADJUST THE TEMPERATURE.	Make a gradual adjustment to the temperature.
GRADUALLY (adv)	Slowly and continuously	GRADUALLY OPEN THE COCKS ON THE RECHARGING UNIT.	
graduated (adj)	GRADUATION (TN)	MEASURE THE DIMENSIONS WITH A METAL RULE THAT HAS GRADUATIONS IN MILLIMETERS.	The measurements are obtained by means of a millimeter-graduated metal rule.
graph (v)	GRAPH (TN)	MAKE A GRAPH OF THE TEST RESULTS.	Graph the test results.
grasp (v)	HOLD (v)	HOLD THE HANDLE.	Grasp the handle.
grave (adj)	DANGEROUS (adj)	HYDRAULIC FLUID IS DANGEROUS FOR YOUR SKIN.	Hydraulic fluid can cause grave skin problems.
grease (v)	LUBRICATE (v)	BEFORE YOU INSTALL THE O-RING, LUBRICATE IT WITH GREASE.	Grease the O-ring before installation.
	GREASE (TN)	APPLY GREASE TO THE FASTENERS.	Grease the fasteners.
great (adj)	LARGE (adj)	AN INDICATION OF A LARGE QUANTITY OF FUEL SHOWS AN INCORRECT ADJUSTMENT OF THE FLOAT SWITCH.	A great amount of fuel indicates incorrect setting of the float switch.
	MORE THAN	IF THE PRESSURE IS MORE THAN 200 PSI, CLOSE THE VALVE.	If the pressure is greater than 200 psi, close the valve.
	VERY (adv)	BE VERY CAREFUL WHEN YOU RELEASE THE PRESSURE.	Take the greatest care when you release the pressure.
greatly (adv)	VERY MUCH	THIS METHOD CAN INCREASE THE RISK OF INJURY VERY MUCH.	This method can greatly increase the risk of injury.
grip (n)	HOLD (v)	HOLD THE JACK BODY TIGHTLY WHILE YOU TURN IT.	Make sure you have a tight grip of the jack body while turning it.

Word (part of speech)	Approved meaning/ ALTERNATIVES	APPROVED EXAMPLE	Not approved example
grip (v)	HOLD (v)	HOLD THE HANDLE LIGHTLY WITH ONE HAND.	Grip the handle lightly with one hand.
GROOVE (n)	A long channel that is not wide	INSTALL THE O-RING IN ITS GROOVE ON THE PISTON.	
GROUND (n)	The surface of the earth	YOU CAN ONLY DO THIS PROCEDURE ON THE GROUND.	
GROUND (v), GROUNDS, GROUNDED, GROUNDED	To connect to the ground or to a large object of zero potential	GROUND THE FUEL TANKS.	
GROUP (n)	Objects that are related	THIS GROUP OF SEALING COMPOUNDS IS ALSO APPLICABLE IN AN EMERGENCY.	
grow (v)	BECOME (v)	IF THE BATTERY BECOMES TOO HOT, DISCONNECT IT.	If the battery grows too hot, disconnect it.
guard (v)	GUARD (TN)	MAKE SURE THAT THE GUARDS ARE INSTALLED ON THE SWITCHES.	Make sure that the switches are guarded.
guide (v)	PUT (v)	PUT THE CABLE THROUGH THE GROMMET.	Guide the cable through the grommet.
	MOVE (v)	WHILE YOU LOWER THE PUMP, MOVE IT ONTO THE MOUNTING BRACKET.	As you lower the pump, guide it onto the mounting bracket.
gush (v)	FLOW (n)	IF YOU TURN THE HANDLE TO “FULL”, THERE WILL BE A STRONG FLOW OF WATER FROM THE OUTLET.	If you turn the valve handle to FULL, water will gush from the outlet.

Word (part of speech)	Approved meaning/ ALTERNATIVES	APPROVED EXAMPLE	Not approved example
halt (v)	STOP (v)	IF YOU HEAR AN UNUSUAL NOISE, STOP THE PROCEDURE.	If you hear an unusual noise, halt the procedure.
halve (v)	HALF (TN)	MAKE SURE THAT THE PITCH DECREASES TO HALF ITS VALUE.	Check that the deviation in pitch is halved.
hamper (v)	PREVENT (v)	STRONG WINDS CAN PREVENT THE CORRECT DEPLOYMENT OF THE ESCAPE SLIDES.	Strong winds can hamper the correct deployment of the escape slides.
handle (v)	MOVE (v)	MOVE THE ACTUATOR WITH A HOIST.	Handle the actuator with a hoist.
	TOUCH (v)	DO NOT TOUCH THE PRINTED CIRCUIT CARD.	Do not handle the printed circuit card.
	CAREFUL (adj) This word can be part of safety sign (rule 8.6).	BE CAREFUL WITH INSTRUMENTS.	Handle instruments with care.
hand-tight (adj)	TIGHTEN ... WITH YOUR HAND	INSTALL THE FILTER CASE AND TIGHTEN IT WITH YOUR HAND.	Install the filter case hand-tight.
hand-tighten (v)	TIGHTEN ... WITH YOUR HAND	TIGHTEN THE SWIVEL NUT WITH YOUR HAND.	Hand-tighten the swivel nut.
HANG (v), HANGS, HUNG, HUNG	To attach or to be attached to something above with no support from below	HANG THE SHACKLE ON THE HOIST.	
		LOWER THE PLATFORM ASSEMBLY UNTIL IT HANGS ON THE SUPPORT FRAME.	
happen (v)	OCCUR (v)	IF AN ACCIDENT OCCURS, GET MEDICAL AID.	If an accident happens, get medical attention.
HARD (adj) (HARDER, HARDEST)	Not easy to cut, not easy to go into or through	IF THE SURFACE IS HARD, USE A CARBIDE DRILL.	

Word (part of speech)	Approved meaning/ ALTERNATIVES	APPROVED EXAMPLE	Not approved example
harden (v)	HARD (adj)	LET THE ADHESIVE BECOME HARD.	Let the adhesive harden.
	CURE (TV)	LET THE BONDING MATERIAL CURE.	Let the bonding material harden.
	Refer to rule 1.12.		
harm (v)	INJURY (n)	BE CAREFUL WHEN YOU CUT LOCKWIRE. IT CAN CAUSE INJURY TO YOUR EYES.	Be careful when you cut lockwire. It can harm your eyes.
	IRRITATION (TN) Refer to rule 1.5.	HYDRAULIC FLUID CAN CAUSE SKIN IRRITATION.	Hydraulic fluid can harm your skin.
harmful (adj)	DANGEROUS (adj)	SOLVENT FUMES ARE DANGEROUS.	Solvent fumes are harmful to your health.
	INJURY (n)	THIS ADHESIVE COMPOUND CAN CAUSE INJURY TO YOUR SKIN.	This adhesive compound is harmful to your skin.
harmonize (v)	AGREE (v)	MAKE THE POSITIONS OF THE VALVES AGREE.	Harmonize the positions of the valves.
	ALIGN (v)	ALIGN THE CENTERLINE ON THE FLUX VALVE.	Harmonize the centerline on the flux valve.
haul (v)	TOW (v)	TOW THE HYDRAULIC CART AWAY FROM THE AIRCRAFT BEFORE YOU START THE ENGINE.	Haul the hydraulic cart away from the aircraft before you start the engine.
	MOVE (v)	MOVE THE HYDRAULIC CART AWAY FROM THE AIRCRAFT BEFORE YOU START THE ENGINE.	Haul the hydraulic cart away from the aircraft before you start the engine.
HAVE (v), HAS, HAD No other forms of this verb.	To possess as a part or quality	THE SYSTEM HAS AN EMERGENCY WARNING DEVICE.	

Word (part of speech)	Approved meaning/ ALTERNATIVES	APPROVED EXAMPLE	Not approved example
have to (v)	MUST (v)	YOU MUST USE EAR PROTECTION WHEN YOU ARE NEAR AN ENGINE THAT IS IN OPERATION.	You have to use ear protection when you are near an engine that is in operation.
hazard (n)	DANGEROUS (adj)	TRICHLORO-ETHYLENE IS DANGEROUS.	Trichloroethylene is a health hazard.
hazardous (adj)	DANGEROUS (adj)	ALL PERSONNEL WHO DO WORK WITH DANGEROUS MATERIALS MUST OBEY THE SAFETY INSTRUCTIONS THAT FOLLOW.	All personnel who work with hazardous materials must obey the following safety instructions.
hazy (adj)	HAZE (TN)	IF THERE IS HAZE, DO NOT DO THE TEST.	If the weather is hazy, do not do the test.
HEAD (n)	The top of something	MAKE SURE THAT THE HEAD OF THE BOLT IS DOWN.	
head (v)	POINT (v)	POINT THE AIRCRAFT INTO THE WIND.	Head the aircraft into the wind.
HEAR (v), HEARS, HEARD, HEARD	To know by sound in the ear	MAKE SURE THAT YOU HEAR THE AUDIO SIGNALS IN ALL THE CREW HEADSETS.	
HEAT (n)	The condition or quality of being hot	USE HEAT TO CURE THE POLYSULPHIDE SEALANT.	
heat (v)	HEAT (n)	USE HEAT TO INCREASE THE TEMPERATURE OF THE OIL TO 40 °C.	Heat the oil to 40 °C.
	HOT (adj)	WHEN THE OIL IS HOT, IT FLOWS FREELY.	When you heat the oil, it flows freely.
HEAVY (adj) (HEAVIER, HEAVIEST)	That has a large mass, weight, or force	A SUPPORT WILL BE NECESSARY BECAUSE THE CONTROL UNIT IS HEAVY.	
HEIGHT (n)	The vertical distance	THE HEIGHT FROM THE GROUND TO THE TOP OF THE FIN CAN CHANGE.	

Word (part of speech)	Approved meaning/ ALTERNATIVES	APPROVED EXAMPLE	Not approved example
heighten (v)	INCREASE (v)	STRONG WINDS WILL INCREASE THE RISK OF DAMAGE.	Strong winds will heighten the risk of damage.
help (n)	AID (n)	GET MEDICAL AID IMMEDIATELY.	You must get medical help immediately.
HELP (v), HELPS, HELPED, HELPED	To make something easier or better	PETROLATUM HELPS TO PREVENT CORROSION OF THE TERMINALS.	
helpful (adj)	HELP (v)	RECORD THE LOCKWIRE POSITIONS. THIS WILL HELP YOU DURING THE ASSEMBLY PROCEDURE.	Record the lockwire positions. This will be helpful during the assembly procedure.
HERE (adv)	In this position	TO DOWNLOAD THE FILE, CLICK HERE.	
hesitation (n)	SMOOTHLY (adv)	MAKE SURE THAT THE SOLENOID OPERATES SMOOTHLY.	Make sure that the solenoid operates without hesitation.
	CORRECTLY (adv)	THE VALVE MUST OPERATE CORRECTLY.	The valve must operate without hesitation.
	IMMEDIATELY (adv)	WHEN YOU OPEN THE CIRCUIT BREAKER, THE LIGHT MUST COME ON IMMEDIATELY.	When you open the circuit breaker, the light must come on without hesitation.
HIGH (adj) (HIGHER, HIGHEST)	That is of large value	USE THE SPECIAL PROTECTION FOR STORAGE IN HIGH TEMPERATURES.	
highly (adv)	VERY (adv)	TOLUENE IS VERY FLAMMABLE.	Toluene is highly flammable.

Word (part of speech)	Approved meaning/ ALTERNATIVES	APPROVED EXAMPLE	Not approved example
hinder (v)	PREVENT (v)	SCRATCHES CAN PREVENT THE FREE MOVEMENT OF THE PISTON IN THE SLEEVE.	Scratches can hinder the movement of the piston in the sleeve.
	DECREASE (v)	A CLOGGED DUCT WILL DECREASE AIRFLOW.	A clogged duct will hinder airflow.
	BLOCKAGE (n)	MAKE SURE THAT THERE IS NO BLOCKAGE IN THE PIPE THAT PREVENTS AIRFLOW.	Make sure that there is nothing in the pipe to hinder airflow.
	CLOGGED (adj)	MAKE SURE THAT THE PIPE IS NOT CLOGGED.	Make sure that there is nothing in the pipe to hinder airflow.
hinge (v)	TURN (v)	THE PANELS TURN ON TWO NYLON STRAPS.	Panels hinge on two nylon straps.
HIT (v), HITS, HIT, HIT	To touch suddenly and with much force	DO NOT HIT THE CARTRIDGE.	
hitch (v)	CONNECT (v)	CONNECT THE TOWING ARM TO THE NOSEWHEEL.	Hitch the towing arm to the nosewheel.
hoist (v)	LIFT (v)	LIFT THE MODULE INTO POSITION.	Hoist the module into position.
hold (n)	HOLD (v)	MAKE SURE THAT YOU HOLD THE ROD TIGHTLY.	Make sure that you have a tight hold on the rod.
HOLD (v), HOLDS, HELD, HELD	1. To continue to have in the hand or grip	HOLD THE ROD TIGHTLY.	
	2. To continue to have in a specified location, position, or condition	HOLD THE AIRSPEED INDICATION AT THE SAME VALUE FOR 2 MINUTES.	

Word (part of speech)	Approved meaning/ ALTERNATIVES	APPROVED EXAMPLE	Not approved example
hold back (v)	PREVENT (v)	AT FULL THRUST, THE BRAKES MUST PREVENT MOVEMENT OF THE AIRCRAFT.	At full thrust, the brakes must hold the aircraft back.
hold off (v)	PREVENT (v)	THE PROTECTIVE PAINT WILL PREVENT CORROSION FOR SIX MONTHS.	The protective paint will hold off corrosion for six months.
HOLE (n)	An empty space in a solid object	MAKE SURE THAT THE HOLES IN THE BUSHING ALIGN WITH THOSE IN THE HOUSING.	
hook (v)	PUT (v)	PUT THE SLING ON THE SHACKLE.	Hook the sling on the shackle.
hook up (v)	CONNECT (v)	CONNECT THE TOWING ARM TO THE NOSE GEAR.	Hook the towing arm up to the nose gear.
HORIZONTAL (adj)	Parallel to the horizon or a baseline	ADJUST THE SLING UNTIL THE FLAP IS HORIZONTAL.	
HORIZONTALLY (adv)	In a horizontal position	THE VALVE IS INSTALLED HORIZONTALLY.	
HOT (adj) (HOTTER, HOTTEST)	At a high temperature	DRY THE SURFACE WITH HOT AIR.	
house (v)	CONTAIN (v)	THE MODULE CONTAINS FIVE COMPONENTS.	The module houses five components.
HOW (adv)	By which manner	PARAGRAPH 7.0 TELLS YOU HOW TO INSTALL THE UNIT.	
however (adv)	BUT (conj)	BUT WE RECOMMEND THIS STEP TO DECREASE THE DRYING TIME.	However, we recommend this step to decrease the drying time.

Word (part of speech)	Approved meaning/ ALTERNATIVES	APPROVED EXAMPLE	Not approved example
huge (adj)	LARGE (adj)	TO REMOVE THE PAINT STRIPPER, FLUSH THE SURFACE WITH LARGE QUANTITIES OF WATER.	To remove the paint stripper, flush the surface with huge quantities of water.
hum (v)	HUM (TN)	ADJUST THE POTENTIOMETER UNTIL YOU DO NOT HEAR A HUM FROM THE LOUDSPEAKER.	Adjust the potentiometer until the loudspeaker no longer hums.
hurt (v)	INJURY (n)	BE CAREFUL WHEN YOU DO THIS PROCEDURE. IF THE UNIT FALLS, IT CAN CAUSE INJURY TO PERSONNEL.	Be careful when you do this procedure. If the unit falls, it can hurt personnel.
HYDRAULIC (adj)	Related to, or operated by, pressurized liquid	CONNECT THE FITTINGS TO THE HYDRAULIC TEST BENCH.	
HYDRAULICALLY (adv)	With hydraulic power	MAKE SURE THAT THE LINES ARE HYDRAULICALLY OPERATED.	

Word (part of speech)	Approved meaning/ ALTERNATIVES	APPROVED EXAMPLE	Not approved example

Blank Page

Word (part of speech)	Approved meaning/ ALTERNATIVES	APPROVED EXAMPLE	Not approved example
identical (adj)	SAME (adj)	THE SAME PROCEDURE IS APPLICABLE TO THE TWO VHF SETS.	The procedure is identical for both VHF sets.
IDENTIFICATION (n)	Something that identifies an object	MAKE SURE THAT THE NAMEPLATE SHOWS THE CORRECT IDENTIFICATION OF THE UNIT.	
IDENTIFY (v), IDENTIFIES, IDENTIFIED, IDENTIFIED	1. To use a specified code or marking to supply with an identity	IDENTIFY THE WIRES WITH SLEEVES OF DIFFERENT COLORS.	
	2. To show an identity For other meanings, use:	A 406 MHz DIGITAL SIGNAL IDENTIFIES THE EMERGENCY LOCATOR TRANSMITTER AND ITS LOCATION.	
	FIND (v)	FIND THE DAMAGED AREAS.	Identify the damaged areas.
idle (v)	IDLE (TN)	OPERATE THE ENGINE AT IDLE FOR 20 MINUTES.	Idle the engine for 20 minutes.
IF (conj)	In the event that, on the condition that, in case of	IF YOU USE A REPLACEMENT FAIRING PLATE, CUT IT UNTIL YOU GET THE CORRECT MATING DIMENSION.	
ignite (v)	IGNITION (TN)	FUEL GOES THROUGH THE ATOMIZERS INTO THE COMBUSTION CHAMBER, WHERE IGNITION OCCURS.	Fuel passes through the atomizers to the combustion chamber, where it is ignited.
IGNORE (v), IGNORES, IGNORED, IGNORED	Not to think about something, not to do something about	IGNORE THE VIBRATION INDICATIONS.	
		CLICK "CANCEL" TO IGNORE ALL CHANGES.	

Word (part of speech)	Approved meaning/ ALTERNATIVES	APPROVED EXAMPLE	Not approved example
ILLUMINATE (v), ILLUMINATES, ILLUMINATED, ILLUMINATED	To make something visible by an external source such as light	MAKE SURE THAT THE STROBE LIGHT ILLUMINATES THE TRACKING PLATES.	
	For lights or lamps, use:		
	COME ON	WHEN THE PRESSURE LIGHT COMES ON, READ THE VALUE ON THE PRESSURE GAUGE.	When the pressure light illuminates, read the value on the pressure gauge.
		MAKE SURE THAT THE LAMP COMES ON.	Ensure the lamp illuminates.
	BE ON	MAKE SURE THAT THE LAMP IS ON.	Ensure the lamp is illuminated.
immediate (adj)	IMMEDIATELY (adv)	MAKE SURE THAT THE SPEEDBRAKE RETRACTS IMMEDIATELY.	Check the speedbrake for immediate retraction.
IMMEDIATELY (adv)	Without a gap of time or space	MAKE SURE THAT THE EMERGENCY LIGHT IMMEDIATELY GOES BACK TO "OFF".	
immerse (v)	SOAK (v)	SOAK THE FILTER IN THE CLEANING SOLUTION FOR 4 HOURS.	Immerse the filter in the cleaning solution for 4 hours.
	PUT FULLY INTO	PUT THE FILTER FULLY INTO THE SOLVENT.	Immerse the filter in solvent.
immobilize (v)	LOCK (v)	LOCK THE PULLEY WITH THE RIGGING PIN.	Immobilize the pulley with the rigging pin.
	PREVENT MOVEMENT	WHEN YOU REPAIR THE UNIT, PREVENT MOVEMENT OF THE ASSEMBLIES.	Immobilize the assemblies when repairing the unit.
impact (n)	HIT (v)	MAKE SURE THAT METAL OBJECTS DO NOT HIT THE MATING SURFACE.	Avoid impact between metal objects and the mating surface.

Word (part of speech)	Approved meaning/ ALTERNATIVES	APPROVED EXAMPLE	Not approved example
impact (v)	HIT (v)	DO NOT LET THE VIBRATOR TOOL HIT THE SURFACE OF THE SEAL.	Do not allow the vibrator tool to impact the seal surface.
	EFFECT (n)	THESE REPAIRS HAVE AN EFFECT ON THE PERFORMANCE OF THE ENGINE.	These repairs impact engine performance.
impair (v)	DECREASE (v)	THE LIFE OF THE UNIT WILL DECREASE IF YOU LET THE UNIT BECOME DIRTY.	The life of the unit is impaired if the unit is allowed to get dirty.
	OPERATE INCORRECTLY	DAMAGE WILL CAUSE THE INSTRUMENT TO OPERATE INCORRECTLY.	Damage will impair operation of the instrument.
	CAUSE DAMAGE	CONTAMINATION OF THE GREASE CAN CAUSE DAMAGE TO THE BEARING.	Contaminants in the grease may impair the bearing.
imperative (adj)	NECESSARY (adj)	REMOVAL OF THE COVER IS NECESSARY.	Removal of the cover is imperative.
	MANDATORY (adj)	THE STEPS THAT FOLLOW ARE MANDATORY.	The following steps are imperative.
	MUST (v)	YOU MUST DO THESE STEPS.	The following steps are imperative.
implement (v)	DO (v)	IF THE TEST DOES NOT GIVE SATISFACTORY RESULTS, DO THE FAULT ISOLATION PROCEDURE.	If the test does not give satisfactory results, implement the fault isolation procedure.
implementation (n)	DO (v)	IT IS MANDATORY TO DO THE PROCEDURE IN THIS SERVICE BULLETIN.	The implementation of the procedure in this Service Bulletin is mandatory.
IMPORTANT (adj)	That has a significant effect	THESE PROCEDURES ARE VERY IMPORTANT.	

Word (part of speech)	Approved meaning/ ALTERNATIVES	APPROVED EXAMPLE	Not approved example
impossible (adj)	NOT POSSIBLE	THUS, IT IS NOT POSSIBLE TO GIVE A STANDARD VALUE.	Thus, it is impossible to give a standard value.
	CANNOT (v)	YOU CANNOT FILL THE CONTAINERS WHILE THEY ARE INSTALLED ON THE AIRCRAFT.	It is impossible to recharge containers while installed on aircraft.
impregnate (v)	SOAK (v)	SOAK THE FILTER ELEMENTS IN OIL.	The filter elements must be impregnated with oil.
impression (n)	THINK (v)	IF YOU THINK THAT A TIRE HAS LOW PRESSURE, DO THE STEPS THAT FOLLOW:	If you have the impression that a tire has low pressure, do the steps that follow:
improper (adj)	INCORRECT (adj)	INCORRECT ADJUSTMENT OF THE STOP BOLTS ON THE RUDDER PEDALS CAN CAUSE DAMAGE.	Improper adjustment of the stop bolts on the rudder pedals can cause damage.
improperly (adv)	INCORRECTLY (adv)	IF THE HINGE IS INCORRECTLY ADJUSTED, YOU CANNOT CLOSE THE PANEL.	If the hinge is improperly adjusted, the panel will not close.
improve (v)	BETTER (adj)	DO NOT ADJUST THE PARTS AGAIN TO MAKE THEIR POSITION BETTER.	Do not readjust the parts to improve their position.
impurity (n)	UNWANTED MATERIAL	USE A FILTER TO REMOVE UNWANTED MATERIAL FROM THE HYDRAULIC OIL.	Filter the hydraulic oil to remove impurities.
	CONTAMINATION (n)	CONTAMINATION CAN CAUSE A FAILURE.	Impurities may cause a failure.
IN (adv)	To, or into a location or position	MAKE SURE THAT THE SWITCHES ARE PUSHED IN.	

Word (part of speech)	Approved meaning/ ALTERNATIVES	APPROVED EXAMPLE	Not approved example
IN (prep)	Function word that shows location, position, condition, time, or limits	LET THE SWITCH STAY IN THE SET POSITION.	
		THE COMPENSATORS ARE IN THE WING BAY.	
		DO NOT PARK THE AIRCRAFT IN STRONG WINDS.	
		MAKE SURE THAT THE LIGHT COMES ON IN 5 SECONDS.	
		MAKE SURE THAT THE CRACKS ARE IN THE LIMITS.	
IN FRONT OF (prep)	In a position ahead of	PUT CHOCKS IN FRONT OF THE WHEELS.	
IN PROGRESS (adv)	Occurring	IF AN INTERACTIVE SESSION IS IN PROGRESS, LET IT CONTINUE UNTIL IT IS FULLY COMPLETED.	
inactive (adj)	OFF (adj)	BEFORE YOU DRAIN THE FUEL, MAKE SURE THAT THE SYSTEM IS OFF.	Before you drain the fuel, make sure that the system is inactive.
	NO EFFECT	IN THIS CONFIGURATION, THE LINE KEY HAS NO EFFECT.	In this configuration, the line key is inactive.
	NOT IN OPERATION	THE AUTOTHRUST SYSTEM IS ON BUT NOT IN OPERATION.	The autothrust system is on but inactive.
	NOT OPERATE	MAKE SURE THAT THE PUMP DOES NOT OPERATE.	Make sure that the pump is inactive.
inadvertent (adj)	ACCIDENTAL (adj)	ACCIDENTAL OPERATION OF THE SYSTEM CAN KILL YOU.	Inadvertent operation of the system could cause fatal injuries.

Word (part of speech)	Approved meaning/ ALTERNATIVES	APPROVED EXAMPLE	Not approved example
inadvertently (adv)	ACCIDENTALLY (adv)	IF YOU ACCIDENTALLY PUT TOO MUCH FUEL IN THE TANK, OPEN THE DRAIN VALVE.	If the tank is inadvertently overfilled, open the drain valve.
INBOARD (adj)	Nearer to the longitudinal axis	REMOVE THE INBOARD FAIRING OF THE FLAP HINGE.	
INBOARD (adv)	In or to a position nearer to the longitudinal axis	THE LANDING GEAR RETRACTS INBOARD INTO THE BAY.	
INBOARD OF (prep)	At a position nearer to the longitudinal axis	PUT THE ACCESS PLATFORM IN POSITION BELOW THE ACCESS PANEL THAT IS INBOARD OF THE FLAP TRACK.	
INCIDENT (n)	An important occurrence that can cause damage or have dangerous results	RECORD ALL INCIDENTS OF WATER FOUND IN THE FUEL.	
incline (n)	SLOPE (n)	YOU CAN ADJUST THE SLOPE OF THE RAMP.	You can adjust the incline of the ramp.
INCLUDE (v), INCLUDES, INCLUDED, INCLUDED	To make, or to be, part of	THIS CHAPTER INCLUDES THE PROCEDURES FOR THE REMOVAL OF THE COMPONENTS.	
including (prep)	THRU (prep)	DO TESTS 4 THRU 8 AGAIN.	Repeat from test 4 up to and including test 8.
	WITH (prep)	SEND THE DEFECTIVE COVER, WITH THE OIL SAMPLES, TO THE REPAIR CENTER.	Return the defective cover, including the oil samples, to the repair center.
incompatible (adj)	NOT COMPATIBLE	THIS SOFTWARE IS NOT COMPATIBLE WITH OTHER OPERATING SYSTEMS.	This software is incompatible with other operating systems.
		THIS COOLANT IS NOT COMPATIBLE WITH ALUMINUM ALLOYS.	This coolant is incompatible with aluminum alloys.

Word (part of speech)	Approved meaning/ ALTERNATIVES	APPROVED EXAMPLE	Not approved example
incomplete (adj)	NOT FULL	IF THE ACTUATOR TRAVEL IS NOT FULL, ADJUST THE RODS AGAIN.	If the actuator travel is incomplete, readjust the rods.
	NOT COMPLETED	IF THE ADJUSTMENT IS NOT COMPLETED, THE VALUES SHOWN ON THE INDICATOR WILL BE INCORRECT.	If the adjustment is incomplete, the values shown on the indicator will be incorrect.
	ALL (adj)	IF YOU DO NOT HAVE ALL THE RESULTS, DO THE TEST AGAIN.	If the results are incomplete, do the test again.
incorporate (v)	INCLUDE (v)	ALL MODELS INCLUDE A MICROPROCESSOR CONTROL UNIT.	All models incorporate a microprocessor control unit.
	HAVE (v)	SOME RELAYS HAVE A POTENTIOMETER.	Some relays incorporate a potentiometer.
incorporation (n)	INCLUDE (v)	THE MANUFACTURER INCLUDED A SECOND DIODE TO INCREASE THE RELIABILITY OF THE MOTOR.	Incorporation of a second diode increased the reliability of the motor.
INCORRECT (adj)	Not correct	IF THE TRIM TAB ADJUSTMENT IS INCORRECT, DO STEPS 7 THRU 10.	
INCORRECTLY (adv)	In an incorrect manner	IF YOU SEE THE RED MARK, THE PART IS INSTALLED INCORRECTLY.	

Word (part of speech)	Approved meaning/ ALTERNATIVES	APPROVED EXAMPLE	Not approved example
increase (n)	INCREASE (v)	THE TEMPERATURE MUST INCREASE.	There must be an increase in temperature.
INCREASE (v), INCREASES, INCREASED, INCREASED	To make or become larger or higher in value	INCREASE THE VANE ANGLE TO 90 DEGREES.	
		MAKE SURE THAT THE TEMPERATURE INCREASES TO 30 °C.	
INCREMENT (n)	One of a sequence of regular operations done one after the other, to increase or decrease an adjustable quantity	ADJUST THE STOP BOLT IN INCREMENTS OF 5 mm.	
incur (v)	CAUSE (v)	REMOVAL OF THE UNIT BEFORE 24 HOURS CAN CAUSE DAMAGE.	Removal of the unit before 24 hours can incur damage.
independent (adj)	INDEPENDENTLY (adv)	THIS UNIT OPERATES INDEPENDENTLY.	The operation of this unit is independent.
INDEPENDENTLY (adv)	Without a relation with	THE EMERGENCY SYSTEM IS CONTROLLED INDEPENDENTLY.	
indicate (v)	SHOW (v)	THE POINTER SHOWS ZERO FEET.	The pointer indicates zero feet.
	IDENTIFY (v)	A 406 MHz DIGITAL SIGNAL IDENTIFIES THE EMERGENCY LOCATOR TRANSMITTER AND ITS LOCATION.	A 406 MHz digital signal indicates the emergency locator transmitter and its location.
INDICATION (n)	Something that is shown	MAKE SURE THAT THE INDICATION IS CORRECT.	
induce (v)	CAUSE (v)	SCRATCHES IN THE WINDSCREEN CAN CAUSE CRACKS.	Scratches in the windscreen may induce cracks.

Word (part of speech)	Approved meaning/ ALTERNATIVES	APPROVED EXAMPLE	Not approved example
ineffective (adj)	UNSERVICEABLE (adj)	IF THE PUMP IS UNSERVICEABLE, REPLACE IT.	If the pump is ineffective, replace it.
	UNSATISFACTORY (adj)	IF THE OPERATION OF THE PUMP IS UNSATISFACTORY, REPLACE IT.	If the operation of the pump is ineffective, replace it.
	NOT OPERATE	IF THE PUMP DOES NOT OPERATE CORRECTLY, REPLACE IT.	If the pump is ineffective, replace it.
inflammable (adj)	FLAMMABLE (adj)	SOLVENT IS VERY FLAMMABLE.	Solvent is highly inflammable.
INFLATE (v), INFLATES, INFLATED, INFLATED	To make or become larger as a result of pressurization by a gas	INFLATE THE TIRES WITH NITROGEN.	
		OPERATION OF THE HANDLE RELEASES THE GAS, AND THE SLIDE-RAFT INFLATES.	
inform (v)	TELL (v)	TELL THE MANUFACTURER ABOUT THE PROBLEM.	Inform the manufacturer about the problem.
INFORMATION (n)	Data collected and made available for a specified function	THIS MANUAL CONTAINS INFORMATION ABOUT MAINTENANCE PROCEDURES.	
ingest (v)	SWALLOW (v)	IF YOU SWALLOW HYDRAULIC FLUID, GET MEDICAL AID IMMEDIATELY.	If you ingest hydraulic fluid, get medical aid immediately.
ingress (n)	GO INTO	IF WATER GOES INTO THE COMPONENT, DAMAGE WILL OCCUR.	Water ingress will damage the component.

Word (part of speech)	Approved meaning/ ALTERNATIVES	APPROVED EXAMPLE	Not approved example
inhale (v)	BREATHE (v)	DO NOT BREATHE THE SOLVENT FUMES.	Do not inhale the solvent fumes.
inhibit (v)	PREVENT (v)	ALODINE PREVENTS CORROSION.	Alodine inhibits corrosion.
INITIAL (adj)	That is related to the start	MOVE THE LEVER BACK TO ITS INITIAL POSITION.	
INITIALLY (adv)	At the start	INITIALLY, THE SPEED INCREASES TO 3000 RPM.	
initiate (v)	START (v)	START THE TEST AT 2000 RPM.	Initiate the test at 2000 rpm.
inject (v)	PUT (v)	PUT OIL INTO THE GEARBOX WITH A SYRINGE.	Inject oil into the gearbox with a syringe.
injection (n)	PUT (v)	PUT THE CORROSION PREVENTIVE OIL INTO THE PUMP WITH A SYRINGE.	The injection of the corrosion preventive oil into the pump must be done with a syringe.
injure (v)	INJURY (n)	MAKE SURE THAT THE END OF THE SAFETY WIRE WILL NOT CAUSE INJURY TO PERSONNEL.	Make sure that the end of the safety wire will not injure personnel.
INJURY (n)	Damage to a person	BEND THE ENDS OF THE LOCKING WIRE TO PREVENT INJURY.	
INNER (adj)	Nearer to the center of an object	THE INNER SURFACE OF THE PART IS BLACK.	
inoperative (adj)	OFF (adj)	BEFORE YOU DRAIN THE FUEL, MAKE SURE THAT THE SYSTEM IS OFF.	Before you drain the fuel, make sure that the system is inoperative.
	NOT OPERATE	MAKE SURE THAT THE DEMISTING SYSTEM DOES NOT OPERATE.	Make sure that the demisting system is inoperative.
INPUT (n)	The data, power, or energy put into equipment or a system	THE UNIT COMPARES THE INPUTS FROM THE TWO SENSORS TO THE COMPUTER.	

Word (part of speech)	Approved meaning/ ALTERNATIVES	APPROVED EXAMPLE	Not approved example
insert (v)	PUT (v)	PUT THE SLEEVE INTO THE OPENING.	Insert the sleeve into the opening.
	ENGAGE (v)	MAKE SURE THAT THE SHAFT ENGAGES IN THE MATING PART.	Check that the shaft is inserted in the mating part.
inside (adj)	INNER (adj)	CLEAN THE INNER SURFACE OF THE CONTAINER.	Clean the inside surface of the container.
inside (n)	INNER (adj)	PAINT THE INNER SURFACE OF THE PUMP ASSEMBLY.	Paint the inside surface of the pump assembly.
inside (prep)	IN (prep)	THERE MUST BE LUBRICANT IN THE HOLE.	There must be lubricant inside the hole.
	INTO (prep)	SAFETY ALL COMPONENTS BEFORE YOU PUT THEM INTO THE FUEL TANK.	Safety all components before you place them inside the fuel tank.
inspect (v)	EXAMINE (v)	EXAMINE ALL THE DRAIN HOLES.	Inspect all the drain holes.
	INSPECTION (n)	DO AN INSPECTION OF THE WORK.	Inspect the work.
INSPECTION (n)	The procedure which compares an object with its standard or specification	USE A VOLTMETER TO DO THIS INSPECTION.	
INSTALL (v), INSTALLS, INSTALLED, INSTALLED	To attach an item in or to a second item	INSTALL THE NEW O-RINGS ON THE SPINDLE.	
INSTALLATION (n)	The procedure which installs an item	REFER TO PARAGRAPH 3 FOR THE INSTALLATION OF THE UNIT.	
instead (adv)	ALTERNATIVE (n)	USE A SPATULA AS AN ALTERNATIVE.	Use a spatula instead.

Word (part of speech)	Approved meaning/ ALTERNATIVES	APPROVED EXAMPLE	Not approved example
instead of (prep)	ALTERNATIVE (n)	YOU CAN USE A CLEAN PLASTIC BAG AS AN ALTERNATIVE TO THE SPECIFIED CONTAINER.	You can use a clean plastic bag instead of the specified container.
INSTRUCTION (n)	A command that you must obey	IF YOU DO NOT OBEY THIS INSTRUCTION, DAMAGE CAN OCCUR.	
INSTRUMENT (n)	An item which measures and/or shows	ALL THE INSTRUMENTS ARE ON THE LEFT QUARTER PANEL.	
insufficient (adj)	NOT SUFFICIENT	IF THIS STEP IS NOT SUFFICIENT TO STOP THE LEAKAGE, REPLACE THE ACTUATOR.	If this step is insufficient to stop the leakage, replace the actuator.
INSULATION (n)	A material that helps to prevent the movement of energy	MAKE SURE THAT YOU DO NOT CAUSE DAMAGE TO THE INSULATION.	
insure (v)	MAKE SURE (v)	MAKE SURE THAT THE WORK BENCH IS CLEAN.	Insure that the work bench is clean.
intact (adj)	NOT DAMAGED	MAKE SURE THAT THE WINDSCREEN IS NOT DAMAGED.	Make sure the windscreen is intact.
integral (adj)	PART (n)	THE PIN IS A PART OF THE PUMP.	The pin is integral with the pump.
INTENSITY (n)	The quantity of strength, force, or energy	OPERATE THE DIMMER SWITCH AND MAKE SURE THAT THE INTENSITY OF THE LIGHT CHANGES.	
INTERCHANGE (v), INTERCHANGES, INTERCHANGED, INTERCHANGED	To put or use each of two things in the place of the other without change	INTERCHANGE THE No. 1 AND No. 2 COMPUTERS. THEN DO THE TEST AGAIN.	

Word (part of speech)	Approved meaning/ ALTERNATIVES	APPROVED EXAMPLE	Not approved example
INTERCHANGE-ABLE (adj)	That you can interchange	RELAYS 2YG AND 8YG ARE FULLY INTERCHANGEABLE ITEMS.	
interconnect (v)	CONNECT (v)	THE PIPES CONNECT SYSTEM A TO SYSTEM B.	The pipes interconnect system A and system B.
INTERFACE (n)	The connection between two systems or components	CLEAN THE INTERFACE BETWEEN THE COMPUTER AND THE SENSOR.	
interfere (v)	HIT (v)	IF YOU INSTALL THE UPPER DRAG STRUT WITH THE HEAD DOWN, IT WILL HIT THE LOWER DRAG STRUT DURING RETRACTION.	Placing the upper drag strut upside down makes it interfere with the lower drag strut.
	CATCH (v)	MAKE SURE THAT THE RODS CANNOT CATCH ON THE FLAPS.	Make sure the rods cannot interfere with the flaps.
	TOUCH (v)	IF THE BRACKET TOUCHES THE PLATE, REMOVE THE UNWANTED MATERIAL WITH A FILE.	Should the bracket interfere with the plate, remove surplus material with a file.
	INTERFERENCE (TN)	MAKE SURE THAT THERE IS NO INTERFERENCE BETWEEN THE RADAR AND THE TACTICAL AIR NAVIGATION SYSTEM (TACAN).	Ensure that the radar does not interfere with the Tactical Air Navigation System (TACAN).

Word (part of speech)	Approved meaning/ ALTERNATIVES	APPROVED EXAMPLE	Not approved example
interference (n)	HIT (v)	IF YOU INSTALL THE UPPER DRAG STRUT WITH THE HEAD DOWN, IT WILL HIT THE LOWER DRAG STRUT DURING RETRACTION.	Placing the upper drag strut upside down causes interference with the lower drag strut.
	CATCH (v)	MAKE SURE THAT THE RODS CANNOT CATCH ON THE FLAPS.	Make sure that there is no interference between the rods and the flaps.
	TOUCH (v)	MAKE SURE THAT THE LOCKING WIRE DOES NOT TOUCH THE LEVER.	Ensure that there is no interference between the locking wire and the lever.
	This word can be a technical name or part of a technical name.	MAKE SURE THAT YOU DO NOT HEAR INTERFERENCE IN THE 400 Hz BAND.	
		INTERFERENCE BOLTS ATTACH THE STRAP AND THE BOTTOM SKIN PANEL TO THE TRIFORM MEMBER.	
interior (n)	INTERNALLY (adv)	EXAMINE THE TANK INTERNALLY.	Examine the interior of the tank.
	INTERNAL (adj)	APPLY A SMALL QUANTITY OF LUBRICANT TO THE INTERNAL SIDE OF THE VALVE HOUSING.	Apply a small quantity of lubricant to the interior of the valve housing.
intermediate (adj)	MIDDLE (adj)	PUT THE LEVER IN THE MIDDLE POSITION.	Put the lever in intermediate position.
	BETWEEN (prep) Give the limits.	WHEN THE INPUT LEVER IS BETWEEN 5 AND 10 mm, MEASURE THE PISTON EXTENSION.	When the input lever is in an intermediate position, measure the piston extension.

Word (part of speech)	Approved meaning/ ALTERNATIVES	APPROVED EXAMPLE	Not approved example
INTERMITTENT (adj)	That stops and starts again, not continuous	THIS CAUSES INTERMITTENT OPERATION OF THE WARNING LIGHTS.	
INTERMITTENTLY (adv)	In an intermittent manner	THE LIGHT COMES ON INTERMITTENTLY.	
INTERNAL (adj)	Of, or on, the inner side and inner space	MAKE SURE THAT FUEL COMES OUT OF THE INTERNAL EJECTORS.	
INTERNALLY (adv)	Of, or on, the inner side and inner space	USE A BORESCOPE TO EXAMINE THE COMBUSTOR INTERNALLY.	
interrupt (v)	STOP (v)	DO NOT STOP THE TEST.	Do not interrupt the test.
intersect (v)	INTERSECTION (TN)	DRILL A HOLE AT THE INTERSECTION OF THE TWO LINES.	Drill a hole where the two lines intersect.
INTERVAL (n)	A distance or gap between two points in space or time	DO THE TEST AT FIVE-MINUTE INTERVALS.	
INTO (prep)	1. Movement to something and access to it	IF AN OBJECT FALLS INTO THE OPENING, DAMAGE TO THE ENGINE CAN OCCUR.	
	2. Function word that shows change, condition, result	THE PARTS OF THE STRUCTURE ARE DIVIDED INTO THREE GROUPS.	
investigate (v)	INVESTIGATION (n)	DO AN INVESTIGATION TO FIND THE CAUSE OF THE FUEL LEAKAGE.	Investigate the cause of the fuel leakage.
INVESTIGATION (n)	An official and accurate enquiry or research	THE TASK TEAM STARTED THE INVESTIGATION TO FIND THE CAUSE OF THE ACCIDENT.	

Word (part of speech)	Approved meaning/ ALTERNATIVES	APPROVED EXAMPLE	Not approved example
inward (adj)	INNER (adj)	MEASURE THE INNER DIAMETER OF THE HOLE.	Measure the inward diameter of the hole.
inward (adv)	IN (adv)	MOVE THE DOOR IN AND REARWARD.	Move the door inward and rearward.
IRREGULAR (adj)	Not regular	IF THE SIGNALS ARE IRREGULAR, REPLACE THE UNIT.	
irregularity (n)	DAMAGE (n)	THERE MUST BE NO DAMAGE TO THE SURFACE FINISH.	There must be no irregularities in the surface finish.
	DEFECT (TN)	THERE MUST BE NO DEFECTS IN THE SURFACE FINISH.	There must be no irregularities in the surface finish.
	IRREGULAR (adj)	IF YOU GET IRREGULAR RESULTS, DO THE TEST AGAIN.	If you get irregularities in the results, do the test again.
	UNUSUAL (adj)	IF YOU GET UNUSUAL RESULTS, DO THE TEST AGAIN.	If you get irregularities in the results, do the test again.
IRREGULARLY (adv)	Not regularly	IF THE UNIT RECEIVES THE SIGNALS IRREGULARLY, REPLACE IT.	
ISOLATE (v), ISOLATES, ISOLATED, ISOLATED	To prevent a supply or connection to	ISOLATE THE ELECTRICAL POWER SUPPLY.	
isolation (n)	ISOLATE (v)	ISOLATE SYSTEM No. 1 TO DO THE LEAKAGE CHECKS.	Isolation of system No. 1 is necessary to do the leakage checks.
IT (pron)	That thing	CAREFULLY MOVE THE DEFLECTOR UNIT DOWN UNTIL IT ENGAGES.	
ITEM (n)	A thing that is determined or specified	LUBRICATE THESE ITEMS WITH GREASE.	
ITS (adj)	Belonging to a thing	REMOVE THE UNIT AND ITS WIRING.	

Word (part of speech)	Approved meaning/ ALTERNATIVES	APPROVED EXAMPLE	Not approved example
jack (v)	JACK (TN)	DO NOT LIFT THE AIRCRAFT ON JACKS IF THE WIND SPEED IS MORE THAN 20 MPH.	Do not jack the aircraft if the wind speed exceeds 20 mph.
jam (v)	FREELY (adv)	MAKE SURE THAT THE VALVE SPINDLE MOVES FREELY.	Check that the valve spindle does not bind or jam.
job (n)	WORK (n)	DO THIS WORK IN A CLEAN AREA.	Do this job in a clean area.
	TASK (n)	DO THIS TASK IN A CLEAN AREA.	Do this job in a clean area.
join (v)	BOND (v)	USE ADHESIVE No. 35 TO BOND METAL TO METAL.	Use adhesive No. 35 to join metal to metal.
	ATTACH (v)	ATTACH SMOKE DETECTOR 1 TO SMOKE DETECTOR 2.	Join smoke detectors 1 and 2.
	CONNECT (v)	CONNECT THE HOSE TO THE PIPE WITH A CLIP.	Join the hose to the pipe with a clip.
JOINT (n)	The area at which two surfaces or edges touch or are attached	SEAL THE JOINT WITH SEALING COMPOUND.	
just (adv)	IMMEDIATELY (adv)	KEEP THE RING IMMEDIATELY ABOVE THE JACK COLLARS.	Keep the ring just above the jack collars.

Blank Page

Word (part of speech)	Approved meaning/ ALTERNATIVES	APPROVED EXAMPLE	Not approved example
KEEP (v), KEEPS, KEPT, KEPT	To continue to have or hold	KEEP THE COMPOUND IN A CLOSED CONTAINER.	
key (v)	REFER (v)	THE ILLUSTRATION REFERS TO EACH STEP.	The illustration will be keyed to each of the steps.
	KEY (TN)	INSTALL THE KEY BETWEEN THE PULLEY AND THE SHAFT.	The pulley must be keyed to the shaft.
KILL (v), KILLS, KILLED, KILLED	To cause death	HIGH VOLTAGES CAN KILL YOU.	
kind (n)	TYPE (n)	YOU CAN USE TWO TYPES OF SEALING COMPOUND.	Two kinds of sealing compound can be used.
kink (v)	KINK (TN)	DO NOT MAKE KINKS IN THE CABLES.	Do not kink the cables.
knock (v)	HIT (v)	HIT THE CLAMP ON THE RIGHT SIDE WITH A HAMMER.	Knock the clamp on the right side with a hammer.
KNOW (v), KNOWS, KNEW, KNOWN	To be sure of data, to have data ready to use	IF YOU KNOW THE CLEARANCE, YOU CAN CALCULATE THE THICKNESS OF THE WASHER.	

Blank Page

Word (part of speech)	Approved meaning/ ALTERNATIVES	APPROVED EXAMPLE	Not approved example
label (v)	IDENTIFY (v)	IDENTIFY THE COMPONENT WITH A CODE TO HELP YOU INSTALL IT AGAIN CORRECTLY.	Label the component with a code to help you install it again correctly.
	LABEL (TN)	PUT A LABEL ON THE RELAY ASSEMBLY.	Label the relay assembly.
lack (n)	NOT SUFFICIENT	DAMAGE CAN OCCUR TO THE PUMP IF THERE IS NOT SUFFICIENT OIL IN THE RESERVOIR.	Damage can occur to the pump if there is a lack of oil in the reservoir.
LAMINATED (adj)	Made of laminations	THIS PANEL IS MADE OF LAMINATED CARBON FIBER.	
LAMINATION (n)	Layers of one or more materials bonded together	IF THE LAMINATION IS DAMAGED, REPLACE THE PANEL.	
land (v)	LANDING (TN)	AFTER A HEAVY LANDING, DO THE HEAVY LANDING INSPECTION.	If the aircraft landed heavily, do the Heavy Landing Inspection.
LARGE (adj) (LARGER, LARGEST)	More than average in dimension, quantity, or capacity	CLEAN YOUR SKIN WITH A LARGE QUANTITY OF CLEAN WATER.	
LAST (adj)	That comes at the end	IMMEDIATELY AFTER THE LAST FLIGHT OF THE DAY, INSTALL ALL COVERS.	
LAST (adv)	After other persons or things	FILL THE OUTER TANKS LAST.	
last (v)	CONTINUE (v)	THE SIGNALS CONTINUE FOR 0.8 SECONDS.	The signals last 0.8 seconds.
LATCH (v), LATCHES, LATCHED, LATCHED	To hold something in position with a latching device	MAKE SURE THAT YOU LATCH THE DOOR CORRECTLY.	
		MAKE SURE THAT THE DOOR LATCHES CORRECTLY.	

Word (part of speech)	Approved meaning/ ALTERNATIVES	APPROVED EXAMPLE	Not approved example
later (adj)	SUBSEQUENT (adj)	A SUBSEQUENT REPAIR WILL BE FOR THE NEW CONFIGURATION.	A later repair will be for the new configuration.
	THEN (adv)	DO THIS STEP FIRST. THEN, DISCONNECT THE CABLES.	Do this step first. At a later time, disconnect the cables.
later (adv)	SUBSEQUENTLY (adv)	MAKE SURE THAT THE TOOL IS SUBSEQUENTLY AVAILABLE FOR THE INSTALLATION PROCEDURE.	Make sure that the tool is available later for the installation procedure.
	WHEN (conj) Give the condition or time, if possible.	DO STEPS (3) THRU (5). DO STEP (6) WHEN THE AILERON MOVEMENT STOPS.	Do steps (3) to (5). Do step (6) later.
	AFTER (prep)	THE BAR MOVES DOWN AFTER 20 SECONDS.	The bar moves down later.
LATERAL (adj)	Related to or on the side	ADJUST THE LATERAL LEVEL.	
LATERALLY (adv)	In a lateral direction	PUSH THE UNIT OUT, AND THEN MOVE IT LATERALLY.	
lay (v)	PUT (v)	PUT COVERS ON THE FLOOR.	Lay covers on the floor.
LAYER (n)	A material applied to a surface or put on, or below, a second material	APPLY A THIN LAYER OF SEALING COMPOUND ON THE BARE METAL.	
lead (v)	PUT (v)	PUT THE TELEFLEX CABLE THROUGH THE BULKHEAD.	Lead the Teleflex cable through the bulkhead.
LEAK (n)	A crack, gap or hole which lets fluid or light go into or come out of something	EXAMINE THE SWIVEL COUPLINGS FOR LEAKS.	

Word (part of speech)	Approved meaning/ ALTERNATIVES	APPROVED EXAMPLE	Not approved example
leak (v)	LEAK (n)	IF THERE ARE LEAKS IN THE SYSTEM, REPAIR THEM.	If the system is leaking, repair it.
LEAKAGE (n)	The quantity of material that goes through a leak	MEASURE THE LEAKAGE FROM THE OUTLET PORT.	
least (adv)	MINIMUM (adj)	REPAIR THE AREA WITH THE MINIMUM DAMAGE FIRST.	Repair the least damaged area first.
least (at least) (adv)	MINIMUM (adj)	THE MINIMUM TIME BETWEEN CHECKS IS TWO MINUTES.	The time between checks is at least two minutes.
	MINIMUM (n)	DO STEPS 13 THRU 16 FOR A MINIMUM OF THREE TIMES.	Do steps 13 thru 16 at least three times.
leave (v)	KEEP (v)	KEEP THE VENT VALVES OPEN.	Leave the vent valves open.
	LET .... STAY (v)	LET THE VENT VALVES STAY OPEN.	Leave the vent valves open.
	GO (v)	IF FUEL SPILLS ON THE GROUND, IMMEDIATELY GO TO A SAFE AREA.	If fuel spills on the ground, immediately leave the contaminated area.
LEFT (adj)	On the west side when you look north	INSTALL THE CONTROL PANEL IN THE LEFT CONSOLE.	
left-hand (adj)	LEFT (adj)	SET THE LEFT SWITCH TO "ON".	Set the left-hand switch to ON.
legible (adj)	READ (v)	MAKE SURE THAT YOU CAN READ THE IDENTIFICATION PLATES.	Make certain that the identification plates are legible.
LENGTH (n)	1. The distance from one end to the other end	MEASURE THE LENGTH OF THE JACK RAM.	
	2. The longer or longest of the three dimensions of an object	THE DIMENSIONS OF THE UNIT ARE: - LENGTH 300 mm - WIDTH 90 mm - THICKNESS 60 mm.	

Word (part of speech)	Approved meaning/ ALTERNATIVES	APPROVED EXAMPLE	Not approved example
lengthen (v)	INCREASE (v)	DO NOT TRY TO INCREASE THE LENGTH OF THIS ROD.	Do not try to lengthen this rod.
LESS (adj) No other forms of this adjective.	Of smaller dimension, value, quantity, volume, or number	THE FLOW RATE MUST NOT BE LESS THAN 4.0 GPM.	
less (prep)	WITHOUT (prep)	THE JUNCTION BOX IS SUPPLIED WITHOUT THE TEST LEADS.	The junction box is supplied less the test leads.
LET (v), LETS, LET, LET	To give opportunity	LET THE ADHESIVE BECOME TACKY.	
lethal (adj)	KILL (v)	THE ELECTRICAL DISCHARGE FROM THESE POWER UNITS CAN KILL YOU.	The electrical discharge from these power units can be lethal.
LEVEL (adj)	Horizontal to a known datum	PARK THE VEHICLE ON LEVEL GROUND.	
LEVEL (n)	A horizontal line, plane, surface, or condition	FILL THE TANK TO THE CORRECT LEVEL (REFER TO FIG. 105).	
level (v)	LEVEL (adj)	MAKE THE AIRCRAFT LEVEL.	Level the aircraft.
liberal (adj)	LARGE (adj)	CLEAN THE WINDOW WITH LARGE QUANTITIES OF WATER AND SOAP.	Clean the window with liberal quantities of water and soap.
	THICK (adj)	APPLY A THICK LAYER OF THE COMPOUND.	Apply a liberal amount of the compound.
liberally (adv)	LARGE (adj)	APPLY A LARGE QUANTITY OF THE SOLUTION.	Apply the solution liberally.
	THICK (adj)	APPLY A THICK LAYER OF THE COMPOUND.	Apply the compound liberally.
lie (v)	BE (v)	THE OTHER PART OF THE CLIP IS ALONG THE BARREL.	The other part of the clip lies along the barrel.

Word (part of speech)	Approved meaning/ ALTERNATIVES	APPROVED EXAMPLE	Not approved example
LIFE (n)	The time during which you can use an item or object	THE LIFE OF THE CARTRIDGE IN THE FIRE EXTINGUISHER IS 3 YEARS.	
LIFT (v), LIFTS, LIFTED, LIFTED	To move something up	LIFT THE COVER CAREFULLY.	
LIGHT (adj) (LIGHTER, LIGHTEST)	That has a small mass, weight, or force For other meanings, use:	APPLY A LIGHT FORCE ON THE RUDDER PEDALS.	
	THIN (adj)	APPLY A THIN LAYER OF SEALANT ON THE SEALING SURFACES OF THE CONTAINER.	Apply a light layer of sealant on the sealing surfaces of the container.
	SMALL (adj)	APPLY A SMALL QUANTITY OF LUBRICANT ON EACH O-RING BEFORE YOU INSTALL IT.	Apply a light quantity of lubricant on each O-ring before its installation.
light (v)	COME ON	MAKE SURE THAT THE FLUID INDICATOR LIGHT COMES ON.	Ensure that the fluid indicator light lights.
LIGHTING (n)	That which gives light to	A DIMMER UNIT CONTROLS THE PANEL LIGHTING.	
LIGHTLY (adv)	In a light manner	LIGHTLY RUB THE DEFECTIVE AREA WITH WET ABRASIVE PAPER.	
LIMIT (n)	A specified maximum or minimum quantity, number, time, or distance	IF THE CLEARANCES ARE NOT IN THE LIMITS GIVEN IN FIG. 4, REFER TO REPAIR SCHEME No. 2.	

Word (part of speech)	Approved meaning/ ALTERNATIVES	APPROVED EXAMPLE	Not approved example
limitation (n)	LIMIT (n)	WHEN YOU LIFT THE AIRCRAFT ON JACKS, KEEP THE CENTER OF GRAVITY BETWEEN THESE LIMITS:	When you lift the aircraft on jacks, observe these center of gravity limitations:
limited (adj)	SMALL (adj)	THERE IS ONLY A SMALL NUMBER OF REPAIRS THAT YOU CAN DO.	There is only a limited number of repairs that you can do.
LINEAR (adj)	In a straight line	MAKE SURE THAT THE RESULTS ARE LINEAR.	
LINEARLY (adv)	In a straight line	INCREASE THE PRESSURE LINEARLY.	
link (v)	CONNECT (v)	CONNECT THE CABLES.	Link the cables.
	ATTACH (v)	ATTACH THE HOIST TO THE POWER UNIT.	Link the hoist to the power unit.
LIQUID (adj)	That has the properties of a liquid	DURING SERVICING, LIQUID OXYGEN FLOWS THROUGH THE STABILIZING CONTAINER.	
LIQUID (n)	A material that is not a gas or a solid	THE CONVERTER CHANGES THE LIQUID INTO A GAS.	
list (v)	RECORD (v)	RECORD THE TEST RESULTS.	List the test results.
	SHOW (v)	TABLE 1 SHOWS ALTERNATIVE ADHESIVES THAT YOU CAN USE.	Table 1 lists alternative adhesives that you can use.
	LIST (TN)	MAKE A LIST OF THE PART NUMBERS IN NUMERICAL SEQUENCE.	List the part numbers in numerical sequence.
LISTEN (v), LISTENS, LISTENED, LISTENED	To use your ears to hear or find	LISTEN FOR THE SIGNAL.	

Word (part of speech)	Approved meaning/ ALTERNATIVES	APPROVED EXAMPLE	Not approved example
little (adj)	SMALL (adj)	THE DIAMETER OF THE TUBE IS TOO SMALL.	The diameter of the tube is too little.
	SHORT (adj)	INSTALL A SHORT LENGTH OF NEW TUBE.	Install a little length of new tube.
	NOT SUFFICIENT	IF THE PRESSURE IS NOT SUFFICIENT, THE TEST WILL STOP.	If the pressure is too little, the test will stop.
little (a little) (adj)	SMALL QUANTITY	ADD A SMALL QUANTITY OF DISINFECTANT TO THE SOLUTION.	Add a little disinfectant to the solution.
little (a little) (adv)	SMALL (adj)	AFTER YOU REMOVE THE NUT, THE RIB CAN MOVE DOWN A SMALL DISTANCE.	After you remove the nut, the rib can move down a little.
LIVE (adj)	That includes explosive material For electrical systems, use:	SOME MAINTENANCE TASKS ARE NOT PERMITTED ON AIRCRAFT THAT HAVE LIVE AMMUNITION.	
	ENERGIZE (v)	THE SYSTEM IS ENERGIZED.	The system has live electrical power.
load (v)	INSTALL (v)	INSTALL THE SPRING IN THE CARTRIDGE.	Load the spring in the cartridge.
	LOAD (TN)	PUT THE LOAD ON THE SPRING BALANCE.	Load the spring balance.
LOCAL (adj)	Applicable to a given location or position	SET THE "LAT" KNOB TO THE LOCAL LATITUDE.	
LOCALLY (adv)	Applicable to a given location or position	YOU CAN ALSO DO THIS TEST WITH A LOCALLY MADE TOOL.	

Word (part of speech)	Approved meaning/ ALTERNATIVES	APPROVED EXAMPLE	Not approved example
locate (v)	FIND (v)	USE TALCUM POWDER TO FIND EXTERNAL LEAKS.	Use talcum powder to locate external leaks.
	ENGAGE (v)	ENGAGE THE GEAR WHEEL ON THE SHAFT.	Locate the gear wheel on the shaft.
	PUT (v)	USE THE MARK TO HELP YOU PUT THE CLAMP IN ITS CORRECT POSITION.	Use the mark to locate the clamp.
LOCATION (n)	A place or site that you can find	FIGURES 1 THRU 4 SHOW THE LOCATIONS OF THE WALKWAYS.	
LOCK (v), LOCKS, LOCKED, LOCKED	To attach something, or hold it in position with a locking device	LOCK THE SWITCH AND REMOVE THE KEY.	
		MAKE SURE THAT THE SWITCH LOCKS BEFORE YOU CONTINUE.	
log (v)	RECORD (v)	RECORD THE COMPASS CORRECTIONS ON THE CARD.	Log the compass corrections on the card.
LONG (adj) (LONGER, LONGEST)	That has large length or duration	USE A TUBE WHICH IS 3.3 FEET LONG.	
long (as long as) (conj)	WHILE (conj)	WHILE THE DC SUPPLY IS ON, THE FAN MUST OPERATE.	As long as the DC supply is ON, the fan must operate.
long (no longer) (adv)	MAXIMUM LENGTH	USE A DRILL BIT THAT HAS A MAXIMUM LENGTH OF 45 mm.	Use a drill bit no longer than 45 mm.
	NOT... AT THIS TIME	THIS PROCEDURE IS NOT APPLICABLE AT THIS TIME.	This procedure is no longer applicable.
	UNTIL...NOT	ADJUST THE POTENTIOMETER UNTIL YOU DO NOT HEAR A HUM FROM THE LOUDSPEAKER.	Adjust the potentiometer until the loudspeaker no longer hums.

Word (part of speech)	Approved meaning/ ALTERNATIVES	APPROVED EXAMPLE	Not approved example
LONGITUDINAL (adj)	Related to the longitudinal axis	ADJUST THE LONGITUDINAL TRIM SYSTEM.	
LONGITUDINALLY (adv)	In a longitudinal direction	MOVE THE STAND LONGITUDINALLY.	
LOOK (v), LOOKS, LOOKED, LOOKED	To use your eyes to see or find	LOOK THROUGH THE INSPECTION WINDOW.	
LOOP (n)	A circular shape made in a flexible material	MAKE A LOOP OF WIRE AROUND THE UNIT.	
loop (v)	LOOP (n)	MAKE A LOOP OF WIRE AROUND THE UNIT.	Loop the wire around the unit.
LOOSE (adj)	1. On, in, but not attached	MAKE SURE THAT YOU REMOVE ALL TOOLS, EQUIPMENT, AND LOOSE OBJECTS FROM THE AIR INTAKE.	
	2. Not tight	DO A CHECK FOR LOOSE FASTENERS AND TIGHTEN THEM, IF NECESSARY.	
	For othcr meanings, use:		
	TENSION (TN)	MAKE SURE THAT THERE IS TENSION IN THE CABLES.	Make sure that the cables are not loose.
LOOSELY (adv)	Not tightly	MAKE SURE THAT THE PARTS ARE LOOSELY ASSEMBLED.	
LOOSEN (v), LOOSENS, LOOSENED, LOOSENED	To cause to be not tight	LOOSEN THE SCREWS THAT ATTACH THE CLAMP TO THE SUPPORT.	
lose (v)	DECREASE (v)	THE EFFECT OF THE SOLVENT DECREASES QUICKLY.	The solvent loses its effectiveness quickly.

Word (part of speech)	Approved meaning/ ALTERNATIVES	APPROVED EXAMPLE	Not approved example
lost (adj)	DECREASE (v)	IF THE PRESSURE DECREASES QUICKLY, THE SHEAR PINS IN THE LATCH WILL BREAK.	If the pressure is lost rapidly, the shear pins in the latch will break.
	NOT FIND	IT IS POSSIBLE THAT YOU CANNOT FIND THE RUBBER BUSHINGS.	The rubber bushings may be lost.
LOW (adj) (LOWER, LOWEST)	That is of small value	THE SWITCH SENSES LOW PRESSURE IN THE DISTRIBUTION CIRCUIT.	
LOWER (v), LOWERS, LOWERED, LOWERED	To move something down	LOWER THE ANTENNA TO GET ACCESS TO THE CABLE CONNECTOR.	
lowermost (adj)	BOTTOM (adj)	THE BOTTOM MARK ON THE BOARD IS RED.	The lowermost mark on the board is red.
LUBRICATE (v), LUBRICATES, LUBRICATED, LUBRICATED	To apply lubricant	LUBRICATE THE O-RINGS WITH HYDRAULIC FLUID (NATO CODE H-515).	
lukewarm (adj)	WARM (adj)	SOAK THE AREA WITH WARM WATER.	Soak the area with lukewarm water.

Word (part of speech)	Approved meaning/ ALTERNATIVES	APPROVED EXAMPLE	Not approved example
MAGNETIC (adj)	Related to the properties of a magnet	REMOVE THE MAGNETIC DRAIN PLUG FROM THE OIL TANK.	
MAGNETICALLY (adv)	In a magnetic manner	RELEASE THE MAGNETICALLY LATCHED SWITCHES.	
magnetized (adj)	MAGNETIC (adj)	DO NOT USE MAGNETIC TOOLS TO ADJUST THE COMPASS.	Do not use magnetized tools to adjust the compass.
main (adj)	PRIMARY (adj)	THE PRIMARY CAUSE OF VALVE FAILURE IS CONTAMINATION OF THE HYDRAULIC FLUID.	The main cause of valve failure is contamination of hydraulic fluid.
maintain (v)	KEEP (v)	KEEP THE FLUID TEMPERATURE AT 70 °F.	Maintain the fluid temperature at 70 °F.
	HOLD (v)	TURN THE ROTATING TUBE 75 DEGREES. THEN HOLD THE TUBE IN THIS POSITION.	Turn the rotating tube 75 degrees. Then maintain the tube in this position.
	MAINTENANCE (n)	DO MAINTENANCE ON THE FUEL SYSTEM.	Maintain the fuel system.
MAINTENANCE (n)	The servicing and/or the repair of something to keep it in the correct condition	DO NOT REFUEL THE AIRCRAFT DURING MAINTENANCE.	
major (adj)	PRIMARY (adj)	THE PRIMARY FUNCTION OF THE POWER UNIT IS TO SUPPLY ELECTRICAL POWER.	The major function of the power unit is to supply electrical power.

Word (part of speech)	Approved meaning/ ALTERNATIVES	APPROVED EXAMPLE	Not approved example
MAKE (v), MAKES, MADE, MADE	To manufacture, to cause to occur or to become	MAKE A RECESS FOR THE STUDS OF 5 mm.	
		MAKE THE CLOTH MOIST WITH WATER.	
		TILT THE FITTING TO MAKE THE BARREL NUT FALL OUT OF IT.	
make certain (v)	MAKE SURE (v)	MAKE SURE THAT THE VALVE IS OPEN.	Make certain that the valve is open.
MAKE SURE (v), MAKES SURE, MADE SURE, MADE SURE	To verify and/or do the necessary steps	MAKE SURE THAT NO LEAK DETECTION FLUID GOES INTO THE OXYGEN SYSTEM.	
maladjusted (adj)	INCORRECTLY ADJUSTED	IF THE WARNING LIGHT IS ON WHEN THE DOOR IS CLOSED, THE MICROSWITCH IS INCORRECTLY ADJUSTED.	If the warning light is on when the door is closed, the microswitch is maladjusted.
MALFUNCTION (n)	Incorrect operation	REFER TO TABLE 104 TO CORRECT THE MALFUNCTION.	
MANDATORY (adj)	That which you must do, use, or obey	THIS SERVICE BULLETIN IS MANDATORY.	
manipulate (v)	MOVE (v)	MOVE THE FLEXIBLE PIPE TO GET ACCESS TO THE BOTTOM BOLTS.	Manipulate the flexible pipe to get access to the bottom bolts.
	ADJUST (v)	ADJUST THE FLEXIBLE PIPE TO GET ACCESS TO THE BOTTOM BOLTS.	Manipulate the flexible pipe to get access to the bottom bolts.
MANUAL (adj)	Operated with the hands	USE THE MANUAL CONTROLS IF THE AUTOMATIC CONTROLS DO NOT OPERATE.	

Word (part of speech)	Approved meaning/ ALTERNATIVES	APPROVED EXAMPLE	Not approved example
MANUAL (n)	A publication that supplies data related to a special subject	THIS MANUAL IDENTIFIES AND SUPPLIES THE PERFORMANCE DATA.	
MANUALLY (adv)	With your hands	MOVE THE LEVER MANUALLY.	
manufacture (v)	MAKE (v)	YOU CAN MAKE THE CLEARING TOOL LOCALLY.	The clearing tool can be manufactured locally.
MANY (adj) (MORE, MOST)	Of large number If possible, give the range, number, or quantity.	CORROSION CAN HAVE MANY CAUSES.	
MARK (n)	Something that you make or is made to show an identification, location, or direction	THE RED MARKS SHOW A MAXIMUM STEERING ANGLE OF 35 DEGREES.	
mark (v)	IDENTIFY (v)	IDENTIFY THE COMPONENT WITH A CODE TO HELP YOU INSTALL IT AGAIN CORRECTLY.	Mark the component with a code to help you install it again correctly.
	MARK (n)	PUT MARKS ON THE OUTER TUBE TO SHOW THE CLAMP LOCATIONS.	Mark the clamp locations on the outer tube.
mask (v)	APPLY (v)	APPLY MASKING TAPE TO THE TRANSPARENT PANEL.	Mask the transparent panel with masking tape.
MASS (n)	The quantity of matter that is in an object and relates to its weight. The equivalent of inertia	STRUCTURAL OR EQUIPMENT MODIFICATIONS CAN CHANGE THE BASIC MASS AND THE CENTER-OF-GRAVITY COORDINATES.	

Word (part of speech)	Approved meaning/ ALTERNATIVES	APPROVED EXAMPLE	Not approved example
match (v)	ALIGN (v)	FOLD THE GIRT ONTO THE CONTAINER UNTIL IT IS ALIGNED WITH THE VELCRO TAPE.	Fold the girt over the container so that the Velcro tape matches.
	AGREE (v)	MAKE SURE THAT THE SEAL AGREES WITH THE SHAPE OF THE PANEL.	Ensure the seal matches the panel shape.
mate (v)	CONNECT (v)	CONNECT THE PLUG TO THE RECEPTACLE.	Mate the plug to the receptacle.
	ENGAGE (v)	ENGAGE THE SPLINES ON THE SHAFT WITH THOSE IN THE GEARBOX.	Mate the splines on the shaft with those in the gearbox.
	TOGETHER (adv)	PUT THE SMOOTH SURFACES TOGETHER.	Mate the smooth surfaces.
MATERIAL (n)	1. A substance or something from which a thing is made	BEFORE YOU INSTALL THE SEAL, REMOVE THE UNWANTED MATERIAL.	
	2. Something made to do a task	YOU CAN USE EQUIVALENT ALTERNATIVES FOR THESE MATERIALS.	
MATING (adj)	Made to touch	APPLY GREASE ON THE MATING SURFACES OF THE STARTER GENERATOR.	
MATT (or MATTE) (adj)	Not glossy	APPLY MATT PAINT ON THE PANEL.	
matter (n)	MATERIAL (n)	REMOVE THE GLOSSY MATERIAL FROM THE SURFACE OF THE BLADE.	Remove the glossy matter from the surface of the blade.
MAXIMUM (adj)	Related to the largest dimension, quantity, number, or value	THE MAXIMUM VALUE THAT THE VOLTMETER SHOWS IS 199.9 V.	
MAXIMUM (n)	The largest dimension, quantity, number, or value	MOVE THE LEVER A MAXIMUM OF 45 DEG.	

Word (part of speech)	Approved meaning/ ALTERNATIVES	APPROVED EXAMPLE	Not approved example
may (v)	CAN (v)	IF YOU USE INCORRECT EQUIPMENT, YOU CAN CAUSE DAMAGE TO THE VANES.	The vanes may be damaged by using incorrect equipment.
	POSSIBLY (adv)	IF YOU CANNOT ENGAGE THE ROD AND THE PISTON, THE LENGTH OF THE NEW ROD IS POSSIBLY INCORRECT.	If you cannot engage the rod and the piston, the length of the new rod may be incorrect.
MEASURE (v), MEASURES, MEASURED, MEASURED	To find the dimensions, capacity, or quantity of something	MEASURE THE LOAD.	
MECHANICAL (adj)	Related to or operated with a mechanism	THE STANDBY SYSTEM IS MECHANICAL.	
MECHANICALLY (adv)	In a mechanical manner	OPERATE THE HORIZONTAL STABILIZER MECHANICALLY.	
MECHANISM (n)	An assembly of mechanical parts that are made to operate together	USE WHITE SPIRIT TO CLEAN A HINGE MECHANISM THAT HAS A GREASE NIPPLE.	
MEDICAL (adj)	Related to medicine	IF YOU GET HYDRAULIC FLUID IN YOUR EYES, FLUSH THEM WITH WATER AND GET MEDICAL AID.	
medium (adj)	MODERATE (adj)	APPLY MODERATE PRESSURE.	A medium amount of pressure must be applied.
meet (v)	ENGAGE (v)	PUSH THE PIN IN UNTIL IT ENGAGES IN THE RETAINING CLIP.	Push the pin in until it meets with the retaining clip.
	TOUCH (v)	THE ROD MUST TOUCH THE CENTER OF THE STRIP.	The rod should meet the center of the strip.

Word (part of speech)	Approved meaning/ ALTERNATIVES	APPROVED EXAMPLE	Not approved example
MELT (v), MELTS, MELTED, MELTED	To change or to cause something to change from solid to liquid because of heat	THE FUSIBLE PLUGS MUST MELT IN 20 SECONDS OR LESS.	
		DO NOT USE A FLAME TO MELT THE ICE.	
mention (v)	GIVE (v)	THE REPAIR LIMITS ARE NOT GIVEN BECAUSE SPECIAL EQUIPMENT IS NECESSARY.	The repair limits are not mentioned as special equipment is required.
mesh (v)	ENGAGE (v)	IF THE GEARS DO NOT ENGAGE CORRECTLY, ADD A SHIM.	If the gears do not mesh correctly, add a shim.
metallic (adj)	METAL (TN)	EXAMINE THE FILTER FOR METAL PARTICLES.	Examine the filter for metallic particles.
meteorological (adj)	WEATHER (n)	DO THE TEST IN GOOD WEATHER.	Carry out the test in good meteorological conditions.
METHOD (n)	A special type of procedure used to get a specified result	THIS PROCEDURE GIVES TWO METHODS FOR THE TEMPORARY REPAIR OF THE WATER TUBES.	
mid (adj)	MIDDLE (adj)	PUT THE THROTTLE LEVER IN THE MIDDLE POSITION.	Place the throttle lever in mid position.
MIDDLE (adj)	At a point that is between two other points	REMOVE THE MIDDLE BELLCRANK UNIT.	
MIDDLE (n)	A point that is at or near the center	LIFT THE SEAT TO THE MIDDLE OF THE TRAVEL.	
mild (adj)	WEAK (adj)	USE A WEAK SOAP SOLUTION TO CLEAN THE SIDEWALL PANEL.	Use a mild soap solution to clean the sidewall panel.
minimize (v)	MINIMUM (n)	USE A TRESTLE TO KEEP THE LOADS ON THE WING TO A MINIMUM.	Use a trestle to minimize the loads on the wing.

Word (part of speech)	Approved meaning/ ALTERNATIVES	APPROVED EXAMPLE	Not approved example
MINIMUM (adj)	Related to the smallest dimension, quantity, number, or value	MAKE SURE THAT THE FLUID IN THE HYDRAULIC RESERVOIR IS ABOVE THE MINIMUM LEVEL.	
MINIMUM (n)	The smallest dimension, quantity, number, or value	LET THE PAINT DRY FOR A MINIMUM OF 1 HOUR.	
minor (adj)	SMALL (adj)	YOU CAN IGNORE SMALL DIFFERENCES IN TEMPERATURE.	Minor differences in temperature can be ignored.
MINUS (prep)	With the subtraction of	INCREASE THE PRESSURE IN THE SHOCK ABSORBER TO 85.5 BAR, PLUS OR MINUS 2.7 BAR.	
miscellaneous (adj)	OTHER (adj)	REMOVE THE TOOLS AND THE OTHER RELATED ITEMS.	Remove the tools and the related miscellaneous items.
	ALL (adj)	REMOVE ALL EQUIPMENT.	Remove miscellaneous equipment.
misrigged (adj)	INCORRECTLY ADJUSTED	IF THE MARKS ARE NOT ALIGNED, THE LINKAGE IS INCORRECTLY ADJUSTED.	If the marks are not aligned, the linkage is misrigged.
MISSING (adj)	That is not there	IF SEALS ARE MISSING, INSTALL NEW SEALS.	
mix (n)	MIXTURE (n)	THE POTLIFE OF THE MIXTURE IS APPROXIMATELY 4 HOURS.	The potlife of the mix is approximately 4 hours.
MIX (v), MIXES, MIXED, MIXED	1. To put together two or more materials to become one	MIX THE COMPOUND.	
	2. To become combined	IF METHANE AND AIR MIX, THERE IS A RISK OF EXPLOSION.	
MIXTURE (n)	The result that you get when you mix materials	DO NOT USE THIS MIXTURE WHEN THE TEMPERATURE IS LESS THAN 15 °C.	

Word (part of speech)	Approved meaning/ ALTERNATIVES	APPROVED EXAMPLE	Not approved example
MOBILE (adj)	That is made to move easily	CONNECT THE MOBILE GROUND POWER UNIT.	
MODE (n)	A special condition of operation	MAKE SURE THAT THE "HEADING SELECT" MODE OPERATES.	
MODERATE (adj)	Between low and high values	APPLY A MODERATE LOAD ON THE BRAKE PEDALS.	
MODERATELY (adv)	In a moderate manner	YOU CAN APPLY A SECOND LAYER WHEN THE SURFACE IS MODERATELY DRY.	
modify (v)	CHANGE (v)	SERVICE BULLETIN NO. 105 CHANGES THE BOLTS THAT ATTACH THE POWER CONTROL UNIT.	Service Bulletin No. 105 modifies the power control unit attachment bolts.
MOIST (adj)	Moderately wet	CLEAN THE PANEL WITH A MOIST CLOTH.	
moisten (v)	MOIST (adj)	MAKE THE CLOTH MOIST WITH SOLVENT.	Moisten the cloth with solvent.
MOMENTARILY (adv)	For a short time	STOP THE PUMP MOMENTARILY. THEN START IT AGAIN.	
momentary (adj)	MOMENTARILY (adv)	MAKE SURE THAT THE SPEED INCREASES MOMENTARILY.	Check for momentary increase in speed.
MONITOR (v), MONITORS, MONITORED, MONITORED	To do a check at something for a period of time to find if there is a change	MONITOR THE INDICATORS ON THE OVERHEAD PANEL.	
		HEAT SENSORS MONITOR THE AMBIENT TEMPERATURE.	
MOOR (v), MOORS, MOORED, MOORED	To hold something in position with ropes and cables	IN BAD WEATHER CONDITIONS, MOOR THE BOAT.	
MORE (adj)	Refer to MANY (adj) and MUCH (adj)		

Word (part of speech)	Approved meaning/ ALTERNATIVES	APPROVED EXAMPLE	Not approved example
MORE (adv)	By a larger dimension, value, quantity, number, or degree	IF IT IS NECESSARY TO DECREASE THE CABLE TENSION MORE, USE THE TURNBUCKLE.	
MOST (adj)	Refer to MANY (adj) and MUCH (adj)		
motion (n)	MOVEMENT (n)	MAKE SURE THAT THE STABILIZER MOVEMENT CONTINUES.	Check that the stabilizer motion continues.
motor (v)	DRY-MOTOR (TV)	DRY-MOTOR THE ENGINE.	Motor the engine without fuel.
	WET-MOTOR (TV)	WET-MOTOR THE ENGINE.	Motor the engine without ignition.
	MOTORING (TN)	DO A MOTORING PROCEDURE.	Motor the engine.
mount (v)	INSTALL (v)	INSTALL THE MANIFOLD.	Mount the manifold.
	ATTACH (v)	ATTACH THE SEATS TO THE SEAT TRACKS.	Mount the seats on the seat tracks.
	PUT (v)	PUT THE WHEEL ON THE AXLE.	Mount the wheel on the axle.
MOVABLE (adj)	That can move if necessary	TURN THE MOVABLE HALF-COLLAR OF THE SUPPORT.	
MOVE (v), MOVES, MOVED, MOVED	To change or cause to change position or location	MAKE SURE THAT THE POINTER MOVES FREELY.	
		MOVE THE VALVE UNTIL IT ALIGNS WITH THE ACTUATOR SHAFT.	
MOVEMENT (n)	A change of position or location	SUDDEN OPERATION OF THE GYRO CAN CAUSE UNWANTED MOVEMENT OF THE HORIZONTAL STABILIZER.	

Word (part of speech)	Approved meaning/ ALTERNATIVES	APPROVED EXAMPLE	Not approved example
MUCH (adj) (MORE, MOST)	Of relatively large dimension, value, quantity, or volume	DO NOT APPLY TOO MUCH PRESSURE.	
MUCH (adv)	To a great degree or extent	BE CAREFUL THAT YOU DO NOT TURN THE INNER RACES TOO MUCH.	
MULTIPLY (v), MULTIPLIES, MULTIPLIED, MULTIPLIED	To use multiplication to get a result For other meanings, use:	MULTIPLY THE INDICATOR VALUE BY THE SCALE VALUE.	
	INCREASE (v)	THE NUMBER OF CRACKS CAN INCREASE.	Cracks can multiply.
MUST (v)	Helping verb that shows obligation	THE BAT 1 SWITCH MUST BE IN THE ON POSITION.	

Word (part of speech)	Approved meaning/ ALTERNATIVES	APPROVED EXAMPLE	Not approved example
NAME (n)	The identification you give to a person or an object	PUSH THE LINE KEY ADJACENT TO THE NAME OF THE SYSTEM.	
		ENTER YOUR NAME AND YOUR PASSWORD.	
NEAR (adj) (NEARER, NEAREST)	At or to a relatively short distance	PUT THE TOOL ON THE NEAR FACE OF THE COMPONENT.	
NEAR (prep)	Approaching in space or condition	BALANCE THE ELEVATOR NEAR ITS MAXIMUM LIMIT.	
		THE FLIGHT DATA RECORDER IS NEAR THE DOOR.	
nearly (adv)	ALMOST (adv)	MOVE THE POINTER UNTIL IT IS ALMOST IN THE GREEN BAND.	Move the pointer until nearly in the green band.
NECESSARY (adj)	That must be	TWO HOURS ARE NECESSARY TO COMPLETE THE PROCEDURE.	
need (v)	NECESSARY (adj)	IT IS NOT NECESSARY TO REPLACE THE BACKING RINGS.	The backing rings do not need to be replaced.
neither (adj)	TWO (TN)...NOT	THE TWO UNITS MUST NOT OPERATE.	Neither unit must operate.
neither (pron)	TWO (TN)...NOT	THE TWO UNITS MUST NOT OPERATE.	Neither of the units must operate.
NEUTRAL (adj)	1. That has no effect	MAKE SURE THAT THE CONTROLS ARE IN THE NEUTRAL POSITION.	
	2. That is not acid or alkaline	CLEAN THE SCREEN WITH A NEUTRAL SOLUTION.	

Word (part of speech)	Approved meaning/ ALTERNATIVES	APPROVED EXAMPLE	Not approved example
neutralize (v)	NEUTRAL (adj)	MAKE THE ELECTROLYTE NEUTRAL WITH VINEGAR OR BORIC ACID.	Neutralize the electrolyte with vinegar or boric acid.
never (adv)	DO NOT	DO NOT LET THE LEVEL IN THE TANK DECREASE TO LESS THAN HALF.	Never allow the level in the tank to drop below half.
NEW (adj) (NEWER, NEWEST)	Not used before	INSTALL A NEW GASKET ON THE FLANGE.	
NEXT (adj)	The first occurrence that follows immediately after in time or sequence.	REMOVE THE UNIT BEFORE THE NEXT FLIGHT.	
next (adv)	THEN (adv)	HOLD THE FLANGE IN ITS POSITION. THEN, INSTALL THE BOLT.	Hold the flange in its position. Next, install the bolt.
next to (prep)	ADJACENT TO (prep)	THE PUMP IS ADJACENT TO THE MIDDLE RIB.	The pump is next to the middle rib.
NO (adj)	Not any	MAKE SURE THAT THERE IS NO FUEL LEAKAGE.	
NO (adv)	Function word that shows the negative answer to a question	DOES THE LIGHT COME ON? YES OR NO?	
NOISE (n)	An unwanted sound	WHILE THE ENGINE CONTINUES TO TURN, LISTEN FOR UNUSUAL NOISES.	
noisy (adj)	NOISE (n)	IF THERE IS NOISE WHEN THE PUMP OPERATES, EXAMINE THE FILTER FOR METAL PARTICLES.	If pump operation is noisy, check the pump filter for metal particles.
NONE (pron)	Not one, not anyone	IF NONE OF THE BRACKETS ARE DAMAGED, CONTINUE THE INSPECTION.	

Word (part of speech)	Approved meaning/ ALTERNATIVES	APPROVED EXAMPLE	Not approved example
nor (conj)	NOT (adv)	IN THESE CONDITIONS, THE MECHANICAL AND ELECTRICAL PITCH TRIM SYSTEMS WILL NOT OPERATE.	In these conditions, neither the mechanical nor the electrical pitch trim systems will operate.
normal (adj)	USUAL (adj)	THE ARMED POSITION IS THE USUAL POSITION OF THE SWITCH.	The ARMED position is the normal position of the switch.
	CORRECT (adj)	THE CORRECT PRESSURE FOR OPERATION IS 3000 PSI.	Normal operating pressure is 3000 psi.
normally (adv)	USUALLY (adv)	THE SWITCH IS USUALLY IN THE ARMED POSITION.	The switch is normally in the ARMED position.
	CORRECTLY (adv)	IF THE SYSTEM DOES NOT OPERATE CORRECTLY, REFER TO THE FAULT ISOLATION PROCEDURE.	If the system does not operate normally, refer to the fault isolation procedure.
NOSE (n)	The front end or part, a part that protrudes	PULL THE TRANSPARENT PLASTIC COLLAR AWAY FROM THE NOSE OF THE ELECTRICAL LATCH.	
NOT (adv)	Adverb of negation	DO NOT LET THE PRESSURE INCREASE TO MORE THAN 3000 PSI.	
NOTCH (n)	An area in the shape of a "V"	MAKE SURE THAT THE BOGIE ROLLERS ARE ALIGNED WITH THE NOTCHES IN THE TRACK.	

Word (part of speech)	Approved meaning/ ALTERNATIVES	APPROVED EXAMPLE	Not approved example
note (v)	RECORD (v)	RECORD DIMENSION "A".	Note dimension A.
notify (v)	TELL (v)	IF YOU FIND CRACKS, TELL THE CHIEF INSPECTOR.	If cracks are found, notify the chief inspector.
	WRITE (v)	IF THE UNIT DOES NOT OPERATE CORRECTLY, WRITE TO THE MANUFACTURER.	If the unit does not operate correctly, notify the manufacturer.
now (adv)	AT THIS TIME	DO NOT TIGHTEN THE NUTS AT THIS TIME.	Do not tighten the nuts now.
NUMBER (n)	1. A symbol or word that identifies an integer and/or decimal	NEGATIVE NUMBERS SHOW THAT THE PRESSURE DECREASES.	
	2. An amount that you can count or calculate	IF THE NUMBER OF FAILURES IS MORE THAN THREE, REPLACE THE UNIT.	

Word (part of speech)	Approved meaning/ ALTERNATIVES	APPROVED EXAMPLE	Not approved example
OBEY (v), OBEYS, OBEYED, OBEYED	To do that which the procedures or instructions tell you	OBEY THE PRECAUTIONS THAT FOLLOW.	
OBJECT (n)	Something that you can see or touch	MAKE SURE THAT THERE ARE NO LOOSE OBJECTS IN THE AIR INTAKE.	
observe (v)	MONITOR (v)	MONITOR THE RADIO ALTIMETER INDICATOR.	Observe the radio altimeter indicator.
	SEE (v)	IF YOU SEE OIL IN THE DRAIN, STOP THE TEST.	If you observe oil in the drain, stop the test.
	OBEY (v)	OBEY ALL SAFETY PRECAUTIONS.	Observe all safety precautions.
obstruct (v)	CATCH (v)	MAKE SURE THAT THE DRAG STRUTS DO NOT CATCH ON THE LANDING GEAR.	Make sure that the drag struts do not obstruct the landing gear.
	PREVENT (v)	ADJUST THE LEVER ARMS IF THEY PREVENT MOVEMENT.	Adjust the lever arms if they obstruct movement.
	BLOCKAGE (n)	REMOVE THE DIRT THAT CAUSES THE BLOCKAGE IN THE TUBE.	Remove the dirt obstructing the tube.
obstructed (adj)	BLOCKED (adj)	IF THE DRAIN HOLE IS BLOCKED, REMOVE THE UNWANTED MATERIAL.	If the drain hole is obstructed, remove the unwanted material.
	CLOGGED (adj)	IF THE DRAIN HOLE IS CLOGGED, CLEAN IT.	If the drain hole is obstructed, clean it.
obstruction (n)	BLOCKAGE (n)	EXAMINE THE DRAIN HOLES FOR BLOCKAGE.	Examine the drain holes for obstruction.
obtain (v)	GET (v)	ADD 0.05 mm TO GET THE CORRECT CLEARANCE FOR A STANDARD SPHERICAL COUPLING.	Add 0.05 mm to obtain the correct clearance for a standard spherical coupling.

Word (part of speech)	Approved meaning/ ALTERNATIVES	APPROVED EXAMPLE	Not approved example
OCCUR (v), OCCURS OCCURRED	To be found or to take place, to come to be	A SUDDEN MOVEMENT OF THE CONTROL SURFACES CAN OCCUR.	
		IF THE SYMPTOMS OCCUR AGAIN, DO A SYSTEM TEST.	
OF (prep)	Function word that shows: from, belonging to, having, or containing	REMOVE THE SPLIT PINS FROM THE BOLTS IN THE BOTTOM ENDS OF THE CONTROL RODS.	
OFF (adj)	Not in operation	MAKE SURE THAT THE BATTERIES ARE OFF.	
OFF (adv)	1. Not in operation	MAKE SURE THAT THE WARNING LIGHT GOES OFF.	
	2. Not in contact	THE HANDLE MUST COME OFF EASILY.	
OFF (prep)	Not in contact with	MAKE SURE THAT THE TIRE IS FULLY OFF THE GROUND.	
often (adv)	FREQUENTLY (adv)	IF A FAILURE OCCURS FREQUENTLY, DO THE SYSTEM TEST.	If a failure occurs often, do the system test.
oil (v)	LUBRICATE (v)	LUBRICATE THE LINKAGE WITH OIL.	Oil the linkage.
	OIL (TN)	PUT OIL ON THE LINKAGE.	Oil the linkage.
oil-free (adj)	OIL (TN)	USE A SOLUTION THAT CONTAINS NO OIL.	Use an oil-free solution.
old (adj)	REMAINING (adj)	REMOVE THE REMAINING SEALANT.	Remove the old sealant.
	USED (adj)	DRAIN THE USED OIL INTO A CONTAINER.	Drain the old oil into a container.
	EXPIRED (adj)	MAKE SURE THAT THE SHELF LIFE OF THE MATERIAL IS NOT EXPIRED.	Make sure that the material is not too old.

Word (part of speech)	Approved meaning/ ALTERNATIVES	APPROVED EXAMPLE	Not approved example
omit (v)	DO NOT DO	DO NOT DO STEPS 3 THRU 5.	Omit steps 3 to 5.
	IGNORE (v)	IGNORE STEPS 3 THRU 5.	Omit steps 3 to 5.
ON (adj)	In operation	MAKE SURE THAT THE BATTERIES ARE ON.	
ON (adv)	1. In operation	MAKE SURE THAT THE WARNING LIGHT COMES ON.	
	2. In contact	MAKE SURE THAT THE LID GOES ON EASILY.	
ON (prep)	Function word that shows contact, support, direction	PUT THE ASSEMBLED PARTS ON THE SUPPORT FITTING.	
once (adv)	ONE (TN) TIME	PUSH THE SWITCH ONE TIME.	Press the switch once.
once (conj)	WHEN (conj)	WHEN THE UNIT IS ON ITS MOUNTING, LOCK IT.	Once the unit is on its mounting, lock it.
ONE (pron)	That refers to a single person or object You can also use this word as a technical name.	IF THE LABEL IS NOT THERE, ATTACH A NEW ONE.	
ONLY (adj)	Nothing more or different	THIS IS THE ONLY APPROVED CONFIGURATION.	
ONLY (adv)	Exclusively	TO TIGHTEN THE NUTS, USE ONLY A TORQUE WRENCH.	
ONTO (prep)	To a position on or upon	WHEN YOU LOWER THE PUMP, MOVE IT ONTO THE MOUNTING BRACKETS.	
opaque (adj)	NOT TRANSPARENT	CLEAN THE PLASTIC SURFACES THAT ARE NOT TRANSPARENT.	Clean the opaque plastic surfaces.
OPEN (adj)	That lets something go in or out	PUT A PLUG ON THE OPEN END OF EACH PIPE.	

Word (part of speech)	Approved meaning/ ALTERNATIVES	APPROVED EXAMPLE	Not approved example
OPEN (v), OPENS, OPENED, OPENED	1. To move or cause to move from the closed position	OPEN THE PRESSURE VALVE.	
		WHEN THE RELIEF VALVE OPENS, THE PRESSURE DECREASES.	
	2. To operate a circuit breaker to interrupt an electrical circuit	OPEN THE CIRCUIT BREAKER.	
OPENING (n)	An aperture or hole in a surface	MAKE SURE THAT THE OPENING IS CLEAR.	
operable (adj)	OPERATE (v)	MAKE SURE THAT THE VALVE CAN OPERATE.	Check that the valve is operable.
		MAKE SURE THAT YOU CAN OPERATE THE VALVE.	Check that the valve is operable.
OPERATE (v), OPERATES, OPERATED, OPERATED	To put, keep, or be in action	OPERATE THE PLUNGER TO DRAIN THE SYSTEM.	
		DO NOT GO NEAR THE ENGINE WHEN IT OPERATES.	
OPERATION (n)	That which occurs when something operates or is operated	THIS PROCEDURE IS IMPORTANT FOR THE CORRECT OPERATION OF THE VALVE.	
operational (adj)	SERVICEABLE (adj)	MAKE SURE THAT THE OTHER MICROSWITCH IS SERVICEABLE.	Check that the other microswitch is operational.
	OPERATE (v)	MAKE SURE THAT THE GROUND-PROXIMITY WARNING SYSTEM CAN OPERATE.	Check that the ground-proximity warning system is operational.
opportunity (n)	Be specific if possible.	REPLACE THE UNIT AT THE NEXT CHECK.	Replace the unit at the first opportunity.

Word (part of speech)	Approved meaning/ ALTERNATIVES	APPROVED EXAMPLE	Not approved example
OPPOSITE (adj)	1. In the other direction	THE RESET DEVICE OPERATES IN THE OPPOSITE DIRECTION.	
	2. Across from	THE BOLTS ARE ON THE OPPOSITE SIDE OF THE FLANGE.	
OPPOSITE (prep)	Across from	THE UNIT IS IN THE WHEEL BAY OPPOSITE THE LANDING GEAR.	
option (n)	POSSIBLE (adj)	IF POSSIBLE, DO THE INSPECTION IN DAYLIGHT.	If you have the option, do the inspection in daylight.
OPTIONAL (adj)	Not mandatory	THIS TOOL IS OPTIONAL.	
OR (conj)	Function word that shows a choice	DO NOT TOUCH THE HIGH TENSION LEADS OR THE IGNITER PLUGS.	
	Do not use this word with the meaning of otherwise. Use a different construction.	MAKE SURE THAT THE L-SHAPED SEAL STAYS BONDED. IF THE SEAL DOES NOT STAY BONDED, A LEAK CAN OCCUR.	Make sure that the L-shaped seal stays bonded or a leak can occur.
order (n)	SEQUENCE (n)	DO THE WEIGHING PROCEDURE IN THIS SEQUENCE:	Do the weighing procedure in the following order:
order (v)	TELL (v)	TELL ALL PERSONNEL TO GO OUT OF THE MAINTENANCE AREA.	Order all personnel out of the maintenance area.
	ORDER (TN)	SEND AN ORDER FOR THE SPECIFIED PAINT DIRECTLY TO THE MANUFACTURER.	Order the required paint directly from the manufacturer.
orient (v)	POINT (v)	MAKE SURE THAT THE BOLT HEADS POINT TO THE REAR.	The bolt heads must be oriented toward the rear.
origin (n)	SOURCE (n)	THE SOURCE OF THE PROBLEM IS IN THE FUEL SYSTEM.	The origin of the problem is in the fuel system.

Word (part of speech)	Approved meaning/ ALTERNATIVES	APPROVED EXAMPLE	Not approved example
original (adj)	INITIAL (adj)	THE SLIDE GOES BACK TO ITS INITIAL POSITION.	The slide returns to the original position.
originate (v)	SOURCE (n)	IF THE VALVE IS THE SOURCE OF THE FUEL LEAK, REPLACE THE VALVE SEALS.	If the fuel leak originates at the valve, replace the valve seals.
OTHER (adj)	Not the same as that given before	CONNECT ONE OF THE STUDS TO THE GROUND AND THE OTHER STUDS TO THE PINS.	
OTHER (pron)	That which is not the same as that given before	CONNECT ONE PLUG, THEN CONNECT THE OTHER.	
otherwise (adv)	IF ...NOT	MAKE SURE THAT THE RODS ARE ADJUSTED CORRECTLY. IF THE ADJUSTMENT IS NOT CORRECT, ADJUST THEM AS FOLLOWS:	Check that the rods are adjusted correctly, otherwise adjust them as follows:
	DIFFERENTLY (adv)	UNLESS THE INSTRUCTIONS TELL YOU DIFFERENTLY, DO REPAIR SCHEME No. 4.	Unless the instructions tell you otherwise, do repair scheme No. 4.
OUT (adj)	Away from, or not in, a location or position	IF THE INDICATOR BUTTON IS OUT, REPLACE THE FILTER.	
OUT (adv)	To a position away from the inside	WHEN THE INDICATOR BUTTON COMES OUT, REPLACE THE FILTER.	
OUT OF (prep)	1. Through or from the inside to the outside	TILT THE FITTING TO MAKE THE BARREL NUT FALL OUT OF IT.	
	2. Not in	KEEP THE COMPOUND OUT OF THE SUNLIGHT.	

Word (part of speech)	Approved meaning/ ALTERNATIVES	APPROVED EXAMPLE	Not approved example
OUTBOARD (adj)	Farther from the longitudinal axis of something	MEASURE THE MOVEMENT OF THE INBOARD AND OUTBOARD FLAPS.	
OUTBOARD (adv)	In, or to, a position farther from the longitudinal axis of something	MOVE THE JACK OUTBOARD TO DISENGAGE IT FROM THE JACKING POINT.	
OUTBOARD OF (prep)	At a position farther from the longitudinal axis	PUT THE ACCESS PLATFORM IN A POSITION THAT IS OUTBOARD OF THE FLAP TRACK.	
OUTDOORS (adv)	Not in a building	DO THIS WORK OUTDOORS.	
OUTER (adj)	Located on the outside or farther from the center of an object	APPLY SEALANT ON THE OUTER SURFACE.	
outline (v)	GIVE (v)	THIS SECTION GIVES YOU THE PROCEDURES THAT ARE NECESSARY TO FIND THE FAULTS.	This section outlines the procedures needed to find the faults.
OUTPUT (n)	The data, power, energy that comes out of equipment or a system	THE IMPEDANCE MATCHING TRANSFORMERS SUPPLY 60 W OUTPUT TO THE LOUDSPEAKERS.	
outside (adj)	EXTERNAL (adj)	THE EXTERNAL TEMPERATURE IS SHOWN ON THE CENTRAL DISPLAY UNIT.	The outside temperature is indicated on the central display unit.
	OUTER (adj)	CLEAN THE OUTER SURFACE WITH WARM WATER.	Clean the outside surface with warm water.

Word (part of speech)	Approved meaning/ ALTERNATIVES	APPROVED EXAMPLE	Not approved example
outside (adv)	OUTDOORS (adv)	DO THIS PROCEDURE OUTDOORS.	Do this procedure outside.
outside (n)	EXTERNAL (adj)	FOLD THE EDGING AGAINST THE EXTERNAL SIDE OF THE INSULATION BLANKET.	Fold the edging against the outside of the insulation blankets.
	OUTER (adj)	FOLD THE EDGING AGAINST THE OUTER SIDE OF THE INSULATION BLANKET.	Fold the edging against the outside of the insulation blankets.
outside (prep)	NEAR (prep)	WHEN YOU DO WORK IN A FUEL TANK, A SECOND PERSON MUST BE NEAR THE TANK AT ALL TIMES.	When you do work in a fuel tank, a second person must be outside the tank at all times.
outwards (adv)	OUT (adv)	MOVE THE DOOR OUT AND FORWARD.	Move the door outwards and forwards.
OVER (prep)	To indicate a position higher than or above land or water Use this word only for movement above land or water. Not permitted for other meanings. Refer to the word "over (prep)" below.	BEFORE FLIGHT OPERATIONS OVER WATER AT LOW ALTITUDES, APPLY CORROSION-PREVENTIVE FLUID TO THE DOOR HINGES OF THE LANDING GEAR.	
over (prep)	ABOVE (prep)	MAKE SURE THAT THE HYDRAULIC FLUID LEVEL IS NOT ABOVE THE "FULL" MARK.	Make sure the hydraulic fluid level is not over the FULL mark.
	ALONG (prep)	APPLY THE WEIGHT EQUALLY ALONG THE STABILIZER SPAN.	The weight must be evenly spread over the stabilizer span.
	ON (prep)	INSTALL THE STOP SLEEVE ON THE SLIDING MEMBER.	Install the stop sleeve over the sliding member.
	MORE THAN	DO NOT INSTALL SHIMS WITH A THICKNESS MORE THAN 0.1 mm.	Do not install shims over 0.1 mm thick.

Word (part of speech)	Approved meaning/ ALTERNATIVES	APPROVED EXAMPLE	Not approved example
OVERBOARD (adv)	Out of a vessel	RELEASE THE AIR OVERBOARD.	
overfill (v)	TOO MUCH	DO NOT PUT TOO MUCH FLUID IN THE LUBRICATION HOLE.	Do not overfill the lubrication hole.
overfull (adj)	TOO FULL	IF THE TANK IS TOO FULL, YOU CANNOT PRESSURIZE IT CORRECTLY.	If the tank is overfull, you cannot pressurize it correctly.
overhaul (v)	OVERHAUL (TN)	DO AN OVERHAUL OF THE No. 2 ENGINE.	Overhaul the No. 2 engine.
overheat (v)	TOO HOT	IF YOU PRESSURIZE THE OXYGEN BOTTLE QUICKLY, IT WILL BECOME TOO HOT.	Rapid charging overheats the oxygen bottle.
OVERLAP (n)	The area in which a part of one surface is on a part of a second surface	REPLACE THE PART IF THE OVERLAP IS MORE THAN 0.01 mm.	
overlap (v)	OVERLAP (n)	MAKE AN OVERLAP OF 10 mm.	Overlap the surfaces by 10 mm.
OVERRIDE (v), OVERRIDES, OVERRODE, OVERRIDDEN	To prevent the automatic operation of a part or system	MANUALLY OVERRIDE THE START SEQUENCE.	
overtighten (v)	TIGHTEN TOO MUCH	DO NOT TIGHTEN THE FITTINGS TOO MUCH.	Do not overtighten the fittings.
	TOO TIGHT	DO NOT MAKE THE FITTINGS TOO TIGHT.	Do not overtighten the fittings.
	TOO TIGHTLY	DO NOT INSTALL THE FITTINGS TOO TIGHTLY.	Do not overtighten the fittings.

Blank Page

Word (part of speech)	Approved meaning/ ALTERNATIVES	APPROVED EXAMPLE	Not approved example
pack (v)	PUT (v)	PUT THE ASSEMBLY INTO THE BOX.	Pack the assembly into the box.
	FILL (v)	FILL THE GROOVE WITH GREASE.	Pack grease into the groove.
PAINT (v), PAINTS, PAINTED, PAINTED	To apply paint to something	PAINT ALL THE SURFACES.	
PAIR (n)	Two objects that are the same or almost the same, and/or that you use together	MEASURE THE DISTANCE BETWEEN EACH PAIR OF AXLES.	
PARALLEL (adj)	Along lines that stay a constant distance apart at all points	MAKE SURE THAT THE TURNBUCKLE IS PARALLEL TO THE AXIS OF THE AIR OUTLET.	
PARK (v), PARKS, PARKED, PARKED	To stop a vehicle and to let it stay in one position on the ground	PARK THE CAR.	
PART (n)	1. A constituent of a machine or other equipment	REPLACE THE DAMAGED PARTS.	
	2. A piece or section of a whole	REFER TO PART 2 FOR THE APPLICABLE PROCEDURE.	
part (v)	DISCONNECT (v)	DISCONNECT THE DUCTING.	Part the ducting.
partial (adj)	NOT FULLY	IF THE FLAPS DO NOT FULLY EXTEND, DO THE TEST AGAIN.	If only partial extension of the flap occurs, do the test again.
partially (adv)	NOT FULLY	IF THE FLAPS DO NOT FULLY EXTEND, DO THE TEST AGAIN.	If the flaps only partially extend, do the test again.
PARTICLE (n)	A very small piece of material	IF YOU FIND METAL PARTICLES IN THE DISASSEMBLED PUMP, FIND THEIR SOURCE AND REPAIR THE DEFECTIVE PART.	

Word (part of speech)	Approved meaning/ ALTERNATIVES	APPROVED EXAMPLE	Not approved example
particular (adj)	ONLY APPLICABLE	THIS PROCEDURE IS ONLY APPLICABLE TO TYPE A PARTS.	This procedure is particular to type A parts.
	VERY (adv)	THIS STEP IS VERY IMPORTANT.	This step is of particular importance.
particularly (adv)	VERY (adv)	THIS STEP IS VERY IMPORTANT.	This step is particularly important.
partly (adv)	NOT FULLY	IF THE VALVE DOES NOT FULLY OPEN, EXAMINE THE POSITIONS OF THE END STOPS ON THE ACTUATOR.	If the valve opens only partly, examine the positions of the end stops on the actuator.
pass (v)	GO (v)	MAKE SURE THAT THE CABLE GOES ONTO THE PULLEY.	Make sure that the cable passes over the pulley.
	THROUGH (prep)	PUT THE ANTENNA CABLE THROUGH THE TOP SLOTS.	Pass the antenna cable through the top slots.
	GIVE (v)	GIVE ONE END OF THE CABLE TO A SECOND PERSON.	Pass one end of the cable to another person.
passage (n)	FLOW (n)	THE FLOW OF FLUID THROUGH THE VALVE MUST BE CONTINUOUS.	The passage of fluid through the valve must be continuous.
	HOLE (n)	CAREFULLY PUT THE TUBE THROUGH THE HOLE.	Carefully put the tube through the passage.
	OPENING (n)	CAREFULLY PUT THE TUBE THROUGH THE OPENING.	Carefully put the tube through the passage.
past (prep)	AROUND (prep)	IF YOU INSTALL THE SEAL INCORRECTLY, LEAKS CAN OCCUR AROUND THE VALVE.	If you install the seal incorrectly, leaks can occur past the valve.
	THROUGH (prep)	IF YOU INSTALL THE SEAL INCORRECTLY, THE FLUID WILL GO THROUGH IT.	An incorrect installation of the seal will let fluid go past it.

Word (part of speech)	Approved meaning/ ALTERNATIVES	APPROVED EXAMPLE	Not approved example
PASTE (n)	A wet, flexible mixture or compound	MIX THE MATERIALS UNTIL THEY MAKE A PASTE.	
PATCH (n)	A piece of material that you use to repair a surface or hole	BOND THE PATCH TO THE TUBE WITH ADHESIVE.	
penetrate (v)	GO THROUGH	IF THE SCRATCH DOES NOT GO THROUGH THE PROTECTIVE LAYER, THE UNIT IS SERVICEABLE.	If the scratch does not penetrate the protective layer, the unit is serviceable.
	GO INTO	MAKE SURE THAT THE HYDRAULIC FLUID DOES NOT GO INTO THE RECEPTACLES.	Make sure that the hydraulic fluid does not penetrate the receptacles.
people (n)	PERSONS (n)	KEEP ALL PERSONS OUT OF THE WORK AREA.	Keep all people out of the work area.
	PERSONNEL (n)	TELL YOUR PERSONNEL ABOUT THESE NEW REGULATIONS.	Tell your people about these new regulations.
per (prep)	FOR EACH	DO NOT USE MORE THAN TWO WASHERS FOR EACH BOLT.	Do not use more than two washers per bolt.
	REFER (v)	DO THE LEAK TEST: (REFER TO PARAGRAPH 7.0).	Do the leak test as per paragraph 7.0.
perforated (adj)	HOLE (n)	MAKE SURE THAT THERE ARE NO HOLES IN THE MEMBRANE.	Make sure that the membrane is not perforated.
perform (v)	DO (v)	DO THE LEAK TEST.	Perform the leak test.
	You can use a more specific command verb.	MEASURE THE VOLTAGE BETWEEN PINS A AND B.	Perform the voltage measurement between pins A and B.

Word (part of speech)	Approved meaning/ ALTERNATIVES	APPROVED EXAMPLE	Not approved example
PERFORMANCE (n)	The ability of a part, a mechanism, or system to do its necessary function when you compare it with a standard or specification For other meanings, use:	SOME REPAIRS CAN DECREASE THE PERFORMANCE OF THE ENGINE.	
	DO (v)	DO THE TEST. THEN GO TO PARAGRAPH C.	After performance of the test, go to paragraph C.
PERIOD (n)	A quantity of time	THE EMERGENCY LOCATOR TRANSMITTER HAS A STANDBY PERIOD OF 30 SECONDS.	
periodically (adv)	INTERVAL (n) Give the value.	EXAMINE THE SERVO CONTROLS AT INTERVALS OF 400 FLIGHT HOURS.	Examine the servo controls periodically.
PERMANENT (adj)	Without a limit in time	MAKE A PERMANENT REPAIR.	
PERMANENTLY (adv)	In a permanent manner	THE "NO SMOKING" SIGNS WILL STAY ON PERMANENTLY DURING THE FLIGHT.	
permissible (adj)	PERMITTED (adj)	EQUIVALENT MATERIALS ARE ALSO PERMITTED.	Equivalent materials are also permissible.
permit (v)	LET (v)	DO NOT LET THE FLUID TOUCH YOU.	Do not permit the fluid to touch you.
PERMITTED (adj)	Allowed	A MAXIMUM OF TWO WASHERS IS PERMITTED FOR EACH BOLT.	
PERPENDICULAR (adj)	At a 90 degree angle to a given line or plane	MAKE SURE THAT THE TOOL AND FASTENER ARE PERPENDICULAR TO THE WORK SURFACE.	

Word (part of speech)	Approved meaning/ ALTERNATIVES	APPROVED EXAMPLE	Not approved example
persist (v)	CONTINUE (v)	IF THE FAULT MESSAGE CONTINUES, PUSH THE AUDIO CANCEL SWITCH ON THE AVIONICS TEST PANEL.	If the fault message persists, push the AUDIO CANCEL switch on the avionics test panel.
persistent (adj)	CONTINUOUS (adj)	IF YOU HEAR CONTINUOUS BACKGROUND NOISE, FIND THE CAUSE.	If you hear persistent background noise, find the cause.
PERSON (n)	A human being	MAKE SURE THAT THERE IS A PERSON NEAR THE TANK.	
PERSONNEL (n)	Persons employed in a group or organization	INJURY TO PERSONNEL AND/OR DAMAGE TO EQUIPMENT CAN OCCUR.	
picket (v)	MOOR (v)	IF STRONG WINDS OCCUR, MOOR THE AIRCRAFT.	Picket the aircraft in strong winds.
PIECE (n)	A quantity or segment of a whole that you can use on its own, or that can operate as an individual item	REMOVE ALL PIECES OF THE BROKEN SEAL.	
pierced (adj)	HOLE (n)	MAKE SURE THAT THERE ARE NO HOLES IN THE MEMBRANE.	Make sure that the membrane is not pierced.
placard (v)	PUT (v) (A PLACARD [TN])	PUT A PLACARD ON THE CONTROLS BEFORE YOU START THE TEST.	Placard the controls before you start the test.
place (n)	POSITION (n)	MAKE SURE THAT THE COVERS ARE IN POSITION.	Ensure that covers are in place.
	AREA (n)	REPAIR THE DAMAGED AREAS.	Repair the damaged places.
place (v)	PUT (v)	PUT THE MANIFOLD IN POSITION.	Place the manifold in position.

Word (part of speech)	Approved meaning/ ALTERNATIVES	APPROVED EXAMPLE	Not approved example
PLAY (n)	Free movement, the length or dimension of this movement	THE PLAY BETWEEN THE TWO GEARS MUST BE LESS THAN 0.05 mm.	
plot (v)	MAKE (v) (A PLOT [TN])	MAKE A PLOT OF THESE POINTS.	Plot these points.
plug (v)	CONNECT (v)	CONNECT THE MICROPHONE TO THE JACK.	Plug the microphone into the jack.
	SEAL (v) (WITH A PLUG [TN] OR PLUGS [TN])	SEAL THE TUBE WITH A PLUG.	Plug the tube.
PLUS (prep)	With the addition of	INCREASE THE PRESSURE IN THE SHOCK ABSORBER TO 85.5 BAR, PLUS OR MINUS 2.7 BAR.	
PNEUMATIC (adj)	Related to, or operated by gas pressure	INSTALL THE PNEUMATIC CONTROLLER.	
PNEUMATICALLY (adv)	With pneumatic power	SOME VALVES OPERATE PNEUMATICALLY.	
POINT (n)	1. A sharp or tapered end	PUT THE POINT OF THE TOOL IN THE SLOT.	
	2. An accurate location	ON THE FLANGE, MEASURE THE DISTANCE BETWEEN POINTS "A" AND "B".	
POINT (v), POINTS, POINTED, POINTED	1. To show the position or direction of	MAKE SURE THAT THE ARROW ON THE CHECK VALVE POINTS AWAY FROM THE GROUND SUPPLY DUCT.	
	2. To turn something in a specified direction	DO NOT POINT THE COMPRESSED AIR OUTLET AT YOUR SKIN.	
POISONOUS (adj)	That contains poison	DO NOT SWALLOW HYDRAULIC FLUID. IT IS POISONOUS.	

Word (part of speech)	Approved meaning/ ALTERNATIVES	APPROVED EXAMPLE	Not approved example
POLISH (v), POLISHES, POLISHED, POLISHED	To make smooth or shiny	POLISH THE WINDOW PANE WITH A SOFT CLOTH.	
poor (adj)	UNSATISFACTORY (adj)	IF SYSTEM PERFORMANCE IS UNSATISFACTORY, DO A TEST OF CIRCUIT A.	If system performance is poor, test circuit A.
	DEFECTIVE (adj)	DEFECTIVE CONTACTS CAN CAUSE A FAILURE OF THE UNIT.	Poor contacts can cause a failure of the unit.
	UNSERVICEABLE (adj)	IF THE PART IS UNSERVICEABLE, REPLACE IT.	If the condition of the part is poor, replace it.
pop (v)	OPEN (v)	IF THE CIRCUIT BREAKER OPENS, DE-ENERGIZE THE CIRCUIT.	If the circuit breaker pops, de-energize the circuit.
	OUT (adv)	IF THE INDICATOR BUTTON CAME OUT, REPLACE THE FILTER.	If the indicator button popped out, replace the filter.
port (adj)	LEFT (adj)	REMOVE THE WING TIP BRAKE FROM THE LEFT WING.	Remove the wing tip brake from the port wing.
portion (n)	PIECE (n)	REMOVE ALL THE PIECES OF THE DAMAGED SEAL.	Remove all portions of the damaged seal.
	PART (n)	ISOLATE THE DEFECTIVE PART OF THE CIRCUIT.	Isolate the defective portion of the circuit.
POSITION (n)	The attitude or setting of something that you can adjust, or the place or site where you put something	SET THE SWITCH TO THE CORRECT POSITION.	
		MAKE SURE THAT THE WARNING NOTICES ARE IN POSITION BEFORE YOU START THE TASK.	

Word (part of speech)	Approved meaning/ ALTERNATIVES	APPROVED EXAMPLE	Not approved example
position (v)	PUT (v)	PUT THE BRACKET ON THE FRAME.	Position the bracket on the frame.
	SET (v)	SET THE SWITCH TO "ON".	Position the switch to ON.
POSSIBLE (adj)	That can occur	IF POSSIBLE, PUT THE RIGGING PIN IN THE HOLE.	
POSSIBLY (adv)	That can occur	IF YOU CANNOT ENGAGE THE ROD AND THE PISTON, THE LENGTH OF THE NEW ROD IS POSSIBLY INCORRECT.	
potential (adj)	POSSIBLE (adj)	PUT BARRIERS AROUND ALL THE POSSIBLE DANGER AREAS.	Put barriers around all the potential danger areas.
pour (v)	PUT (v)	PUT THE MIXTURE INTO THE CYLINDER.	Pour the mixture into the cylinder.
power (v)	SUPPLY (v)	THE HYDRAULIC SYSTEM SUPPLIES POWER TO THE ACTUATORS.	The hydraulic system powers the actuators.
practice (n)	PROCEDURE (n)	IT IS STANDARD PROCEDURE TO TIGHTEN THE BOLTS IN THIS SEQUENCE.	It is common practice to tighten the bolts in this sequence.
PRECAUTION (n)	That which you do to prevent injury and/or damage	OBEY THESE PRECAUTIONS.	
precautionary (adj)	PRECAUTION (n)	OBEY THESE PRECAUTIONS.	Take these precautionary measures.
precede (v)	BEFORE (conj)	BEFORE YOU APPLY THE TAPE, CLEAN THE SURFACE	Application of the tape must be preceded by cleaning of the surface.
precise (adj)	ACCURATE (adj)	THE GYRO ADJUSTMENT MUST BE ACCURATE.	The gyro requires precise adjustment.
precisely (adv)	ACCURATELY (adv)	ADJUST THE GYRO ACCURATELY.	The gyro must be adjusted precisely.

Word (part of speech)	Approved meaning/ ALTERNATIVES	APPROVED EXAMPLE	Not approved example
PRECISION (n)	The quality of being accurate	THE TEMPERATURE OF THE OIL HAS AN EFFECT ON THE PRECISION OF THE OIL LEVEL INDICATOR.	
preferable (adj)	RECOMMEND (v)	IT IS POSSIBLE TO MIX APPROVED ENGINE OILS OF THE SAME TYPE. BUT WE RECOMMEND THAT YOU ALWAYS USE THE SAME TYPE OF ENGINE OIL FROM THE SAME MANUFACTURER.	It is possible to mix approved engine oils of the same type, but it is always preferable to use the same type of engine oil from the same manufacturer.
preferably (adv)	RECOMMEND (v)	WE RECOMMEND THAT YOU REFUEL THE VEHICLE OUTDOORS.	Preferably, refueling should be done outdoors.
preheat (v)	HEAT (n)	BEFORE YOU START THIS PROCEDURE, APPLY HEAT TO THE AREA WITH A HEAT LAMP.	Preheat the area with a heat lamp.
preload (v)	LOAD (TN)	BEFORE YOU START THIS PROCEDURE, APPLY A LOAD OF 10 LBS. TO THE SPRING.	Preload the spring to 10 lbs.
premature (adj)	BEFORE (conj)	IF THE VALVE OPENS BEFORE THE SET PRESSURE, STOP THE TEST.	If the valve opening is premature, stop the test.
prematurely (adv)	BEFORE (conj)	IF THE VALVE OPENS BEFORE THE SET PRESSURE, STOP THE TEST.	If the valve opens prematurely, stop the test.
preparation (n)	PREPARE (v)	PREPARE THE MIXTURE 20 MINUTES BEFORE YOU USE IT.	Mixture preparation must be accomplished 20 minutes before use.

Word (part of speech)	Approved meaning/ ALTERNATIVES	APPROVED EXAMPLE	Not approved example
PREPARE (v), PREPARES, PREPARED, PREPARED	To make or become ready	PREPARE THE TUBE FOR THE LEAK TEST.	
		PREPARE FOR ELECTRICAL BONDING.	
prescribed (adj)	SPECIFIED (adj)	FILL THE HYDRAULIC TANK WITH THE SPECIFIED HYDRAULIC FLUID.	Fill the hydraulic tank with the prescribed hydraulic fluid.
presence (n)	BE (v)	THIS INSPECTION IS NECESSARY WHEN THERE ARE DEFECTS.	This inspection is necessary in the presence of defects.
present (adj)	BE (v)	IF THERE IS A VOLTAGE AT PIN C OF CONNECTOR F7, THE UNIT IS UNSERVICEABLE.	If a voltage is present at pin C of connector F7, the unit is faulty.
present (v)	GIVE (v)	THIS SECTION GIVES THE APPLICABLE TORQUE VALUES.	The applicable torque values are presented in this section.
	SHOW (v)	THE TABLE THAT FOLLOWS SHOWS THE TEST DATA.	The test data is presented in the following table.
preserve (v)	PRESERVATION (TN)	THIS PROCEDURE IS FOR THE PRESERVATION OF STEEL CABLES.	This procedure will preserve the steel cables.
preset (v)	SET (v)	SET THE TENSION INDICATOR TO ZERO.	Preset the tension indicator to zero.
press (v)	PUSH (v)	PUSH AND HOLD THE TEST BUTTON.	Press and hold the TEST button.
PRESSURIZATION (n)	The procedure that supplies or increases the pressure	CONTINUE THE PRESSURIZATION FOR 10 MINUTES.	
PRESSURIZE (v), PRESSURIZES, PRESSURIZED, PRESSURIZED	To supply pressure	PRESSURIZE THE HYDRAULIC SYSTEM.	

Word (part of speech)	Approved meaning/ ALTERNATIVES	APPROVED EXAMPLE	Not approved example
PREVENT (v), PREVENTS, PREVENTED, PREVENTED	To make sure that something does not occur	TO PREVENT DAMAGE TO THE ENGINE, IMMEDIATELY INSTALL PROTECTIVE COVERS ON ALL OPENINGS.	
prevent (from) (v)	LET (v)	DO NOT LET THE DOOR OPEN.	Prevent the door from opening.
preventive (adj)	PREVENT (v)	THIS PREVENTS CORROSION.	This is a corrosion preventive measure.
previous (adj)	BEFORE (conj)	REMOVE THE PLATE BEFORE YOU ADJUST THE CABLE.	Adjustment of the cable requires previous removal of the plate.
previously (adv)	BEFORE (conj)	IF THE CONTROL WAS ADJUSTED BEFORE YOU STARTED THIS PROCEDURE, DO NOT DO THE VALVE TEST.	If the control was adjusted previously, omit the valve test.
primarily (adv)	PRIMARY (adj)	THE PRIMARY FUNCTION OF THE HYDRAULIC SYSTEM IS TO SUPPLY THE PRESSURE TO OPERATE THE FLIGHT CONTROLS.	The hydraulic system primarily supplies the pressure to operate the flight controls.
PRIMARY (adj)	First in importance	THE PRIMARY FUNCTION OF THE POWER UNIT IS TO SUPPLY ELECTRICAL POWER.	
prime (v)	FILL (v)	FILL THE FILLER HOSE.	Prime the filler hose.
	PUT (v)	PUT FUEL IN THE FUEL PUMP.	Prime the fuel pump.
principal (adj)	PRIMARY (adj)	THIS IS THE PRIMARY FUNCTION OF THE SYSTEM.	This is the principal function of the system.
prior to (prep)	BEFORE (conj)	DRAIN THE SYSTEM BEFORE YOU DISCONNECT THE COMPONENTS.	Drain the system prior to disconnecting the components.

Word (part of speech)	Approved meaning/ ALTERNATIVES	APPROVED EXAMPLE	Not approved example
priority (n)	IMPORTANT (adj)	THIS PROCEDURE IS VERY IMPORTANT.	This procedure has high priority.
	BEFORE (prep)	FILL THE INNER TANK BEFORE YOU FILL THE OUTER TANK.	The inner tank filling procedure has priority over the outer tank filling procedure.
	FIRST (adv)	DURING THE PROCEDURE THAT FOLLOWS, REPAIR THE UPPER FLANGE FIRST.	During the following repair procedure, the upper flange has priority.
probability (n)	RISK (n)	OBEY THE SAFETY PRECAUTIONS TO DECREASE THE RISK OF FIRE.	Obey the safety precautions to decrease the probability of fire.
probable (adj)	POSSIBLE (adj)	IF YOU DO NOT SAFETY THE DOOR, IT IS POSSIBLE THAT STRONG WINDS WILL CAUSE DAMAGE TO THE STRUCTURE.	If you do not safety the door, damage to the structure due to strong winds is probable.
	RISK (n)	IF YOU DO NOT SAFETY THE DOOR, THERE IS A RISK THAT STRONG WINDS WILL CAUSE DAMAGE TO THE STRUCTURE.	If you do not safety the door, damage to the structure due to strong winds is probable.
probe (v)	FIND (v) (WITH A PROBE [TN])	USE A PROBE TO FIND THE BLOCKAGE IN THE TUBE.	Probe the tube to find the blockage.
PROBLEM (n)	Something that is difficult and for which you must find the correct answer	IF YOU FIND A PROBLEM, REFER TO THE TROUBLESHOOTING MANUAL.	
PROCEDURE (n)	Steps that are in a sequence	THESE MAINTENANCE PROCEDURES ARE APPLICABLE TO THE MAIN LANDING GEAR ONLY.	
proceed (v)	CONTINUE (v)	CONTINUE WITH THE ELECTRICAL TEST.	Proceed with the electrical test.

Word (part of speech)	Approved meaning/ ALTERNATIVES	APPROVED EXAMPLE	Not approved example
process (in the process of) (prep)	DURING (prep)	DURING THIS REPAIR, APPLY SUFFICIENT HEAT TO THE UNIT.	Apply sufficient heat to the unit in the process of doing this repair.
	WHILE (conj)	WHILE YOU DO THIS REPAIR, APPLY SUFFICIENT HEAT TO THE UNIT.	Apply sufficient heat to the unit in the process of doing this repair.
process (n)	PROCEDURE (n)	THE ASSEMBLY PROCEDURES ARE ALMOST THE SAME.	The assembly processes are almost the same.
	This word is a technical name when it means a sequence of changes that occur and cause a result.	CORROSION IS AN ELECTROLYTIC PROCESS.	
process (v)	PUT (v)	PUT THE PARTS IN AN ELECTROLYTIC BATH AND DO THE ELECTROPLATING PROCEDURE.	Process the parts in an electrolytic bath.
produce (v)	CAUSE (v)	ELECTRICAL EQUIPMENT CAN CAUSE INTERFERENCE IN TELEPHONE COMMUNICATION.	Electrical equipment can produce interference in telephone communication.
	GIVE (v)	A USED SOLUTION WILL NOT GIVE YOU THE CORRECT RESULTS.	An old solution does not produce the correct results.
	MAKE (v)	THE LEVER MAKES A CLICK WHEN IT ENGAGES.	The lever produces a click when engaged.
	SUPPLY (v)	THE POWER UNIT SUPPLIES THE SYSTEM WITH A VOLTAGE OF 28 VDC.	The power unit produces a voltage of 28 VDC.
product (n)	MATERIAL (n)	WASH PRIMER IS A DANGEROUS MATERIAL.	Wash primer is a dangerous product.

Word (part of speech)	Approved meaning/ ALTERNATIVES	APPROVED EXAMPLE	Not approved example
profile (n)	CONTOUR (n)	MAKE SURE THAT THE CONTOUR OF THE TEMPLATE IS THE SAME.	Make sure that the profile of the template is the same.
program (n)	SEQUENCE (n)	DISASSEMBLE THE ROTOR IN THIS SEQUENCE:	Disassemble the rotor according to the following program:
progress (n)	CONTINUE (v) Also refer to IN PROGRESS (adv).	MAKE SURE THAT THE TEST CONTINUES SATISFACTORILY.	Monitor the progress of the test.
progress (v)	CONTINUE (v)	YOU CAN CONTINUE THE TEST.	You can progress with the test.
progressive (adj)	GRADUALLY (adv)	GRADUALLY TORQUE THE NUT.	Apply progressive torque to the nut.
progressively (adv)	GRADUALLY (adv)	TORQUE THE BOLTS GRADUALLY.	Torque the bolts progressively.
prohibit (v)	PREVENT (v)	PUT A WARNING NOTICE ON THE PANEL TO PREVENT OPERATION OF THE CONTROLS.	Display a warning notice on the panel prohibiting the operation of the controls.
	TELL (NOT TO) (v)	TELL THE PERSONNEL NOT TO USE ASBESTOS PARTS.	Prohibit the use of asbestos parts.
prolonged (adj)	LONG (adj)	OPERATION OF THE MOTOR FOR LONG PERIODS CAN CAUSE DAMAGE TO IT.	Prolonged motor operation can cause damage.
prompt (adj)	IMMEDIATELY (adv)	IF YOU SWALLOW SOLVENT, GET MEDICAL AID IMMEDIATELY.	Prompt action is required if you swallow solvent.

Word (part of speech)	Approved meaning/ ALTERNATIVES	APPROVED EXAMPLE	Not approved example
propagation (n)	INCREASE (v)	IF THE DIMENSION OF A CRACK INCREASES, THE PART CAN BREAK.	Crack propagation can cause the part to break.
	Give the dimensions or limits.	IF THE DEPTH OF A CRACK IS MORE THAN 0.2 mm, REJECT THE PART.	Crack propagation can cause the rejection of the part.
		IF A CRACK IS LARGER THAN 0.2 mm, REJECT THE PART.	Crack propagation can cause the rejection of the part.
proper (adj)	CORRECT (adj)	DO A CHECK FOR THE CORRECT OPERATION OF THE UNIT.	Check the unit for proper operation.
properly (adv)	CORRECTLY (adv)	CLOSE THE DOORS CORRECTLY.	Close the doors properly.
PROPERTY (n)	A characteristic or attribute of an object	THE MAGNETIC PROPERTIES OF THIS METAL DECREASE AT HIGH TEMPERATURES.	
PROPORTION (n)	The ratio of something to something else	MAKE SURE THAT YOU ALWAYS MIX CLEANING AGENTS WITH COLD WATER, AND IN THE CORRECT PROPORTIONS.	
	For other meanings, use:		
	RELATION (n)	THE TEMPERATURE DECREASES IN RELATION TO THE ALTITUDE.	The temperature decreases in proportion to the altitude.
proportional (adj)	RELATION (n)	THE CENTER OF GRAVITY MOVES IN RELATION TO THE LOADS ON THE WINGS.	Movement of the center of gravity is proportional to the loads on the wings.
	PROPORTION (n)	MIX THE PIGMENT AND SOLVENT IN THE CORRECT PROPORTIONS.	Mix the correct quantity of pigment proportional to the quantity of solvent.

Word (part of speech)	Approved meaning/ ALTERNATIVES	APPROVED EXAMPLE	Not approved example
proportionally (adv)	RELATION (n)	THE CABLE TENSION CHANGES IN RELATION TO THE TEMPERATURE.	The cable tension changes proportionally with temperature.
protect (v)	PREVENT (v)	USE CELLOPHANE TO PREVENT DAMAGE TO THE SURFACE.	Use cellophane to protect the surface.
	PROTECTION (n)	MAKE SURE THAT ALL SHARP EDGES HAVE VINYL TAPE PROTECTION.	Make sure that all sharp edges are protected with vinyl tape.
PROTECTION (n)	Something to prevent injury, damage, or failure	APPLY TWO LAYERS OF SEALING COMPOUND ON THE SURFACES THAT HAVE NO PROTECTION.	
PROTECTIVE (adj)	That gives protection	REMOVE THE PROTECTIVE LAYER.	
protrude (v)	ABOVE (prep)	THE TOPS OF THE BOLT HEADS MUST NOT BE MORE THAN 5 mm ABOVE THE PLATE.	The tops of the bolt heads must not protrude from the plate more than 5 mm.
	OUT (adv)	REPLACE THE FILTER ELEMENT IF THE CLOGGING INDICATOR IS OUT.	Replace the filter element if the clogging indicator protrudes.
provide (v)	GIVE (v)	THIS SECTION GIVES THE PROCEDURES FOR THE TEST.	This section provides the procedures for the test.
	SUPPLY (v)	THE HYDRAULIC PRESSURE SYSTEM SUPPLIES PRESSURE FOR THE OPERATION OF THE FLIGHT CONTROLS.	The hydraulic pressure system provides pressure for operating the flight controls.
provided (that) (conj)	IF (conj)	IF A REPLACEMENT IS AVAILABLE, REPLACE THE DAMAGED FAIRING.	Provided that a replacement is available, replace the damaged fairing.

Word (part of speech)	Approved meaning/ ALTERNATIVES	APPROVED EXAMPLE	Not approved example
providing (that) (conj)	IF (conj)	IF A NEW COMPONENT IS INSTALLED, YOU CAN CONTINUE THE OPERATION OF THE MACHINE.	Providing that a new component is installed, you can continue the operation of the machine.
proximity (n)	NEAR (prep)	DO NOT DO MAINTENANCE PROCEDURES NEAR AN ENGINE THAT IS IN OPERATION.	Do not do maintenance procedures in proximity to a running engine.
PULL (v), PULLS, PULLED, PULLED	To use a force on something to move it toward the source of the force Use this word together with a preposition or an adverb to show direction.	PULL THE CABLE DOWN AND AWAY FROM THE BRACKET.	
pump (v)	PUMP (TN)	WHEN YOU REMOVE OIL FROM THE TANK, OPEN THE VALVE BEFORE YOU START THE PUMP.	Before you pump oil from the tank, open the valve.
puncture (v)	HOLE (n)	MAKE SURE THAT YOU DO NOT MAKE A HOLE IN THE BLADDER.	Take care not to puncture the bladder.
purge (v)	REMOVE (v)	REMOVE UNWANTED MATERIAL FROM THE SYSTEM.	Purge the system.
purify (v)	CLEAN (v)	CLEAN THE WATER.	Purify the water.
purpose (n)	FUNCTION (n)	THE FUNCTION OF THIS SPECIAL TOOL IS TO REMOVE THE BUSHING.	The purpose of this special tool is to extract the bushing.
	DO (v)	BE CAREFUL WHEN YOU OPEN THE COWLS TO DO A MAINTENANCE TASK.	Be careful when you open the cowls for maintenance purposes.

Word (part of speech)	Approved meaning/ ALTERNATIVES	APPROVED EXAMPLE	Not approved example
PUSH (v), PUSHES, PUSHED, PUSHED	1. To apply a force to something to move it away from the source of the force	PUSH THE INSERT DOWN UNTIL IT TOUCHES THE BOTTOM OF THE HOLE.	
	2. To move with a force against something Use this word together with a preposition or an adverb to show direction.	THE SPRING UNIT PUSHES AGAINST THE BRAKE PEDAL.	
PUT (v), PUTS, PUT, PUT	To cause something to move or to be in a specified position or condition	PUT THE ADAPTER IN POSITION AGAINST ITS SUPPORT.	
		PUT THE LEVER BACK TO ITS INITIAL POSITION.	
PUT ON (v), PUTS ON, PUT ON, PUT ON	To cover your skin or face with clothing, or other items that give protection	PUT ON SAFETY GOGGLES AND PROTECTIVE CLOTHING.	

Word (part of speech)	Approved meaning/ ALTERNATIVES	APPROVED EXAMPLE	Not approved example
qualified (adj)	APPROVED (adj)	AN APPROVED PERSON MUST DO THE IMPORTANT CHECKS.	A qualified person must do the important checks.
QUALITY (n)	Condition, property, or type	THE CLEANING SOLVENT MUST BE OF GOOD QUALITY.	
QUANTITY (n)	A specified amount or number	APPLY A SMALL QUANTITY OF OIL ON THE THREADS OF THE BOLTS.	
	For an amount that you can count, use NUMBER (n).	MAKE SURE THAT YOU HAVE THE CORRECT NUMBER OF SHIMS AVAILABLE.	
quarterly (adv)	INTERVAL (n)	SEND REPORTS TO THE AUTHORITIES AT INTERVALS OF THREE MONTHS.	Send reports to the authorities quarterly.
questionable (adj)	NOT SURE	IF YOU ARE NOT SURE OF THEIR CONDITION, REJECT THE PARTS.	Reject the parts that are in a questionable condition.
quick (adj)	QUICKLY (adv)	MAKE SURE THAT THE ROD DOES NOT MOVE QUICKLY.	Make sure that there is no quick movement of the rod.
QUICKLY (adv)	In a fast manner	MAKE SURE THAT THE PRESSURE DECREASES QUICKLY.	
quit (v)	GO (v)	BEFORE YOU START THE ENGINE, MAKE SURE THAT ALL PERSONS GO AWAY FROM THE DANGER AREA.	Before you start the engine, make sure that all persons quit the danger area.
	STOP (v)	IF THE RED LIGHT COMES ON, STOP THE REFUEL PROCEDURE.	If the red light comes on, quit the refuel procedure.
quote (v)	GIVE (v)	THE TABLE GIVES THE MAXIMUM TOLERANCE FOR THE HOLE.	The table quotes the maximum tolerance for the hole.

Blank Page

Word (part of speech)	Approved meaning/ ALTERNATIVES	APPROVED EXAMPLE	Not approved example
RADIAL (adj)	Along a radius	DO A CHECK FOR RADIAL CUTS, CRACKS, OR SPLITS.	
RADIALLY (adv)	Along a radius	THE FAN BLADES ARE INSTALLED RADIALLY.	
RADIOACTIVE (adj)	Related to radioactivity	DO A CHECK FOR RADIOACTIVE CONTAMINATION.	
raise (v)	LIFT (v)	DO NOT LIFT THE COVER MORE THAN NECESSARY.	Do not raise the cover more than necessary.
	INCREASE (v)	INCREASE THE TEMPERATURE BY 10 DEG.	Raise the temperature 10 deg.
	RETRACT (v)	RETRACT THE LANDING GEAR.	Raise the landing gear.
RANDOM (adj)	Not regular	DO RANDOM CHECKS OF THE TIRE PRESSURE DURING THIS PROCEDURE.	
RANDOMLY (adv)	Not regularly	DO THESE CHECKS RANDOMLY.	
RANGE (n)	The limits within which something operates	THE AUDIO SPECTRUM IS THE FREQUENCY RANGE AT WHICH HUMANS CAN HEAR.	
rapid (adj)	FAST (adj)	HYDRAULIC FLUID CAUSES FAST DETERIORATION OF PAINTED SURFACES.	Hydraulic fluid causes rapid deterioration of painted surfaces.
rapidly (adv)	QUICKLY (adv)	QUICKLY TURN THE AILERON CONTROL WHEEL CLOCKWISE.	Rapidly turn the aileron control wheel clockwise.
RATE (n)	A measurement of how frequently or how quickly something occurs	MAKE SURE THAT THE FLOW RATE IS CORRECT.	

Word (part of speech)	Approved meaning/ ALTERNATIVES	APPROVED EXAMPLE	Not approved example
re- (prefix)	In general, use the word AGAIN (adv) or BACK (adv) with the basic word to replace words that have "re-" as a prefix.		
reach (v)	GET (v)	WHEN YOU GET THE CORRECT PRESSURE, CLOSE THE VALVE.	When the correct pressure is reached, close the valve.
	TOUCH (v)	MAKE SURE THAT THE EXTENSION BAR TOUCHES THE TARGET PLATE.	Make sure that the extension bar reaches the target plate.
	BE (v)	WHEN THE FLAPS ARE AT THE 30 DEGREE POSITION, INSTALL THE RIGGING PIN.	When the flaps reach the 30 degree position, install the rigging pin.
reactivate (v)	OPERATE (v)	OPERATE THE UNIT AGAIN.	Reactivate the unit.
	START (v)	START THE SYSTEM AGAIN.	Reactivate the system.
READ (v), READS, READ, READ	To come to know information with the eyes or electronically	READ THE INDICATORS ON THE CONTROL PANEL.	
reading (n)	INDICATION (n)	MAKE SURE THAT THE EXHAUST FROM THE GROUND CART DOES NOT CAUSE AN INCORRECT INDICATION.	Avoid an erroneous reading caused by exhaust from ground cart.
readjust (v)	ADJUST (v)	ADJUST THE LINKAGE AGAIN AFTER THE INSPECTION.	Readjust the linkage after the inspection.
ready (adj)	PREPARE (v)	MAKE SURE THAT THE CONTROL SYSTEM IS PREPARED FOR OPERATION.	Make sure that the control system is in an operational ready state.
ready (v)	PREPARE (v)	PREPARE THE REMOVAL KIT FOR THE REMOVAL OF THE STRUT.	Ready the removal kit for the removal of the strut.

Word (part of speech)	Approved meaning/ ALTERNATIVES	APPROVED EXAMPLE	Not approved example
real (adj)	AGREE (v)	MAKE SURE THAT THE INDICATION ON THE GAUGE AGREES WITH THE QUANTITY THAT IS IN THE TANK.	Make sure that the gauge shows the real quantity that is in the tank.
REAR (adj)	Away from the person who looks or from a reference point	INSTALL THE TWO BOLTS IN THE REAR FITTINGS.	
REAR (n)	The rear part or rear surface	THE CIRCUIT BREAKERS ARE INSTALLED ON THE REAR OF THE UNIT.	
rear of (prep)	AFT OF	THE PUMP IS INSTALLED ON RIB 3, AFT OF THE FRONT SPAR.	The pump is located on rib 3, rear of the front spar.
	BEHIND (prep)	THE PUMP IS INSTALLED ON RIB 3, BEHIND THE FRONT SPAR.	The pump is located on rib 3, rear of the front spar.
rearmost (adj)	REAR (n)	YOU CANNOT REMOVE THE HALF CLAMP NEAREST TO THE REAR.	The rearmost half clamp is not removable.
REARWARD (adv)	In the direction of the rear	PULL THE LEVER REARWARD.	
reason (n)	CAUSE (n)	FIND THE CAUSE OF THE INCORRECT INDICATION.	Find the reason for the wrong indication.
	BECAUSE OF (prep)	BECAUSE OF THE POSSIBLE EFFECTS ON YOUR HEALTH, USE ONLY POTABLE WATER.	For health reasons, only use potable water.
reassemble (v)	ASSEMBLE (v)	ASSEMBLE THE ACTUATOR AGAIN.	Reassemble the actuator.
rebuild (v)	ASSEMBLE (v)	ASSEMBLE THE UNIT AGAIN. USE NEW PARTS IF NECESSARY.	Rebuild the unit if necessary.

Word (part of speech)	Approved meaning/ ALTERNATIVES	APPROVED EXAMPLE	Not approved example
RECEIVE (v), RECEIVES, RECEIVED, RECEIVED	To get energy, material, or a signal from a different source	MAKE SURE THAT THE CAPTAIN'S BOOMSET RECEIVES THE SIGNALS CORRECTLY.	
RECESS (n)	An indentation in a surface	WHEN YOU CLEAN THE SURFACES, REMOVE THE STAINS FROM THE RECESSES.	
recess (v)	RECESS (n)	DO NOT MAKE A RECESS OF MORE THAN 5 mm FOR THE STUDS.	Do not recess studs more than 5 mm.
		THE FLANGE IS IN A RECESS IN THE FUSELAGE.	The flange is recessed into the fuselage.
recharge (v)	CHARGE (v)	CHARGE THE BATTERY AGAIN.	Recharge the battery.
	FILL (v)	FILL THE FLUID RESERVOIR.	Recharge the fluid reservoir.
	PRESSURIZE (v)	PRESSURIZE THE ACCUMULATOR.	Recharge the accumulator.
recheck (v)	MEASURE (v)	MEASURE THE DISTANCE BETWEEN THE FACES AGAIN.	Recheck the distance between the faces.
	EXAMINE (v)	EXAMINE THE CASTING FOR CORROSION AGAIN.	Recheck the casting for corrosion.
	CHECK (n)	DO A LEAKAGE CHECK OF THE FILTER ASSEMBLY AGAIN.	Recheck the filter assembly for leakage.
reclaim (v)	REPAIR (v)	REPAIR THE DAMAGED FABRIC AREA.	Reclaim the damaged fabric area.
recleat (v)	ATTACH (v) (TO A CLEAT [TN])	ATTACH THE TIEDOWN CORD TO A CLEAT.	Recleat the tiedown cord.
reclose (v)	CLOSE (v)	AFTER SERVICING, CLOSE THE COMPARTMENT AGAIN.	Reclose the compartment after servicing.

Word (part of speech)	Approved meaning/ ALTERNATIVES	APPROVED EXAMPLE	Not approved example
recoil (v)	MOVE BACK	BE CAREFUL WHEN THE SLIDE MOVES BACK.	Be careful when the slide recoils.
	WIND (v)	WIND THE PROBE WIRE BACK ON THE SPOOL.	Recoil the probe wire back on the spool.
	RECOIL (TN)	MOVE THE BARREL OF THE GUN TO THE RECOIL POSITION.	Recoil the gun barrel.
RECOMMEND (v), RECOMMENDS, RECOMMENDED, RECOMMENDED	To advise that which is best	WE RECOMMEND THAT YOU DO THIS PROCEDURE IN A DEDICATED AREA.	
recondition (v)	REPAIR (v)	REPAIR THE SEATS.	Recondition the seats.
	OVERHAUL (TN)	DO AN OVERHAUL OF THE ENGINE.	Recondition the engine.
reconstruction (n)	REPAIR (n)	RECORD ALL REPAIRS TO THE SKIN SURFACE.	Record all reconstruction of the skin surface.
	REPAIR (v)	THIS PROCEDURE TELLS YOU HOW TO REPAIR THE SURFACE.	This procedure is for the reconstruction of the surface.
RECORD (v), RECORDS, RECORDED, RECORDED	1. To make notes of and keep data to use subsequently	RECORD THE RESULTS.	
	2. To put data on a storage medium	THE COCKPIT VOICE RECORDER RECORDS SIGNALS FROM FOUR AUDIO INPUTS DURING THE FLIGHT.	
recover (v)	COLLECT (v)	REMOVE THE NUT AND THE BOLT, AND COLLECT THE WASHER.	Remove the nut and the bolt, and recover the washer.
rectify (v)	CORRECT (v)	CORRECT THE DEFECT BEFORE YOU START THE PROCEDURE AGAIN.	Rectify the defect before you restart the procedure.
recur (v)	OCCUR (v)	IF THE SYMPTOMS OCCUR AGAIN, DO A SYSTEM TEST.	If the symptoms recur, do a system test.

Word (part of speech)	Approved meaning/ ALTERNATIVES	APPROVED EXAMPLE	Not approved example
RECYCLE (v), RECYCLES, RECYCLED, RECYCLED	To put a used material through a process to make it possible to use it or its components again For other meanings, use:	MAKE SURE THAT YOU RECYCLE THE REMOVED INSULATION MATERIAL.	
	CYCLE (n)	DO ONE MORE TEST CYCLE OF THE COMPONENT.	Recycle the component through a test.
reduce (v)	DECREASE (v)	SLOWLY DECREASE THE RATE OF DESCENT.	Slowly reduce the rate of descent.
REFER (v), REFERS, REFERRED, REFERRED	1. To tell a person where to find information	REFER TO CHAPTER 28 FOR THE TEST PROCEDURE.	
	2. To give information	BEFORE YOU DO THIS PROCEDURE, ENGAGE THE RAMP SERVICE DOOR SAFETY CONNECTOR PIN (THE PIN THAT HOLDS THE RAMP SERVICE DOOR, REFERRED TO IN THIS PROCEDURE AS THE "SAFETY CONNECTOR PIN").	
reference (n)	REFER (v)	REFER TO CHAPTER 20 FOR THE STANDARD TORQUE VALUES.	Reference is made to Chapter 20 for the standard torque values.
referenced (adj)	GIVEN (adj)	THIS POINT IS GIVEN ON THE GRAPH.	This point is referenced on the graph.
refill (v)	FILL (v)	FILL THE CONTAINER AGAIN.	Refill the container.
refit (v)	INSTALL (v)	INSTALL THE LINKAGE AGAIN.	Refit the linkage.
	REPAIR (v)	ON THE SUBMARINE, REPAIR ALL DAMAGE AND REPLACE ALL WORN PARTS.	Refit the submarine.

Word (part of speech)	Approved meaning/ ALTERNATIVES	APPROVED EXAMPLE	Not approved example
reflect (v)	REFLECTION (n)	THE RETICLE IMAGE MAKES A REFLECTION ON THE BOTTOM SURFACE OF THE GLASS.	The reticle image reflects on the bottom surface of the glass.
REFLECTION (n)	Something that occurs when energy comes against a surface which sends it back	A CLEAN SURFACE GIVES A BETTER REFLECTION.	
REFUEL (v), REFUELS, REFUELED, REFUELED	To supply with fuel	REFUEL THE AIRCRAFT.	
register (v)	SHOW (v)	ADJUST THE "SET +40" CONTROL UNTIL THE POINTER SHOWS +40.	Adjust the "SET +40" control until the pointer registers +40.
regrease (v)	APPLY (v)	APPLY GREASE TO THE ROD AGAIN.	Regrease the rod.
	MORE (adj)	PUT MORE GREASE ON THE JOINT UNTIL YOU CAN MOVE IT.	Regrease the joint until you can move it.
REGULAR (adj)	At specified or equal intervals	THE COMPUTER GIVES REGULAR INPUTS TO THE CONTROL SYSTEM.	
REGULARLY (adv)	In a regular manner	IF THE FAILURE OCCURS REGULARLY, DO A SYSTEM TEST.	
regulate (v)	CONTROL (v)	CONTROL THE ELECTRICAL CURRENT.	Regulate the electrical current.
	ADJUST (v)	ADJUST THE TIRE PRESSURE AS NECESSARY.	Regulate the tire pressure as necessary.

Word (part of speech)	Approved meaning/ ALTERNATIVES	APPROVED EXAMPLE	Not approved example
regulation (n)	ADJUSTMENT (n)	THE TEMPERATURE ADJUSTMENT IS AUTOMATIC.	The regulation of temperature is automatic.
	CONTROL (n)	THE TEMPERATURE CONTROL IS AUTOMATIC.	The regulation of temperature is automatic.
	CONTROL (v)	A SENSOR CONTROLS THE TEMPERATURE IN THE COMPARTMENT.	The regulation of temperature in the compartment is effected by a sensor.
reinflate (v)	INFLATE (v)	INFLATE THE TIRE AGAIN.	Reinflate the tire.
reinforce (v)	MAKE … STRONGER	DOUBLERS MAKE THE JOINT STRONGER.	Doublers reinforce the joint.
reinstall (v)	INSTALL (v)	INSTALL THE COVER AGAIN AFTER YOU ADJUST THE UNIT.	Reinstall the cover after you adjust the unit.
reinstallation (n)	INSTALLATION (n)	ONLY APPROVED PERSONNEL CAN DO THE INSTALLATION PROCEDURE.	This reinstallation must be performed by qualified personnel.
REJECT (v), REJECTS, REJECTED, REJECTED	To make a decision that something is unsatisfactory	REJECT THE PARTS THAT ARE DAMAGED.	
rejection (n)	REJECT (v)	REJECT THE COMPONENT IF THE TEST IS NOT SATISFACTORY.	Rejection is mandatory if the test proves the component to be defective.
relate (v)	RELATED (adj)	THE PROCEDURES ARE RELATED TO THE SYSTEM TEST.	The procedures relate to the system test.
RELATED (adj)	That has a relation	CLEAN THE BALLS OF THE ELECTRICAL LATCH AND ITS RELATED SOCKETS.	
RELATION (n)	The connection between two or more things caused by their functions, values, or conditions	THE RELATION BETWEEN THE TWO CONTROL SYSTEMS KEEPS THE AIRCRAFT STABLE.	

Word (part of speech)	Approved meaning/ ALTERNATIVES	APPROVED EXAMPLE	Not approved example
relay (v)	SEND (v)	SEND THE INFORMATION TO PERSON B.	Relay the information to Person B.
RELEASE (v), RELEASES, RELEASED, RELEASED	To make free, to let go	RELEASE THE INSTRUMENT PANEL TO GET ACCESS TO THE CONNECTIONS.	
relevant (adj)	RELATED (adj)	REMOVE THE ROD END FROM THE RELATED PIVOT.	Withdraw the rod end from the relevant pivot.
	THEIR (pron)	INSTALL THE BOLTS IN THEIR HOLES.	Install the bolts in the relevant holes.
	ITS (pron)	INSTALL THE COMPUTER IN ITS RACK.	Install the computer in the relevant rack.
relieve (v)	RELEASE (v)	RELEASE THE TENSION IN THE CABLE.	Relieve the tension in the cable.
remain (v)	STAY (v)	THE FLAGS MUST STAY OUT OF VIEW.	The flags must remain out of view.
REMAINING (adj)	That continues to stay	USE A SOLVENT TO REMOVE THE REMAINING SEALANT.	
remnant (n)	REMAINING (adj)	REMOVE THE REMAINING SEALANT FROM THE FLOOR COVERING.	Remove sealant remnants from the floor covering.
removable (adj)	REMOVE (v)	FIND THE PART OF THE BARREL THAT YOU CAN REMOVE.	Find the removable part of the barrel.
REMOVAL (n)	A procedure which removes an object	THE SUPERVISOR MUST SUPPLY THE APPLICABLE PROCEDURE FOR CORROSION REMOVAL.	
REMOVE (v), REMOVES, REMOVED, REMOVED	To take or move something away from its initial position	REMOVE THE INDICATOR FROM THE PANEL.	
render (v)	MAKE (v)	MAKE THE SYSTEM ELECTRICALLY SAFE.	Render the system electrically safe.

Word (part of speech)	Approved meaning/ ALTERNATIVES	APPROVED EXAMPLE	Not approved example
renew (v)	NEW (adj)	USE NEW ADHESIVE TAPE.	Renew the adhesive tape.
renovate (v)	REPAIR (v)	REPAIR ALL WORN SURFACES.	Renovate all worn surfaces.
	NEW (adj)	APPLY NEW PAINT TO THE CONTROL BOX IF ITS PAINT IS DAMAGED.	Renovate any damaged paint on the control box.
repack (v)	PUT (v)	PUT THE ASSEMBLY BACK IN THE BOX.	Repack the assembly in the box.
	FILL (v)	FILL THE HOLE WITH GREASE AGAIN.	Repack the hole with grease.
repaint (v)	PAINT (v)	PAINT ALL THE SURFACES AGAIN.	Repaint all the surfaces.
REPAIR (n)	The act of repairing or the result when something is repaired	AFTER THE REPAIR, THE SURFACE FINISH OF THE REPAIRED AREA MUST BE BETTER THAN 0.8 MICRONS.	
REPAIR (v), REPAIRS, REPAIRED, REPAIRED	To make an item serviceable	REPAIR THE DEFECTIVE WIRING.	
repairable (adj)	REPAIR (v)	IT IS POSSIBLE TO REPAIR THIS DAMAGE.	This damage is repairable.
repeat (v)	AGAIN (adv)	DO STEPS (10) THRU (14) AGAIN.	Repeat steps (10) to (14).
repeated (adj)	AGAIN AND AGAIN	IF YOU HEAR SIGNALS AGAIN AND AGAIN, DO A CHECK OF THE "AP ENGAGE" BUTTON ON THE FLIGHT MODE PANEL.	If repeated signals are heard, check the AP ENGAGE button on the flight mode panel.
repeatedly (adv)	AGAIN AND AGAIN	IF THE VOLTAGE DECREASES AGAIN AND AGAIN, ADJUST THE VOLTAGE REGULATOR.	If voltage drops repeatedly, adjust the voltage regulator.

Word (part of speech)	Approved meaning/ ALTERNATIVES	APPROVED EXAMPLE	Not approved example
REPLACE (v), REPLACES, REPLACED, REPLACED	To remove an item and to install a new or serviceable item of the same type	REPLACE THE RELAY.	
REPLACEMENT (n)	The item you install when you replace an item	MAKE SURE THAT THE REPLACEMENT HAS THE CORRECT DIMENSIONS.	
replenish (v)	FILL (v)	FILL THE ACCUMULATOR WITH OIL.	Replenish the accumulator with oil.
	ADD (v)	ADD OIL TO THE ACCUMULATOR.	Replenish the oil in the accumulator.
	REFUEL (v)	REFUEL THE AIRCRAFT.	Replenish the aircraft with fuel.
REPORT (n)	The information that is recorded about a subject or occurrence	IF THERE IS A REPORT OF DAMAGE, DO AN ACCURATE INSPECTION.	
reposition (v)	MOVE (v)	MOVE THE LEVER BACK TO ITS INITIAL POSITION.	Reposition the lever.
	PUT (v)	PUT THE LEVER BACK TO ITS INITIAL POSITION.	Reposition the lever.
	SET (v)	SET THE SWITCH BACK TO THE OFF POSITION.	Reposition the switch to OFF.
represent (v)	SHOW (v)	THE SUCTION SHOWS AS AN ALTIMETER INDICATION.	The suction is represented by an altimeter reading.
repressurize (v)	PRESSURIZE (v)	PRESSURIZE THE TANK AGAIN.	Repressurize the tank.
request (n)	TELL (v)	TELL THE FIRE SERVICE TO HELP YOU.	Make a request for external fire support.
	WRITE (v)	IF AID IS NECESSARY, WRITE TO THE MANUFACTURER.	Any request for assistance should be addressed to the manufacturer.

Word (part of speech)	Approved meaning/ ALTERNATIVES	APPROVED EXAMPLE	Not approved example
request (v)	TELL (v)	TELL THE FIRE SERVICE TO HELP YOU.	Request external fire support.
	WRITE (v)	WRITE TO THE MANUFACTURER FOR REPAIR INSTRUCTIONS.	Request repair instructions from the manufacturer.
require (v)	NECESSARY (adj)	IF NECESSARY, INSTALL CLEAN FILTERS.	Install clean filters if required.
reseal (v)	SEAL (v)	SEAL THE CONTAINER AGAIN.	Reseal the container.
reset (v)	SET (v)	SET THE INDICATOR TO A NEW POSITION.	Reset the indicator to a new position.
residual (adj)	REMAINING (adj)	REMOVE ALL THE REMAINING GREASE.	Remove all residual grease.
residue (n)	REMAINING (adj)	REMOVE THE REMAINING GREASE.	Remove the grease residues.
RESISTANT (adj)	That will help to prevent something	THE MATERIAL OF THE SEATS IS RESISTANT TO FIRE.	
respective (adj)	RELATED (adj)	ATTACH THE CABLES TO THEIR RELATED SHACKLES.	Attach the cables to their respective shackles.
	CORRECT (adj)	INSTALL THE BOLTS IN THEIR CORRECT POSITIONS.	Install the bolts in their respective positions.
respectively (adv)	RELATED (adj)	ENGINES 1 AND 2 SUPPLY HOT AIR TO THEIR RELATED WINGS.	Engines 1 and 2 supply the left and right wings respectively with hot air.
respond (v)	RESULT (n)	IF THE INSTRUMENT DOES NOT GIVE A RESULT, DO A TEST.	If the instrument fails to respond, do a test.
rest (n)	REMAINING (adj)	THE REMAINING INSTRUCTIONS ARE APPLICABLE ONLY TO AIRCRAFT WITH FUSELAGE TANKS.	The rest of these instructions only apply to aircraft with fuselage tanks.

Word (part of speech)	Approved meaning/ ALTERNATIVES	APPROVED EXAMPLE	Not approved example
rest (v)	BE (v)	THE AIRCRAFT MUST BE ON ITS WHEELS.	The aircraft must rest on its wheels.
	PUT (v)	PUT THE REMOVED PART ON A CLEAN SURFACE.	Rest the removed part on a clean surface.
restart (v)	START (v)	START THE UNIT AGAIN.	Restart the unit.
restore (v)	REPAIR (v)	DO NOT REPAIR THE CHROMIC ACID ANODIZING ON THE SKIN PANEL.	Do not restore the chromic acid anodizing on the skin panel.
	PUT (v)	PUT THE PINS BACK IN THEIR LOCATION.	Restore the pins to their location.
restrain (v)	HOLD (v)	HOLD THE HOSES IN POSITION.	Restrain the hoses in position.
	STOP (v)	STOP THE FLUID FLOW.	Restrain all fluid flow.
	PREVENT (v)	THE STOP PREVENTS THE MOVEMENT OF THE ROD.	The stop restrains the movement of the rod.
	LIMIT (n)	THE VALVE KEEPS THE AIRFLOW IN LIMITS.	The valve restrains the airflow.
restrict (v)	DECREASE (v)	THE METERING PLUG DECREASES THE FUEL FLOW.	The metering plug restricts the fuel flow.
	PREVENT (v)	THE STOP PREVENTS THE MOVEMENT OF THE ROD.	The stop restricts the movement of the rod.
	ONLY (adv)	USE THESE NUTS ONLY FOR THE TEST.	Restrict the use of these nuts to test only.
	LIMIT (n)	THE VALVE KEEPS THE AIRFLOW IN LIMITS.	The valve restricts the airflow.
RESULT (n)	Something that occurs when you do something	IF THESE TESTS DO NOT GIVE THE CORRECT RESULTS, REPLACE THE UNIT.	

Word (part of speech)	Approved meaning/ ALTERNATIVES	APPROVED EXAMPLE	Not approved example
result (v)	CAUSE (v)	AN INCORRECT CONNECTION WILL CAUSE DAMAGE.	An incorrect connection will result in damage.
	RESULT (n)	REPAIR DAMAGE THAT IS THE RESULT OF CORROSION.	Repair any damage resulting from corrosion.
resume (v)	START (v)	IF THERE IS NO CONTAMINATION, START THE OPERATION AGAIN.	If there is no contamination, resume the operation.
	CONTINUE (v)	CONTINUE THE TEST.	Resume the test.
retain (v)	KEEP (v)	KEEP THE PLUG.	Retain the plug.
retorque (v)	TORQUE (v)	TORQUE THE FITTING AGAIN.	Retorque the fitting.
retouch (v)	APPLY (v)	APPLY ENAMEL ON THE AREAS WHERE IT IS MISSING.	Retouch the missing enamel areas.
	PAINT (v)	PAINT THE DAMAGED LETTERS ON THE LABEL.	Retouch the damaged label letters.
RETRACT (v), RETRACTS, RETRACTED, RETRACTED	1. To pull in(to)	RETRACT THE LANDING GEAR.	
	2. To move in(to)	THE NOSE LANDING GEAR RETRACTS INTO THE FUSELAGE.	
retractable (adj)	RETRACT (v)	THE AIRCRAFT HAS A LANDING GEAR THAT RETRACTS.	The aircraft has a retractable landing gear.
RETRACTION (n)	The result when something retracts	DURING THE RETRACTION SEQUENCE, MAKE SURE THAT THE LANDING GEAR DOORS CLOSE.	
return (n)	GO (v)	AFTER THE AILERONS GO BACK TO NEUTRAL, MAKE SURE THAT THEY ARE FLUSH WITH THE FLAPS.	After the return of the ailerons to neutral, check that they are flush with the flaps.

Word (part of speech)	Approved meaning/ ALTERNATIVES	APPROVED EXAMPLE	Not approved example
return (v)	GO (v)	THE PLUNGERS MUST GO BACK TO THE CLOSED POSITION.	Plungers must return to the closed position.
reusable (adj)	USE (v)	YOU CAN USE THE MIXTURE AGAIN.	The mixture is reusable.
reuse (v)	USE (v)	USE THE FABRIC AGAIN IF IT IS NOT DAMAGED.	Reuse the fabric if it is not damaged.
reveal (v)	SHOW (v)	IF THE FUEL SAMPLE SHOWS FREE WATER, DRAIN THE SYSTEM.	If the fuel sample reveals free water, drain the system.
reverse (adj)	OPPOSITE (adj)	MOVE THE CONTROL IN THE OPPOSITE DIRECTION.	Move the control in the reverse direction.
	This word is part of a technical name when it refers to rearward thrust.	MANUALLY TURN THE PROPELLER BLADES TO THE FULL REVERSE POSITION.	
reverse (v)	OPPOSITE (adj)	MOVE THE ROD IN THE OPPOSITE DIRECTION.	Reverse the movement of the rod.
review (n)	INSPECTION (n)	DO AN INSPECTION OF THE DAMAGED AREA.	Do a review of the damaged area.
rewind (v)	WIND (v)	WIND THE CABLE BACK ONTO ITS REEL.	Rewind the cable onto its reel.
rework (v)	Refer to technical verbs, manufacturing process.	DO AN INSPECTION FOR SCRATCHES. THEN, LIGHTLY POLISH ALL DAMAGED AREAS.	Do an inspection for scratches and rework all damage areas.
rig (v)	ADJUST (v)	ADJUST THE SYSTEM.	Rig the system.
RIGHT (adj)	On the east side when you look north	DO A FLOW CHECK OF THE PUMP IN THE RIGHT WING TANK.	
right-hand (adj)	RIGHT (adj)	THE FUEL CONNECTOR IS IN THE RIGHT WING.	The fuel connector is in the right-hand wing.
RIGID (adj)	That cannot easily bend or change shape	REMOVE THE SCREWS FROM THE RIGID COVER.	

Word (part of speech)	Approved meaning/ ALTERNATIVES	APPROVED EXAMPLE	Not approved example
rinse (v)	FLUSH (v)	FLUSH THE FILTER WITH HOT WATER.	Rinse the filter in hot running water.
	REMOVE (v)	REMOVE THE CLEANING AGENT FROM THE SURFACE.	Rinse the cleaning agent from the surface.
	CLEAN (v)	CLEAN THE WINDOW PANE WITH A LARGE QUANTITY OF WATER.	Rinse the window pane with a large quantity of water.
ripped (adj)	DAMAGED (adj)	IF THE SEAT COVER IS DAMAGED, REPLACE IT.	If the seat cover is ripped, replace it.
rise (n)	INCREASE (v)	WHEN A FIRE STARTS, THE TEMPERATURE INCREASES AND CAUSES THE DETECTOR TO OPERATE.	When a fire starts, the temperature rise causes the detector to operate.
rise (v)	INCREASE (v)	USE A PUMP TO FILL THE STRUT WITH FLUID UNTIL THE PRESSURE INCREASES.	Pump fluid into the strut until the pressure rises.
RISK (n)	The possibility that something that is dangerous occurs	OBEY THE PRECAUTIONS TO PREVENT THE RISK OF FIRE.	
rope off (v)	ROPE (TN)	PUT ROPES AROUND THE AREA.	Rope off the area.
rotary (adj)	TURN (v)	TURN THE GEARS TO ALIGN THEM.	Use a rotary movement to align the gears.
rotate (v)	TURN (v)	SLOWLY TURN THE VANE.	Slowly rotate the vane.
rotation (n)	TURN (v)	MAKE SURE THAT THE FAN TURNS FREELY.	Ensure the free rotation of the fan.
rotational (adj)	TURN (v)	TURN THE GEARS TO ALIGN THEM.	Use a rotational movement to align the gears.
ROUGH (adj) (ROUGHER, ROUGHEST)	Not smooth	CLEAN THE ROUGH SURFACES CAREFULLY.	
roughen (v)	ROUGH (adj)	MAKE THE FAYING SURFACES ROUGH.	Roughen the faying surfaces.

Word (part of speech)	Approved meaning/ ALTERNATIVES	APPROVED EXAMPLE	Not approved example
roughness (n)	ROUGH (adj)	MAKE SURE THAT THE AREA OF THE BOND IS SUFFICIENTLY ROUGH.	Make sure the area of the bond has sufficient roughness.
round (adj)	CIRCULAR (adj)	REPAIR THE DAMAGE WITH A CIRCULAR PATCH.	Repair the damage with a round patch.
ROUNDED (adj)	Not angular	WHEN YOU INSTALL THE PART, MAKE SURE THAT THE ROUNDED EDGE IS AGAINST THE STRUCTURE.	
route (n)	ROUTING (n)	MAKE SURE THAT YOU DO NOT CHANGE THE WIRE ROUTING.	Make sure you do not change the wire routes.
route (v)	PUT (v)	PUT THE FUELING HOSE ALONG THE GROUND.	Route the fueling hose along the ground.
	ROUTING (n)	MAKE SURE THAT THE ROUTING OF THE WIRE IS CORRECT.	Make sure that the wiring is correctly routed.
routine (adj)	USUAL (adj)	IF SUCH FAILURES BECOME USUAL, REPLACE THE UNIT.	If such failures become routine, replace the unit.
ROUTING (n)	The specified direction that cables, pipes, wires, and other parts of a system must go along	DO NOT CHANGE THE ROUTING OF THE PIPE.	
ROW (n)	A number of objects in a line	THE PANELS ARE INSTALLED IN ROWS.	
RUB (v), RUBS, RUBBED, RUBBED	To move or cause something to move with pressure and friction along a surface	RUB THE SURFACE WITH A CLEAN CLOTH.	
		MAKE SURE THAT THE COILS OF THE OXYGEN HOSE DO NOT RUB TOGETHER.	
run (v)	OPERATE (v)	OPERATE THE ENGINE AT 100% THRUST.	Run the engine at 100% thrust.
running (adj)	FLUSH (v)	FLUSH THE PIPES WITH WATER.	Clean the pipes with running water.

Blank Page

Word (part of speech)	Approved meaning/ ALTERNATIVES	APPROVED EXAMPLE	Not approved example
SAFE (adj) (SAFER, SAFEST)	Not dangerous, disarmed	MAKE THE AREA SAFE BEFORE YOU CONTINUE THE WORK.	
SAFELY (adv)	In a safe manner	MAKE SURE THAT THE LADDER IS SAFELY ATTACHED.	
SAFETY (n)	A condition that is safe	FOR YOUR SAFETY, ALWAYS USE A PROTECTIVE MASK.	
SAFETY (v), SAFETIES, SAFETIED, SAFETIED	1. To prevent accidental operation	OPEN, TAG, AND SAFETY THE CIRCUIT BREAKERS.	
	2. To make sure that something does not become loose	SAFETY THE NUT WITH LOCKWIRE.	
safety-clip (v)	LOCK (v) (WITH A CLIP [TN] OR CLIPS [TN])	LOCK THE TURNBUCKLE ON THE CONTROL ROD WITH A CLIP.	Safety-clip the turnbuckle on the control rod.
	SAFETY (v) (WITH A CLIP [TN] OR CLIPS [TN])	OPEN, TAG, AND SAFETY THE CIRCUIT BREAKER WITH A CLIP.	Open, tag, and safety-clip the circuit breaker.
safety-wire (v)	LOCK (v) (WITH LOCKWIRE [TN])	LOCK THE COVER TO THE FIRE EXTINGUISHER SWITCH WITH LOCKWIRE.	Safety-wire the cover to the fire extinguisher switch.
	SAFETY (v) (WITH SAFETY WIRE [TN])	SAFETY THE NUT WITH SAFETY WIRE.	Safety-wire the nut.
SAME (adj)	Agrees in all details	THE SAME PROCEDURE IS APPLICABLE TO THE TWO UNITS.	

Word (part of speech)	Approved meaning/ ALTERNATIVES	APPROVED EXAMPLE	Not approved example
SAME (pron)	Agrees in all details	THE ADJUSTMENT OF THE LEFT AND RIGHT ACTUATORS MUST BE THE SAME.	
SAMPLE (n)	A piece or quantity of something that you use for an inspection or test procedure to show that it has specified qualities	GET A SAMPLE OF THE FUEL AND MEASURE ITS SPECIFIC GRAVITY.	
sample (v)	SAMPLE (n)	GET A SAMPLE OF FUEL AND DO A TEST FOR WATER CONTAMINATION.	Sample and test the fuel for water contamination.
SATISFACTORILY (adv)	In a satisfactory manner	MAKE SURE THAT THE MICROSWITCHES OPERATE SATISFACTORILY.	
SATISFACTORY (adj)	Agrees with all that is necessary	A CLEARANCE OF 3 mm IS SATISFACTORY.	
saturate (v)	SOAK (v)	SOAK THE ELEMENT FOR FOUR HOURS.	Saturate the element for four hours.
save (v)	KEEP (v)	KEEP THE O-RINGS FOR THE SUBSEQUENT PROCEDURES.	Save the O-rings for the subsequent procedures.
scan (v)	EXAMINE (v)	EXAMINE THE SURFACE ALONG THE TRAILING EDGE.	Scan the surface along the trailing edge.
	SCAN (TN)	THE ANTENNA DOES A SCAN THROUGH 80 DEG.	Antenna scans through 80 deg.
SCHEDULE (v), SCHEDULES, SCHEDULED, SCHEDULED	To plan something in a time or a sequence	SCHEDULE THE CHECKS OF THE TIRES AT REGULAR INTERVALS.	
scored (adj)	SCORE (TN)	IF THE BEARING RACE HAS SCORES, REPLACE IT.	If the bearing race is scored, replace it.
scrap (v)	DISCARD (v)	DISCARD THE REMOVED SEAL.	Scrap the removed seal.

Word (part of speech)	Approved meaning/ ALTERNATIVES	APPROVED EXAMPLE	Not approved example
scrape (v)	REMOVE (v) (WITH A SCRAPER [TN])	USE A SCRAPER TO REMOVE THE REMAINING SEALANT.	Scrape off the old sealant.
scratched (adj)	SCRATCH (TN)	REPAIR THE SURFACES THAT HAVE SCRATCHES.	Repair the scratched surfaces.
screw (v)	TURN (v)	CONTINUE TO TURN THE END FITTING UNTIL IT IS IN THE CORRECT POSITION.	Continue screwing the end fitting until it is in the correct position.
	ATTACH (v) (WITH A SCREW [TN] OR SCREWS [TN])	ATTACH THE STRAPS TO THE PANELS WITH SCREWS.	Screw the straps to the panels.
scrub (v)	CLEAN (v) (WITH A BRUSH [TN])	CLEAN THE METAL PARTS WITH A BRUSH.	Scrub all metal parts.
SEAL (n)	Something that prevents access or leaks	APPLY SEALANT TO THE FILLET SEALS.	
SEAL (v), SEALS, SEALED, SEALED	To prevent access or leaks	SEAL THE OPENINGS WITH POLYETHYLENE.	
search (v)	EXAMINE (v)	EXAMINE THE SURFACE FOR CRACKS.	Search the surface for cracks.
seat (v)	INSTALL (v)	INSTALL THE O-RING IN ITS GROOVE ON THE PISTON.	Seat the O-ring in its groove on the piston.
	AGAINST (prep)	MAKE SURE THAT THE VALVE IS IN ITS CORRECT POSITION AGAINST THE FRAME.	Make sure that the valve is seated on the frame.
SECONDARY (adj)	Second in importance	DO AN OPERATIONAL CHECK OF THE SECONDARY FLIGHT CONTROLS.	

Word (part of speech)	Approved meaning/ ALTERNATIVES	APPROVED EXAMPLE	Not approved example
secure (adj)	TIGHT (adj)	MAKE SURE THAT THE BOLTS ARE TIGHT.	Make sure that the bolts are secure.
	SAFE (adj)	BEFORE YOU LIFT THE PLATFORM, MAKE SURE THAT IT IS SAFE.	Make sure that the platform is secure before you lift it.
	CORRECTLY (adv)	MAKE SURE THAT THE BRACKETS ARE CORRECTLY ATTACHED.	Check the brackets for secure attachment.
secure (v)	ATTACH (v)	REMOVE THE BOLTS THAT ATTACH THE UNIT.	Remove the bolts that secure the unit.
	SAFETY (v)	SAFETY THE PULLEY GUARD PINS WITH SPLIT PINS.	Secure the pulley guard pins with split pins.
securely (adv)	CORRECTLY (adv)	MAKE SURE THAT THE BRACKETS ARE CORRECTLY ATTACHED.	Make sure that the brackets are securely attached.
	SAFELY (adv)	MAKE SURE THAT THE LADDER IS SAFELY ATTACHED.	Make sure the ladder is securely attached.
	TIGHT (adj)	MAKE SURE THAT THE BOLTS ARE TIGHT.	Make sure that the bolts are securely installed.
security (n)	CORRECTLY (adv)	MAKE SURE THAT THE COMPONENTS ARE CORRECTLY ATTACHED.	Check the security of the components.

Word (part of speech)	Approved meaning/ ALTERNATIVES	APPROVED EXAMPLE	Not approved example
SEE (v), SEES, SAW, SEEN	To know with the eyes	MAKE SURE THAT YOU CAN SEE THE CABLE THROUGH THE INSPECTION HOLE.	
	For other meanings, use:		
	REFER (v)	REFER TO TABLE 8001 FOR THE APPLICABLE TORQUE VALUES.	See Table 8001 for the applicable torque value.
	EXAMINE (v)	EXAMINE THE MEMBRANE FOR HOLES.	Check the membrane to see if there are holes.
	MAKE SURE (v)	MOVE THE TUBE TO MAKE SURE THAT ITS INNER CONNECTION IS TIGHT.	Move the tube to see if its inner connection is tight.
seek (v)	GET (v)	GET MEDICAL AID IMMEDIATELY.	Seek medical attention immediately.
seepage (n)	LEAKAGE (n)	CLEAN THE AREAS WHERE THERE IS HYDRAULIC FLUID LEAKAGE.	Clean the areas where there is hydraulic fluid seepage.
seized (adj)	CATCH (v)	IF THE CONTROL CABLE IS CAUGHT IN THE PULLEY, RELEASE THE CABLE TENSION.	If the control cable is seized in the pulley, release the cable tension.
	MOVE (v)	IF YOU CANNOT MOVE THE BOLTS, APPLY SOME PENETRATING OIL.	If the bolts are seized, apply some penetrating oil.
	TURN (v)	IF THE FLAP CONTROL MOTOR CANNOT TURN, USE THE ALTERNATIVE MODE.	If the flap control motor is seized, use the alternate mode.

Word (part of speech)	Approved meaning/ ALTERNATIVES	APPROVED EXAMPLE	Not approved example
SELECT (v), SELECTS, SELECTED, SELECTED	To make a choice	SELECT THE HYDRAULIC SYSTEM THAT YOU WILL PRESSURIZE.	
		SELECT A LANGUAGE FROM THE MENU.	
	Do not use this word as a synonym for SET (v).	SET THE SWITCH TO "TEST".	Select the switch to TEST.
SELECTION (n)	The action or result of choosing	THE OPERATION OF THE INDICATOR DOES NOT PREVENT THE SELECTION OF SYSTEM 1.	
SEMICIRCULAR (adj)	That has the shape of half a circle	THE VALVE FLAPS ARE SEMICIRCULAR.	
SEND (v), SENDS, SENT, SENT	To cause to go	SEND THE FILTER ELEMENT TO THE MAINTENANCE SHOP FOR THE NECESSARY INSPECTION.	
SENSE (v), SENSES, SENSED, SENSED	To get an input automatically	THE TEMPERATURE BULB SENSES THE EXTERNAL AIR TEMPERATURE.	
SENSITIVE (adj)	That can sense small changes	THE CAPSULE IS SENSITIVE TO PRESSURE CHANGES.	
separable (adj)	DISASSEMBLE (v)	YOU CAN DISASSEMBLE THIS UNIT INTO TWO PARTS.	This unit is separable into two parts.
	DISCONNECT (v)	YOU CAN DISCONNECT THESE LINE FITTINGS.	These line fittings are separable.

Word (part of speech)	Approved meaning/ ALTERNATIVES	APPROVED EXAMPLE	Not approved example
separate (adj)	NOT CONNECTED	THESE TWO TRACKS ARE NOT CONNECTED TO THE OTHER TWO.	These two tracks are separate from the other two.
	ISOLATED (v)	EACH HYDRAULIC SYSTEM IS FULLY ISOLATED.	All hydraulic systems are completely separate.
	NOT ATTACHED	THE CAP IS NOT ATTACHED TO THE COUPLING.	The cap is separate from the coupling.
separate (v)	DISCONNECT (v)	DISCONNECT THE LINE FITTINGS.	Separate the line fittings.
	DIVIDE (v)	YOU CAN DIVIDE THE DRAINS INTO THREE GROUPS.	You can separate the drains into three groups.
SEPARATION (n)	The action or result of separating	SEPARATION OF THESE PARTS IS NOT EASY.	
SEQUENCE (n)	The relation of items that follow one after the other in a list or the relation of steps or events that occur one after the other in time	TIGHTEN THE BOLTS IN THE SEQUENCE THAT IS GIVEN IN FIGURE 3.	
serious (adj)	IMPORTANT (adj)	VIRUS CONTAMINATION IS AN IMPORTANT PROBLEM.	Virus contamination is a serious problem.
	DANGEROUS (adj)	HYDRAULIC FLUID IS DANGEROUS FOR YOUR EYES.	Hydraulic fluid can cause serious eye injury.
serrated (adj)	SERRATION (n)	USE GLOVES WHEN YOU TOUCH PARTS WITH EDGE SERRATIONS.	Use gloves when handling parts with serrated edges.
SERRATION (n)	One or more notches or teeth as in a saw	MAKE MARKS THAT SHOW THE LOCATION OF THE SERRATIONS ON THE PLATES.	
serve (v)	Use an action verb.	THE RADIO MASTER SWITCH CONTROLS THE BUS BAR.	The radio master switch serves to control the bus bar.

Word (part of speech)	Approved meaning/ ALTERNATIVES	APPROVED EXAMPLE	Not approved example
service (v)	SERVICING (n)	DO THE SERVICING OF THE FIRE EXTINGUISHERS.	Service the fire extinguishers.
serviceability (n)	SERVICEABLE (adj)	THE TIRE IS SERVICEABLE IF THE CUTS ARE SMALLER THAN 10 mm.	The serviceability of the tire is not affected by any cut smaller than 10 mm.
SERVICEABLE (adj)	Correct or satisfactory	DO A VISUAL INSPECTION OF THE ADAPTER TO MAKE SURE THAT IT IS SERVICEABLE.	
SERVICING (n)	The steps necessary to prepare something for operation	BEFORE YOU DO SERVICING OF THE HYDRAULIC SYSTEM, ISOLATE IT.	
SET (n)	A group of related items that you (can) use for the same purpose	USE THE SET OF TOOLS THAT IS SUPPLIED WITH THE UNIT.	
SET (v), SETS, SET, SET	To put something into a given adjustment, condition, or mode	SET THE ALTIMETER SCALE TO 1013 mbar.	
setting (n)	ADJUSTMENT (n)	DO NOT CHANGE THE ADJUSTMENT OF THE SWITCHES.	Do not change the setting of the switches.
	POSITION (n)	MAKE SURE THAT THE POSITION OF THE FLAPS IS EIGHT DEG.	Check flaps for eight deg. setting.
	SET (v)	THE MANUFACTURER SETS THE LENGTH OF THE ROD IN THE FACTORY.	Setting of the rod length is achieved in the factory.
settle (v)	COLLECT (v)	LET THE SEDIMENT COLLECT AT THE BOTTOM OF THE TANK.	Let the sediment settle at the bottom of the tank.
	STABLE (adj)	LET THE INDICATIONS BECOME STABLE.	Let the indications settle.
sever (v)	CUT (v)	MAKE SURE THAT YOU DO NOT CUT THE CONTROL CABLE.	Make sure you do not sever the control cable.

Word (part of speech)	Approved meaning/ ALTERNATIVES	APPROVED EXAMPLE	Not approved example
several (adj)	SOME (adj)	SOME FUNCTIONS ARE NOT AVAILABLE.	Several functions are not available.
	If this alternative is not sufficient, give the range, quantity, number, or minimum value.	SOAK THE FILLER CAPS IN DISTILLED WATER FOR A MINIMUM OF EIGHT HOURS.	Soak the filler caps in distilled water for several hours.
severe (adj)	DANGEROUS (adj)	HYDRAULIC FLUID IS DANGEROUS FOR YOUR SKIN.	Hydraulic fluid can cause severe skin problems.
SHAKE (v), SHAKES, SHOOK, SHAKEN	To move or cause to move quickly up and down or from side to side	SHAKE THE CONTAINER.	
		THE CONTROL COLUMN STARTS TO SHAKE BEFORE A STALL OCCURS.	
shall (v)	MUST (v)	HOLES MUST NOT HAVE SHARP EDGES.	Holes shall not have sharp edges.
SHAPE (n)	The contour of an object	THE PATCH MUST BE THE SAME SHAPE AS THE CUTOUT.	
shape (v)	SHAPE (n)	MAKE THE PATCH THE SAME SHAPE AS THE CUTOUT.	Shape the patch to suit the cutout.
SHARP (adj) (SHARPER, SHARPEST)	That can cut or make a hole	MAKE A HOLE WITH A SHARP TOOL.	
sharply (adv)	QUICKLY (adv)	QUICKLY PULL ON THE CABLES TO DISENGAGE THEM.	Sharply pull on the cables to disengage them.
	SUDDENLY (adv)	MOVE THE CONTROL COLUMN FORWARD SUDDENLY.	Move the control column forward sharply.
shear (v)	BREAK (v)	TIGHTEN THE NUT. CONTINUE UNTIL THE NUT BREAKS.	Tighten the nut until it shears.
	CUT (v)	DO NOT CUT THE WIRES.	Do not shear the wires.

Word (part of speech)	Approved meaning/ ALTERNATIVES	APPROVED EXAMPLE	Not approved example
SHEET (n)	A piece of material that is thin in relation to its length and width	IF THE SHEETS HAVE DENTS WITH A DEPTH OF MORE THAN 0.5 mm, REMOVE THESE DENTS.	
shield (v)	PREVENT (v)	PUT ON GOGGLES TO PREVENT INJURY TO YOUR EYES.	Shield your eyes.
	PROTECTION (n)	GIVE THE CONTROL UNIT PROTECTION FROM DAMAGE.	Shield the control unit from damage.
shift (v)	MOVE (v)	MOVE THE LEVER FORWARD.	Shift the lever forward.
SHINY (adj) (SHINIER, SHINIEST)	That can cause a reflection of light	CLEAN THE SHINY AREA OF THE OLEO STRUT.	
SHOCK (n)	The sudden effect of energy on a person or object	SHOCKS CAN CAUSE DAMAGE TO THE GYROSCOPE.	
SHORT (adj) (SHORTER, SHORTEST)	That has small length or duration	ATTACH THE SHORT ARM OF THE BELLCRANK TO THE ROD.	
	Be specific if possible.	AFTER YOU STOP THE POWER SUPPLY, THE TEMPERATURE DECREASES IN A SHORT TIME. THIS TIME MUST NOT BE MORE THAN 30 SECONDS.	
shorten (v)	DECREASE (v)	DO NOT DECREASE THE TEST TIME.	Do not shorten the test time.
should (v)	MUST (v)	PERSONNEL MUST USE PROTECTIVE CLOTHING.	Personnel should wear protective clothing.
	IF (conj)	IF A FAILURE OCCURS, STOP THE TEST.	Should a failure occur, stop the test.

Word (part of speech)	Approved meaning/ ALTERNATIVES	APPROVED EXAMPLE	Not approved example
SHOW (v), SHOWS, SHOWED, SHOWN	1. To cause to be seen	THE INDICATOR SHOWS THE LEVEL OF HYDRAULIC FLUID IN THE RESERVOIR.	
	2. To be in view or come into view	THE SYMBOL SHOWS ON THE SCREEN.	
shut down (v)	STOP (v)	STOP THE ENGINE.	Shut down the engine.
SIDE (n)	1. The specified surface or area of an object	LUBRICATE ONE SIDE OF THE WASHER.	
	2. A location or direction that has a relationship to a center or a line of division	PUSH THE UNIT OUT AND THEN MOVE IT TO THE LEFT SIDE.	
sideways (adv)	SIDE (n)	PUSH THE UNIT OUT AND THEN MOVE IT TO THE LEFT SIDE.	Push the unit out and then move it sideways.
sight (n)	VIEW (n)	TURN THE KNOB TO MAKE THE DISPLAY GO OUT OF VIEW.	Turn the knob to make the display go out of sight.
sight (v)	LOOK (v)	LOOK ALONG THE SKIN.	Sight along the skin.
SIGN (n)	The indication of a possible condition	EXAMINE THE SURFACE FOR SIGNS OF INTERNAL DAMAGE.	
significant (adj)	IMPORTANT (adj)	THIS PROCEDURE IS IMPORTANT FOR THE CORRECT OPERATION OF THE UNIT.	This procedure is significant for the correct operation of the unit.
silence (v)	STOP (v)	STOP THE ALARM BELL.	Silence the alarm bell.
similar (adj)	EQUIVALENT (adj)	USE MATERIAL 11-001 OR AN EQUIVALENT MATERIAL.	Use material 11-001 or a similar material.
	ALMOST THE SAME	THE TWO ITEMS HAVE ALMOST THE SAME SHAPE.	The two items are similar in shape.

Word (part of speech)	Approved meaning/ ALTERNATIVES	APPROVED EXAMPLE	Not approved example
SIMULATE (v), SIMULATES, SIMULATED, SIMULATED	To make a condition that is the same as one that can occur in operation	SIMULATE A WHEEL SPEED OF 80 KNOTS.	
simultaneous (adj)	AT THE SAME TIME	YOU MUST REMOVE THE FILTER ELEMENT AND THE FILTER HOUSING AT THE SAME TIME.	Simultaneous removal of the filter element and the filter housing is mandatory.
simultaneously (adv)	AT THE SAME TIME	DO THESE TWO STEPS AT THE SAME TIME.	Do these two steps simultaneously.
SINCE (conj)	Function word that shows: "from some time in the past until a later time or now" For other meanings, use:	IF IT IS MORE THAN TWO HOURS SINCE YOU MIXED THE COMPOUND, DO NOT USE IT.	
	BECAUSE (conj)	BE CAREFUL WHEN YOU USE ALODINE, BECAUSE IT IS A DANGEROUS MATERIAL.	Since Alodine is a dangerous material, be careful when you use it.
single (adj)	ONE (TN)	ONE CRACK, 0.50 INCH LONG, IS PERMITTED.	A single crack, 0.50 inch long, is allowed.
situated (adj)	INSTALL (v)	THE TRIM CONTROL KNOB IS INSTALLED ON THE CONTROL PEDESTAL.	The trim control knob is situated on the control pedestal.
	BE (v)	THE TRIM CONTROL KNOB IS ON THE CONTROL PEDESTAL.	The trim control knob is situated on the control pedestal.
size (n)	DIMENSION (n)	THESE CASES HAVE DIFFERENT DIMENSIONS.	These cases are of different sizes.
skid (v)	SKID (TN)	THIS SHOWS THAT THE AIRCRAFT IS IN A SKID.	This indicates that the aircraft is skidding.

Word (part of speech)	Approved meaning/ ALTERNATIVES	APPROVED EXAMPLE	Not approved example
slack (adj)	TENSION (TN)	MAKE SURE THAT THE TENSION IN THE CONTROL CABLES IS CORRECT.	Check for slack control cables.
slack (n)	TENSION (TN)	MAKE SURE THAT THE TENSION IN THE CONTROL CABLES IS CORRECT.	Check for slack in the control cables.
slacken (v)	TENSION (TN)	RELEASE THE TENSION IN THE CONTROL CABLES.	Slacken the control cables.
slackness (n)	TENSION (TN)	MAKE SURE THAT THE TENSION IN THE CABLES IS CORRECT.	Make sure there is no slackness in the cables.
slave (v)	CONTROL (v)	THE PILOT'S INDICATORS CONTROL THE COPILOT'S INDICATORS.	The copilot's indicators are slaved to the pilot's indicators.
slide (v)	MOVE (v)	MOVE THE COMPUTER HORIZONTALLY TO PREVENT DAMAGE.	Slide the computer horizontally to avoid damage.
slight (adj)	SMALL (adj)	IF THE LEAKAGE IS SMALL, TORQUE THE CONNECTION TO 4.0 Nm. THEN, DO THE TEST AGAIN.	If slight leakage is found, torque the connection to 4.0 Nm. Then repeat the test.
	LIGHT (adj)	APPLY A LIGHT FORCE TO COMPRESS THE SPRING.	Apply a slight force to compress the spring.
slightly (adv)	SMALL (adj)	MOVE THE ADAPTER A SMALL DISTANCE FORWARD.	Move the adapter slightly forward.

Word (part of speech)	Approved meaning/ ALTERNATIVES	APPROVED EXAMPLE	Not approved example
slip (v)	MOVE (v)	MOVE THE CARRIAGE ALONG THE FLAP TRACK.	Slip the carriage along the flap track.
	REMOVE (v)	REMOVE THE CLAMP FROM THE CABLE.	Slip the clamp from the cable.
	CAUSE TO FALL	IF FUEL OR LUBRICATING OIL SPILLS, CLEAN THE AREA IMMEDIATELY. A WET SURFACE IS DANGEROUS AND CAN CAUSE YOU TO FALL.	If fuel or lubricating oil spills, clean the area immediately. You can slip and fall.
	MOVE ACCIDENTALLY	MAKE SURE THAT THE CLUTCH DOES NOT MOVE ACCIDENTALLY.	Make sure that the clutch does not slip.
SLOPE (n)	A surface that is not level	IF THE AIRCRAFT IS PARKED ON A SLOPE, MAKE SURE THAT WHEEL CHOCKS ARE IN POSITION.	
SLOT (n)	A long opening that is not wide	PUSH THE CABLES THROUGH THE SLOT IN FRAME 8 AND ATTACH THEM TO THE CLIPS.	
SLOW (adj) (SLOWER, SLOWEST)	At low speed	MAKE SURE THAT THE MOVEMENT OF THE ELEVATORS IS SLOW.	
slow down (v)	DECREASE (v)	DECREASE THE SPEED OF THE SCREWING MACHINE TO PREVENT DAMAGE TO FASTENERS.	Slow down the screwing machine to prevent damage to fasteners.
SLOWLY (adv)	In a slow manner	TURN THE KNOB SLOWLY.	
sluggish (adj)	SLOW (adj)	IF THE MOVEMENT OF THE NEEDLE IS TOO SLOW, REPLACE THE SENSOR.	If the movement of the needle is sluggish, replace the sensor.

Word (part of speech)	Approved meaning/ ALTERNATIVES	APPROVED EXAMPLE	Not approved example
SMALL (adj) (SMALLER, SMALLEST)	Less than average in dimension, quantity, quality, or capacity	WHEN YOU REFUEL THE AIRCRAFT, A SMALL QUANTITY OF FUEL COMES OUT OF THE VENT LINE.	
smear (v)	APPLY (v)	APPLY GREASE TO THE TAPER SLEEVE.	Smear grease on the taper sleeve.
SMELL (v), SMELLS, SMELLED, SMELLED	To sense with the nose	IF YOU SMELL OIL IN THE COMPARTMENT, REPLACE THE RECIRCULATION FILTERS.	
SMOKE (n)	A gas that has particles of burned material and that you can see	WHEN THE CONCENTRATION OF SMOKE IS MORE THAN THE LIMIT, AN ALARM OPERATES.	
SMOKE (v), SMOKES, SMOKED, SMOKED	To breathe the fumes of burning material and especially tobacco	DO NOT SMOKE IN THE WORK AREA.	
SMOOTH (adj) (SMOOTHER, SMOOTHEST)	1. That has a continuous uniform surface	MAKE SURE THAT THE PAINTED SURFACE IS SMOOTH.	
	2. That has a constant rate of movement	EXAMINE THE LINKAGE FOR SMOOTH MOVEMENT.	
smooth (v)	SMOOTH (adj)	MAKE THE TAPE SMOOTH.	Smooth the tape out.
SMOOTHLY (adv)	In a smooth manner	MAKE SURE THAT THE REDUCTION GEAR MOVES SMOOTHLY.	
snag (v)	CATCH (v)	THE CLOTH WILL CATCH ON BROKEN WIRES.	The cloth will snag on broken wires.
snap (n)	CLICK (n)	WHEN YOU ATTACH THE SPRING CLIP, MAKE SURE THAT YOU HEAR A CLICK.	Ensure that there is a snap when you fit the spring clip.

Word (part of speech)	Approved meaning/ ALTERNATIVES	APPROVED EXAMPLE	Not approved example
snap (v)	MOVE QUICKLY	THE RETAINING CLIP MUST MOVE QUICKLY INTO ITS POSITION.	The retaining clip must snap into its position.
	BREAK (v)	IF YOU BEND THIS PART TOO FAR, IT WILL BREAK.	If you bend this part too far, it will snap.
so (that) (conj)	UNTIL (prep)	ADJUST THE LENGTH OF THE ROD UNTIL YOU GET A MINIMUM CLEARANCE OF 0.1 in.	Adjust the length of rod so that a minimum gap of 0.1 in. is obtained.
	PREVENT (v)	LOCK THE WHEEL TO PREVENT MOVEMENT.	Lock the wheel so that it does not move.
SOAK (v), SOAKS, SOAKED, SOAKED	1. To put something into a liquid and keep it there until it is fully wet or soft	SOAK THE FILTER IN THE CLEANING SOLUTION FOR 4 HOURS.	
	2. To stay in liquid until fully wet or soft.	THE COMPONENT MUST SOAK IN THE SOLUTION FOR A MINIMUM OF 2 HOURS.	
soapy (adj)	SOAP (TN)	CLEAN THE SURFACE WITH A SOLUTION OF SOAP AND WATER.	Clean the surface using soapy water.
SOFT (adj) (SOFTER, SOFTEST)	Flexible, not hard	RUB THE SURFACE WITH A SOFT, DRY CLOTH.	
soften (v)	MAKE…SOFT	PUT THE TOOLS IN SOLVENT TO MAKE THE COMPOUND SOFT.	Put the tools in solvent to soften the compound.
soiled (adj)	DIRTY (adj)	DISCARD THE DIRTY RAGS.	Scrap the soiled rags.
SOLID (adj)	That has the properties of a solid	THE RAILS ARE MADE FROM SOLID BARS.	
SOLID (n)	Material that is not a gas or a liquid	EXAMINE THE OIL FOR SOLIDS.	
solidify (v)	SOLID (adj)	DO NOT LET THE FOAM BECOME SOLID.	Do not allow the foam to solidify.

Word (part of speech)	Approved meaning/ ALTERNATIVES	APPROVED EXAMPLE	Not approved example
SOLUTION (n)	1. A liquid that includes a dissolved material	PREPARE THE SOLUTION IN A CONTAINER MADE OF CERAMIC MATERIAL.	
	2. The answer to a problem	A BITE TEST WILL GIVE THE SOLUTION.	
solve (v)	SOLUTION (n)	IF THE REMOVAL OF THE VALVE WAS NOT THE SOLUTION TO THE PROBLEM, REPLACE THE UNIT.	If the removal of the valve did not solve the problem, replace the unit.
SOME (adj)	Related to a quantity not specified	SOME MODELS DO NOT HAVE THIS FUNCTION.	
SOME (pron)	Related to a quantity not specified	SOME MODELS HAVE THIS FUNCTION, BUT SOME DO NOT.	
SOMETHING (pron)	A thing that is not determined or specified	IF SOMETHING UNUSUAL OCCURS, DO A BITE TEST.	
soon (adv)	TIME (n)	WHEN YOU STOP THE POWER SUPPLY, THE TEMPERATURE WILL DECREASE IN A SHORT TIME.	When you stop the power supply, the temperature will decrease soon.
	Give the time if possible.	WHEN YOU STOP THE POWER SUPPLY, THE TEMPERATURE WILL DECREASE IN APPROXIMATELY 1 MINUTE.	When you stop the power supply, the temperature will decrease soon.
SOUND (n)	Something that you can hear	INCREASE THE VOLUME OF THE SOUND.	
sound (v)	OPERATE (v)	THE ALARM BELL OPERATES.	The alarm bell sounds.

Word (part of speech)	Approved meaning/ ALTERNATIVES	APPROVED EXAMPLE	Not approved example
SOURCE (n)	1. Something that supplies energy or data	USE A HYDRAULIC SOURCE TO OPERATE THE SWAGING TOOL.	
	2. The point where something starts	WHEN YOU FIND THE SOURCE OF THE LEAKAGE, CLEAN THE SURFACE WITH SOLVENT.	
SPACE (n)	A distance, area, or volume	PUT THE SEALANT INTO THE SPACE BEHIND THE FLANGE OF THE WATER INJECTOR.	
space (v)	DISTANCE (n)	MAKE SURE THAT THERE IS AN EQUAL DISTANCE BETWEEN THE HOLES.	Make sure you space the holes equally.
spacing (n)	DISTANCE (n)	REFER TO FIG. 401 FOR THE DISTANCE BETWEEN RIVETS.	The rivet spacing is given in Fig. 401.
spanwise (adj)	SPAN (TN)	MAKE SURE THAT THERE ARE NO SIGNS OF MOVEMENT ALONG THE SPAN.	Check for signs of spanwise movement.
sparingly (adv)	QUANTITY (n)	APPLY A SMALL QUANTITY OF GREASE.	Apply grease sparingly.
SPARK (n)	A hot, bright particle of a material that burns	DO NOT USE TEST EQUIPMENT THAT CAN CAUSE SPARKS.	
spark (v)	SPARK (n)	PUT ON GOGGLES WHEN YOU USE ELECTRICAL EQUIPMENT WHICH MAKES SPARKS.	Wear goggles when using electrical equipment that sparks.
SPEAK (v), SPEAKS, SPOKE, SPOKEN	To use a voice to make words	SPEAK INTO THE MICROPHONE.	
SPECIAL (adj)	For a specified function	TIGHTEN THE SPECIAL NUT.	

Word (part of speech)	Approved meaning/ ALTERNATIVES	APPROVED EXAMPLE	Not approved example
SPECIALLY (adv)	In a special manner	THIS EXTRACTOR IS SPECIALLY MADE TO REMOVE THESE PINS.	
specific (adj)	SPECIAL (adj)	USE SPECIAL TOOLS TO PREVENT POSSIBLE DAMAGE.	Use specific tools to preclude the possibility of damage.
	SPECIFIED (adj)	THESE AREAS HAVE SPECIFIED DAMAGE LIMITS.	These areas have specific damage limits.
specifically (adv)	SPECIALLY (adv)	THESE MATERIALS ARE SPECIALLY MADE FOR PLASTIC SURFACES.	These materials are specifically designed for plastic surfaces.
SPECIFIED (adj)	Given in, identified in, or related to a specification, regulation, or procedure	INFLATE THE TIRE WITH NITROGEN TO THE SPECIFIED PRESSURE.	
speck (n)	PARTICLE (n)	REMOVE ALL PARTICLES OF DUST FROM THE LENS.	Remove all specks of dust from the lens.
speech (n)	VOICE (n)	IF YOU CANNOT HEAR THE VOICE, REPLACE THE UNIT.	If the speech is not audible, replace the unit.
SPEED (n)	The rate of movement	IF THE WIND SPEED IS MORE THAN 20 KNOTS, DO NOT OPEN THE RADOME.	
SPHERICAL (adj)	That has the shape of a sphere	THE ACCUMULATORS ARE SPHERICAL AND HAVE BLADDERS.	
SPILL (v), SPILLS, SPILLED, SPILLED	To accidentally flow, or cause to flow, out of a container	IF THE ELECTROLYTE SPILLS, REMOVE IT IMMEDIATELY.	
		IF YOU SPILL THE ELECTROLYTE, REMOVE IT IMMEDIATELY.	

Word (part of speech)	Approved meaning/ ALTERNATIVES	APPROVED EXAMPLE	Not approved example
spillage (n)	CONTAMINATION (n)	FIND THE AREA WHERE THERE IS MERCURY CONTAMINATION.	Find the area of the mercury spillage.
	SPILL (v)	IF ACID SPILLS, CLEAN THE AREA.	After acid spillage, clean the area.
spin (v)	TURN … QUICKLY	TURN THE WHEEL QUICKLY WITH YOUR HAND.	Spin the wheel by hand.
splash (v)	GET (v)	IF YOU GET FLUID IN YOUR EYES, GET MEDICAL AID.	If fluid splashes into eyes, get medical aid.
splined (adj)	SPLINE (TN)	THE DRIVE SHAFT HAS SPLINES.	The drive shaft is splined.
split (adj)	DAMAGED (adj)	REPLACE DAMAGED CASINGS.	Split casings must be replaced.
split (v)	DIVIDE (v)	DIVIDE THE ELECTRICAL LOAD BETWEEN THE TWO BUS BARS.	Split the electrical load between the two bus bars.
	BREAK (v)	MAKE SURE THAT THE CASING DOES NOT BREAK WHEN YOU APPLY THE MAXIMUM TEST PRESSURE.	Make sure that the casing does not split when you apply the maximum test pressure.
spot (n)	MARK (n)	A RED MARK ON THE PIN IDENTIFIES THE LUBRICATION POINT.	A red spot on the pin identifies the lubrication point.
	STAIN (TN)	REMOVE ALL STAINS FROM THE CARPET WITH SOAP AND WARM WATER.	Remove any spots from the carpet with soap and warm water.
spot (v)	FIND (v)	IF YOU FIND CORROSION IN THIS AREA, USE REPAIR SCHEME No. 3.	If you spot corrosion in this area, use repair scheme No. 3.
SPRAY (n)	A jet of very small drops of liquid	THE SEALANT IS RESISTANT TO SALT SPRAY.	

Word (part of speech)	Approved meaning/ ALTERNATIVES	APPROVED EXAMPLE	Not approved example
SPRAY (v) SPRAYS, SPRAYED, SPRAYED	To apply as a spray	DO NOT SPRAY DISINFECTANTS DIRECTLY ONTO THE CARPETS OR INSTRUMENT PANELS.	
spread (v)	APPLY (v)	APPLY THE COMPOUND EQUALLY.	Spread compound evenly.
spring (v)	MOVE (v)	MAKE SURE THAT THE PINS DO NOT MOVE OUT OF THE NOTCH.	Make certain that the pins do not spring out of the notch.
spurious (adj)	UNWANTED (adj)	IF THERE ARE UNWANTED INDICATIONS, DO A TEST.	In the event of spurious indications, do a test.
stability (n)	STABLE (adj)	MAKE SURE THAT THE AIRCRAFT IS STABLE BEFORE YOU LIFT IT WITH JACKS.	Before jacking up aircraft make sure that aircraft stability is ensured.
stabilization (n)	STABLE (adj)	MAKE SURE THAT THE OIL PRESSURE DECREASES WHEN THE TEMPERATURE IS STABLE.	Check that oil pressure drops after stabilization of temperature.
stabilize (v)	STABLE (adj)	LET THE V-BARS BECOME STABLE.	Allow V-bars to stabilize.
STABLE (adj)	That does not change or move	MAKE SURE THAT THE PRESSURE INDICATIONS ARE STABLE.	
stage (n)	STEP (n)	DO NOT REMOVE THE PIN DURING THIS STEP.	Do not remove the pin at this stage.
stained (adj)	DIRTY (adj)	IF THE CARPET IS DIRTY, CLEAN IT WITH MATERIAL NO. 8.	If the carpet is stained, clean it with the Material No. 8.
	STAIN (TN)	IF THERE ARE HYDRAULIC FLUID STAINS ON THE COVER, EXAMINE THE LINE COUPLINGS.	If the cover is stained by hydraulic fluid, check line couplings.

Word (part of speech)	Approved meaning/ ALTERNATIVES	APPROVED EXAMPLE	Not approved example
stamp (v)	STAMP (TN)	PUT THE TEST DATE ON THE CYLINDER NECK WITH THE APPLICABLE STAMP.	Stamp the test date on the cylinder neck.
stand (v)	STAY (v)	STAY AWAY FROM SURFACES THAT MOVE.	Do not stand near moving surfaces.
STANDARD (adj)	Related to equipment and procedures that are normally used	IN THIS PROCEDURE, USE STANDARD TOOLS.	
starboard (adj)	RIGHT (adj)	ON THE RIGHT WING, MOVE THE AILERONS UP.	On the starboard wing, move the ailerons up.
START (n)	The beginning of a movement or operation	MOVEMENT CAN BE SLOW AT THE START.	
START (v), STARTS, STARTED, STARTED	1. To begin a procedure, movement, or operation	START THE ENGINE.	
	2. To come into being, activity, or operation	MEASURE THE BEND RADIUS OF THE HOSE FROM THE POINT WHERE THE BEND STARTS.	
state (n)	CONDITION (n)	EXAMINE THE CONDITION OF THE UNIT.	Examine the state of the unit.
state (v)	TELL (v)	UNLESS THE MANUFACTURER'S INSTRUCTIONS TELL YOU DIFFERENTLY, DO THE PROCEDURE THAT FOLLOWS.	Unless otherwise stated by the manufacturer, do the following procedure.
station (v)	PUT (v)	PUT A PERSON NEAR THE FUEL TANK.	Station a person near the fuel tank.
stationary (adj)	STABLE (adj)	WHEN THE PRESSURE IS STABLE, CLOSE THE VALVE.	When the pressure is stationary, close the valve.
	STOP (v)	WHEN THE FLYWEIGHTS STOP, DO THE TEST.	When the flyweights are stationary, do the test.

Word (part of speech)	Approved meaning/ ALTERNATIVES	APPROVED EXAMPLE	Not approved example
STAY (v), STAYS, STAYED, STAYED	To continue to be in a location or condition	IF THE AIRCRAFT MUST STAY IN THE HANGAR, MAKE SURE THAT THE EXIT IS CLEAR.	
steadily (adv)	CONTINUOUSLY (adv)	IF THE TEMPERATURE INCREASES CONTINUOUSLY, STOP THE ENGINE.	If the temperature increases steadily, stop the engine.
steady (adj)	STABLE (adj)	MAKE SURE THAT THE SYSTEM PRESSURE IS STABLE.	Make sure that the system pressure is steady.
STEP (n)	A specified part of a procedure	DO STEPS 13 THRU 16 A MINIMUM OF THREE TIMES.	
stick (v)	ATTACH (v)	ATTACH THE LABEL TO THE SURFACE.	Stick the label to the surface.
	FREELY (adv)	MAKE SURE THAT THE FAN BLADES MOVE FREELY.	Check that the fan blades do not stick.
	CATCH (v)	REMOVE ALL JEWELRY THAT CAN CATCH IN THE MACHINE.	Remove all jewelry that can get stuck in the machine.
sticky (adj)	TACKY (adj)	PUT THE PATCH ON THE SURFACE WHEN THE CEMENT IS TACKY.	Put the patch on the surface when the cement is sticky.
stiff (adj)	MOVE (v)	IF THE PISTON DOES NOT MOVE EASILY IN THE SLEEVE, DO A DIMENSIONAL INSPECTION OF THE PARTS.	If the piston is stiff, check the dimensions of the piston and sleeve.
still (adv)	STAY (v)	MAKE SURE THAT THE "OVHT" LIGHT STAYS ON.	Make sure that the "OVHT" light is still illuminated.
	CONTINUE (v)	MAKE SURE THAT THE ROTOR CONTINUES TO TURN.	Make sure that the rotor still turns.

Word (part of speech)	Approved meaning/ ALTERNATIVES	APPROVED EXAMPLE	Not approved example
stimulate (v)	INCREASE (v)	INCREASE THE AIRFLOW.	Stimulate the airflow.
stir (v)	MIX (v)	MIX THE BONDING MATERIAL WITH A SPATULA.	Stir the bonding material with a spatula.
STOP (v), STOPS, STOPPED, STOPPED	1. To cause the end of a procedure, movement, or an operation	STOP THE ENGINE.	
	2. To come to an end	WHEN THE FLOW STOPS, REMOVE THE DRAIN HOSE.	
store (v)	KEEP (v)	KEEP THE CARTRIDGES IN A SAFETY AREA.	Cartridges should be stored in a safety area.
	CONTAIN (v)	THE FIRE EXTINGUISHER BOTTLES CONTAIN THE AGENT HALON 1301 IN LIQUID CONDITION.	The fire extinguisher bottles store the agent Halon 1301 in liquid condition.
	STORAGE (TN)	WHEN YOU PUT THE UNIT INTO STORAGE, OBEY THE APPLICABLE STORAGE INSTRUCTIONS.	When you store the unit, obey the applicable storage instructions.
STOW (v), STOWS, STOWED, STOWED	To move or cause to move into a specified position of storage.	STOW THE LIFE VEST SAFELY BELOW THE PASSENGER SEAT.	
		MAKE SURE THAT THE THRUST REVERSER STOWS WHEN YOU PUSH THE QUADRANT CONTROL FORWARD.	
STRAIGHT (adj) (STRAIGHTER, STRAIGHTEST)	Without curves or bends	MAKE SURE THAT THE STRAIGHT PART OF THE RIGGING PIN IS SMOOTH.	
straighten (v)	STRAIGHT (adj)	DO NOT TRY TO MAKE THE BENDS IN THE CABLE STRAIGHT.	Do not try to straighten the bends in the cable.

Word (part of speech)	Approved meaning/ ALTERNATIVES	APPROVED EXAMPLE	Not approved example
strain (v)	FILTER (TN)	PUT THE OIL THROUGH A FILTER BEFORE YOU USE IT.	Strain the oil before use.
	STRAIN (TN)	DO NOT PUT STRAIN ON THE BELLOWS.	Do not strain bellows.
strap (v)	ATTACH (v) (WITH STRAPS [TN])	USE A STRAP TO ATTACH THE CABLES TO THE STRUCTURE.	Strap the cables to the structure.
stress (v)	FORCE (TN)	DO NOT PUT TOO MUCH FORCE ON THE DUCTING.	Do not stress the ducting unnecessarily.
strictly (adv)	FULLY (adv)	FULLY OBEY THE SAFETY PRECAUTIONS.	Strictly obey the safety precautions.
strike (v)	HIT (v)	MAKE SURE THAT THE SLING DOES NOT HIT THE FUSELAGE SKIN.	Make certain that the sling does not strike the fuselage skin.
STRIP (n)	A piece of material that is long and thin, but not wide	PUT A STRIP OF ADHESIVE TAPE ALONG THE INSULATION BLANKET.	
strip (v)	REMOVE (v)	REMOVE THE INSULATION FROM THE WIRE.	Strip the insulation from the wire.
STRIPE (n)	A line on a surface which is a different color	THE FLAG HAS RED AND WHITE STRIPES.	
striped (adj)	STRIPE (n)	THE FLAG HAS RED AND WHITE STRIPES.	The flag is striped red and white.
stroke (n)	TRAVEL (n)	TO BLEED THE BRAKING SYSTEM, OPERATE THE HAND PUMP HANDLE THROUGH ITS FULL TRAVEL.	Bleed the braking system by applying full strokes to the hand pump handle.
STRONG (adj) (STRONGER, STRONGEST)	With much strength, power, or concentration	MOOR THE AIRCRAFT CAREFULLY WHEN WINDS ARE STRONG.	
STRUCTURAL (adj)	Related to the structure	STRUCTURAL REPAIRS ARE SPECIFIED IN THE STRUCTURAL REPAIR MANUAL.	

Word (part of speech)	Approved meaning/ ALTERNATIVES	APPROVED EXAMPLE	Not approved example
STRUCTURALLY (adv)	Related to the structure	THE LIST GIVES YOU STRUCTURALLY IMPORTANT ITEMS.	
STRUCTURE (n)	1. A construction	ATTACH THE WIRES TO THE STRUCTURE.	
	2. The arrangement of something	THE INTRODUCTION GIVES YOU THE STRUCTURE OF THE MAINTENANCE MANUAL.	
SUBASSEMBLY (n)	An assembly that is a part of a larger assembly	DO NOT CHANGE THE POSITION OF THE SUBASSEMBLIES.	
subject (v)	APPLY (v)	APPLY SUCTION TO THE SYSTEM.	Subject the system to suction.
submerge (v)	PUT (v)	PUT THE FILTER ELEMENT FULLY INTO THE FLUID.	Submerge the filter element in the fluid.
submit (v)	SEND (v)	SEND THE METAL PARTICLES TO A LABORATORY FOR ANALYSIS.	Submit the metal particles to a laboratory for analysis.
SUBSEQUENT (adj)	Following in an unspecified time or sequence	A SUBSEQUENT REPAIR WILL BE FOR THE NEW CONFIGURATION ONLY.	
	Use NEXT (adj) for the first occurrence that follows immediately in time or sequence.	DO THE NEXT INSPECTION.	Do the immediately subsequent inspection.
SUBSEQUENTLY (adv)	After an unspecified time	MAKE SURE THAT THE TOOL IS SUBSEQUENTLY AVAILABLE FOR THE INSTALLATION PROCEDURE.	
	Use THEN (adv) for an occurrence that follows immediately in time or sequence.	LIFT THE COVER. THEN PULL THE HANDLE.	Lift the cover and subsequently pull the handle.

Word (part of speech)	Approved meaning/ ALTERNATIVES	APPROVED EXAMPLE	Not approved example
subsequent to (prep)	AFTER (conj)	AFTER YOU INSTALL THE COMPONENT, REMOVE THE HANDLING TOOL.	Remove the handling tool subsequent to the installation of the component.
substance (n)	MATERIAL (n)	THE PRIMER CONTAINS A MATERIAL THAT PREVENTS CORROSION.	The primer contains a corrosion-inhibiting substance.
substitute (adj)	EQUIVALENT (adj)	USE AN EQUIVALENT MATERIAL IF THE SPECIFIED ONE IS NOT AVAILABLE.	A substitute material may be used if the specified one is not available.
	ALTERNATIVE (adj)	IF THIS SPECIAL TOOL IS NOT AVAILABLE, USE AN ALTERNATIVE METHOD.	If this special tool is not available, use a substitute method.
substitute (n)	ALTERNATIVE (n)	YOU CAN USE ALTERNATIVES FOR THESE ITEMS.	Equivalent substitutes may be used for these items.
substitute (v)	REPLACE (v)	DO NOT REPLACE THE SELF-LOCKING NUTS WITH PLAIN NUTS.	Do not substitute the self-locking nuts with plain nuts.
SUBTRACT (v), SUBTRACTS, SUBTRACTED, SUBTRACTED	To take from a number or quantity	SUBTRACT COEFFICIENT "A" FROM THE INDICATION.	
successful (adj)	CORRECT (adj)	IF THE TEST RESULTS ARE NOT CORRECT, REPLACE THE UNIT.	If the test is not successful, replace the unit.
	SATISFACTORY (adj)	IF THE TEST RESULTS ARE NOT SATISFACTORY, REPLACE THE UNIT.	If the test is not successful, replace the unit.
SUCH (adj)	Of the same group	WHEN YOU DO WORK ON SUCH SYSTEMS, USE GOGGLES.	
such as	FOR EXAMPLE	WHEN YOU REMOVE THE STUD, USE APPLICABLE TOOLS. FOR EXAMPLE, USE A PLASTIC MALLET AND A PUNCH.	When you remove the stud, use applicable tools such as a plastic mallet and a punch.

Word (part of speech)	Approved meaning/ ALTERNATIVES	APPROVED EXAMPLE	Not approved example
suck (v)	SUCTION (TN)	USE SUCTION TO REMOVE THE UNWANTED MATERIAL.	Suck out the unwanted material.
SUDDEN (adj)	That occurs in a short time	SUDDEN MOVEMENT CAN CAUSE DAMAGE TO THE LIMIT STOPS.	
SUDDENLY (adv)	In a sudden manner	IF THE TEMPERATURE INCREASES SUDDENLY, STOP THE ENGINE.	
SUFFICIENT (adj)	Not less (or more) than necessary	ADJUST THE CLAMP UNTIL THERE IS SUFFICIENT FRICTION TO KEEP IT IN POSITION.	
SUFFICIENTLY (adv)	Not less (or more) than necessary	WHEN THE PAINT IS SUFFICIENTLY SOFT, REMOVE IT WITH A PLASTIC SPATULA.	
suitable (adj)	APPLICABLE (adj)	PUT THE APPLICABLE SCREWDRIVER IN THE SLOT.	Put a suitable screwdriver in the slot.
	CORRECT (adj)	PUT THE CORRECT PLUGS ON THE DISCONNECTED LINES.	Put suitable plugs on the disconnected lines.
suitably (adv)	SUFFICIENTLY (adv)	WHERE THE PAINT IS SUFFICIENTLY SOFT, REMOVE IT WITH A PLASTIC SPATULA.	Where the paint is suitably soft, remove it with a plastic spatula.
SUM (n)	The result when you add	CALCULATE THE SUM OF THE LOADS FOR EACH AREA.	
SUPPLY (n)	Something that is supplied	STOP THE ELECTRICAL POWER SUPPLY.	
SUPPLY (v), SUPPLIES, SUPPLIED, SUPPLIED	To give something that is necessary	SUPPLY ELECTRICAL POWER TO THE AUXILIARY SYSTEM.	

Word (part of speech)	Approved meaning/ ALTERNATIVES	APPROVED EXAMPLE	Not approved example
support (n)	SUPPORT (TN)	PUT A SUPPORT BELOW THE ITEM BEFORE YOU DISCONNECT IT.	Make certain there is adequate support for the item before disconnecting it.
	HOLD (v)	MAKE SURE THAT A PERSON HOLDS THE COMPONENT WHILE YOU DISCONNECT IT.	Make certain that someone provides adequate support during component disconnection.
support (v)	HOLD (v)	HOLD THE SWITCH AND REMOVE THE SCREWS.	Support the switch and remove the screws.
	SUPPORT (TN)	PUT A SUPPORT BELOW THE PUMP, THEN PULL IT OFF THE DRIVE SHAFT.	Support the pump, then pull it off the drive shaft.
SURE (adj)	Certain	IF YOU ARE NOT SURE THAT A PART IS SERVICEABLE, DISCARD IT.	
SURFACE (n)	One or more of the faces of something	RUB THE SURFACE QUICKLY WITH A SOFT, DRY CLOTH.	
surplus (adj)	UNWANTED (adj)	REMOVE THE UNWANTED GREASE.	Wipe off surplus grease.
surrounding (adj)	ADJACENT (adj)	CLEAN THE MOUNTING AND THE ADJACENT AREA.	Clean the mounting and the surrounding area.
	AROUND (prep)	REMOVE ALL SEALANT THAT IS AROUND THE HOLE.	Remove all sealant surrounding the hole.
survey (v)	EXAMINE (v)	EXAMINE THE SURFACE.	Survey the surface.
	INSPECTION (n)	DO AN INSPECTION OF THE SURFACE.	Survey the surface.
suspect (adj)	THINK (v)	REMOVE THE VALVE THAT YOU THINK IS UNSERVICEABLE.	Remove the suspect valve.

Word (part of speech)	Approved meaning/ ALTERNATIVES	APPROVED EXAMPLE	Not approved example
suspect (v)	THINK (v)	IF YOU THINK THAT THE HYDRAULIC PUMP DOES NOT OPERATE CORRECTLY, DO A SYSTEM CHECK.	If you suspect that the hydraulic pump does not operate correctly, do a system check.
suspend (v)	HANG (v)	HANG THE ENGINE IN A FIREPROOF BAY.	Suspend the engine in a fireproof bay.
	STOP (v)	IF THE CABLE CATCHES, STOP THE PROCEDURE.	If the cable catches, suspend all operations.
SWALLOW (v), SWALLOWS, SWALLOWED, SWALLOWED	To take through the mouth and esophagus into the stomach Use this word for safety instructions only.	IF YOU SWALLOW NITRIC ACID, GET MEDICAL AID IMMEDIATELY.	
swap (v)	INTERCHANGE (v)	INTERCHANGE THE COMPUTERS 1 AND 2. THEN DO THE TEST AGAIN.	Swap computers 1 and 2. Then do the test again.
sweep (v)	SCAN (TN)	MAKE A SCAN OF THE FREQUENCY BAND.	Sweep the frequency band.
swing (v)	MOVE (v)	MOVE THE SHOCK ABSORBERS INTO THEIR CORRECT POSITIONS.	Swing the shock absorbers into their correct positions.
	CALIBRATE (v)	YOU CAN CALIBRATE ALL THE COMPASS SYSTEMS AT THE SAME TIME.	All compass systems may be swung simultaneously.
switch (v)	SET (v)	SET THE SELECTOR TO "SYSTEM DISPLAY".	Switch the selector to "SYSTEM DISPLAY".
switch off (v)	STOP (v)	STOP THE UNIT AFTER 5 MIN.	Switch off the unit after 5 min.
	SWITCH (TN)	SET THE NAV 1 SWITCH TO "OFF".	Switch off NAV 1.

Word (part of speech)	Approved meaning/ ALTERNATIVES	APPROVED EXAMPLE	Not approved example
switch on (v)	START (v)	START THE UNIT.	Switch on the unit.
	SWITCH (TN)	SET THE NAV 1 SWITCH TO "ON".	Switch on NAV 1.
SYMBOL (n)	A written sign that identifies a quantity, operation, item, or condition	THE SPECIFICATION GIVES THE SYMBOLS FOR RELAYS, DIODES, AND OTHER PARTS OF THE SYSTEM.	
SYMMETRICAL (adj)	That has or is with symmetry	MAKE SURE THAT THE POSITION OF THE WING FLAPS IS SYMMETRICAL.	
SYMMETRICALLY (adv)	In a symmetrical manner	THE ELEVATORS MUST MOVE SYMMETRICALLY.	
SYMPTOM (n)	A sign of a condition	IF THE SYMPTOMS OCCUR AGAIN, REMOVE THE UNIT.	
synchronize (v)	ADJUST (v)	ADJUST COMPASS SYSTEM 1 UNTIL IT AGREES WITH COMPASS SYSTEM 2.	Synchronize compass system 1 with compass system 2.
SYNCHRONIZED (adj)	1. That operates at the same time	MAKE SURE THAT THE TWO ROTORS ARE SYNCHRONIZED.	
	2. That agrees with	STOP THE TEST WHEN THE TIMERS ARE SYNCHRONIZED.	
SYSTEM (n)	An assembly of related parts to do a specified operation	DO NOT FILL THE SYSTEM DIRECTLY FROM A HIGH-PRESSURE CYLINDER.	

Blank Page

Word (part of speech)	Approved meaning/ ALTERNATIVES	APPROVED EXAMPLE	Not approved example
TACKY (adj) No other forms of this adjective	Sticky when you touch it	WHEN THE COMPOUND IS TACKY, APPLY THE PATCH.	
TAG (v), TAGS, TAGGED, TAGGED	To put a tag on	OPEN, TAG, AND SAFETY THE CIRCUIT BREAKERS.	
take (v)	REMOVE (v)	REMOVE THE FILTER FROM THE SOLVENT.	Take the filter from the solvent.
	READ (v)	LET THE DIAL BECOME STABLE BEFORE YOU READ THE INDICATIONS.	Let the dial become stable before taking readings.
	GET (v)	GET A SAMPLE OF HYDRAULIC FLUID.	Take a sample of hydraulic fluid.
	NECESSARY (adj)	TWO HOURS ARE NECESSARY TO DO THE PROCEDURE.	The procedure takes 2 hours to complete.
take care (v)	MAKE SURE (v)	DURING THIS STEP, MAKE SURE THAT YOU DO NOT CAUSE DAMAGE TO THE SURFACE.	During this step, take care not to cause damage to the surface.
	OBEY (v)	OBEY THE SAFETY PRECAUTIONS WHEN YOU DO WORK WITH HIGH VOLTAGES.	You must take care when you work with high voltages.
	DO NOT (v)	DO NOT INTERCHANGE THE TWO UNITS.	Take care not to interchange the two units.
	CAREFUL (adj)	BE CAREFUL WHEN YOU INSTALL THE ELECTRONIC BOARD.	Take care when installing the electronic board.
talk (v)	SPEAK (v)	SPEAK INTO THE MICROPHONE.	Talk into the microphone.
TAP (v), TAPS, TAPPED, TAPPED	To hit lightly and quickly	TAP THE BOLT WITH A MALLET.	

Word (part of speech)	Approved meaning/ ALTERNATIVES	APPROVED EXAMPLE	Not approved example
tape (v)	TAPE (TN)	PUT THE FOIL IN POSITION AND ATTACH IT WITH ADHESIVE TAPE.	Tape the foil in position.
taped (adj)	RECORD (v)	MAKE SURE THAT YOU CAN HEAR THE RECORDED ANNOUNCEMENTS.	Make sure that you can hear the taped announcements.
TASK (n)	Assigned work or procedure	EACH STEP OF THE TASK HAS AN IDENTIFICATION LETTER.	
taut (adj)	TENSION (TN)	MAKE SURE THAT THE TENSION IN THE CABLES IS CORRECT.	Make sure the cables are taut.
tear (v)	DAMAGE (n)	BE CAREFUL THAT YOU DO NOT CAUSE DAMAGE TO THE BELLOWS.	Take care you do not tear the bellows.
technique (n)	METHOD (n)	THIS METHOD GIVES THE BEST RESULTS.	This technique gives the best results.
TELESCOPIC (adj)	Related to items that retract into each other	INSTALL THE TELESCOPIC DUCTS IN THE WING DE-ICING SYSTEM.	
TELL (v), TELLS, TOLD, TOLD	To give an order or supply information	PUT A WARNING NOTICE IN THE COCKPIT TO TELL PERSONS NOT TO OPERATE THE FLIGHT CONTROLS.	
		THE COMPUTER TELLS THE FUEL CONTROLLER TO OPEN THE FUEL CONTROL VALVE.	
TEMPORARILY (adv)	In a temporary manner	ATTACH THE COVER TEMPORARILY.	
TEMPORARY (adj)	For a short time only, not permanent	THIS INSTALLATION IS ONLY TEMPORARY.	
tension (v)	TENSION (TN)	ADJUST THE TENSION IN THE CABLES TO THE CORRECT VALUE.	Tension cables to the correct value.

Word (part of speech)	Approved meaning/ ALTERNATIVES	APPROVED EXAMPLE	Not approved example
terminate (v)	STOP (v)	STOP THE TEST AFTER 2 SECONDS.	Terminate the test after 2 seconds.
TERTIARY (adj)	Third in importance	THERE ARE THREE CLASSES OF STRUCTURE: PRIMARY, SECONDARY, AND TERTIARY.	
TEST (n)	The procedure where an object or system is operated to make sure that its performance and/or function is correct	DISCONNECT ALL THE SYSTEMS WHICH ARE NOT NECESSARY FOR THE TEST.	
test (v)	TEST (n)	DO A FUNCTIONAL TEST OF THE WARNING SYSTEM.	Functionally test warning system.
THAN (conj)	Function word that you use with comparative adjectives or adverbs	MAKE SURE THAT THE TOTAL ANGLE IS SMALLER THAN 20 DEG.	
		CORROSION CAN OCCUR MORE EASILY IN THIS AREA THAN IN OTHER AREAS.	
THAT (conj)	Function word that starts a subordinate clause	MAKE SURE THAT THE SURFACE IS CLEAN.	
THAT (pron)	Function word that shows the person or thing referred to	EXAMINE THE UNIT TO SEE IF THAT IS THE PROBLEM.	
THE (art)	Function word: definite article	REMOVE THE SEVEN FLEXIBLE HOSES.	
THEIR (adj)	Belonging to persons or things	WRITE TO THE MANUFACTURERS FOR THEIR REPAIR SCHEMES.	
THEM (pron)	Function word that shows the persons or things referred to	IF THE CANOPIES ARE DIRTY, CLEAN THEM.	
THEN (adv)	Immediately after in time or sequence	LIFT THE COVER. THEN PULL THE HANDLE.	

Word (part of speech)	Approved meaning/ ALTERNATIVES	APPROVED EXAMPLE	Not approved example
THERE (adv)	In that position	IF THE LABEL IS NOT THERE, ATTACH A NEW ONE.	
THERE (pron)	Function word that starts a sentence or a subordinate clause	THERE ARE THREE JACKING POINTS ON THE FUSELAGE.	
		MAKE SURE THAT THERE ARE NO LOOSE OBJECTS.	
therefore (adv)	THUS (adv)	THIS TORQUE SHAFT CANNOT MOVE INBOARD. THUS, BEFORE YOU REMOVE IT, REMOVE THE OUTBOARD TORQUE SHAFT.	This torque shaft cannot move inboard. Therefore, before you remove it, remove the outboard torque shaft.
	AS A RESULT	THIS TORQUE SHAFT CANNOT MOVE INBOARD. AS A RESULT, BEFORE YOU REMOVE IT, REMOVE THE OUTBOARD TORQUE SHAFT.	This torque shaft cannot move inboard. Therefore, before you remove it, remove the outboard torque shaft.
THESE (adj)	Refers to specified persons or things	REMOVE THESE PARTS.	
THESE (pron)	Function word that shows the persons or things referred to	THESE ARE SPECIAL TOOLS.	
THEY (pron)	Function word that shows the persons or things referred to	DENTS ARE NOT PERMITTED IF THEY PREVENT CORRECT OPERATION OF THE PARTS.	
THICK (adj) (THICKER, THICKEST)	That has a large thickness	APPLY A THICK LAYER OF GREASE ON THE BARE METAL.	
THICKNESS (n)	The smallest of the three dimensions of an object	THE THICKNESS OF PRIMER LAYER MUST BE BETWEEN 0.05 mm AND 0.06 mm.	

Word (part of speech)	Approved meaning/ ALTERNATIVES	APPROVED EXAMPLE	Not approved example
THIN (adj) (THINNER, THINNEST)	That has a small thickness	APPLY A THIN LAYER OF SEALING COMPOUND ON THE BARE METAL.	
thin (v)	LIQUID (adj)	ADD SOLVENT TO MAKE THE PAINT MORE LIQUID.	Add solvent to thin the paint.
THINK (v), THINKS, THOUGHT, THOUGHT	To have an opinion	IF YOU THINK THAT THE WATER WILL FREEZE, ADD ANTIFREEZE.	
THIS (adj)	Refers to a specified person or thing	DO THIS TEST AT FULL VOLTAGE.	
THIS (pron)	Function word that shows the person or thing referred to	REFER TO TESTING AND FAULT ISOLATION. THIS TELLS YOU THE PARTS TO REPLACE.	
thorough (adj)	FULL (adj)	DO A FULL CHECK OF THE HYDRAULIC SYSTEM.	Do a thorough check of the hydraulic system.
thoroughly (adv)	FULLY (adv)	DRY THE SURFACE FULLY.	Dry the surface thoroughly.
THOSE (pron)	Function word that shows the persons or things referred to	WHEN YOU EXAMINE THE PARTS, DISCARD THOSE THAT ARE DAMAGED.	
thread (v)	PUT (v)	PUT TAPE ON THE ROLLER.	Thread tape on the roller.
	TURN (v)	TURN THE LOCK NUT ON THE FITTING.	Thread the lock nut on the fitting.
threaded (adj)	THREAD (TN)	PUT THE END OF THE ROD THAT HAS A THREAD IN THE FITTING.	Put the threaded end of the rod in the fitting.
THROUGH (adv)	Function word that shows movement into one end or side and out of the other	IF FUMES COME THROUGH, DO THE EMERGENCY PROCEDURE.	
THROUGH (prep)	Function word that shows movement into one end or side and out of the other	MAKE SURE THAT THE FLUID THAT GOES THROUGH THE TUBE HAS NO BUBBLES.	

Word (part of speech)	Approved meaning/ ALTERNATIVES	APPROVED EXAMPLE	Not approved example
throughout (prep)	DURING (prep)	MAKE SURE THAT THE DOORS ARE CLOSED DURING THE PRESSURIZATION TEST.	Make sure that the doors are closed throughout the pressurization test.
THRU (prep)	From a point in a sequence to another	DO STEPS 4 THRU 10 AGAIN.	
THUS (adv)	For that reason	THIS BOILER CONTROLLER IS APPLICABLE TO ALL SYSTEMS. THUS, IT CAN ALSO CONTROL THERMAL OIL HEATERS.	
	For other meanings, use:		
	FOLLOW (v)	CALCULATE THE THICKNESS OF THE SHIM AS FOLLOWS: T=X-Y.	The thickness of the shim is thus calculated: T=X-Y.
tie (v)	ATTACH (v)	ATTACH THE THREADING CORD TO THE NEW CABLE.	Tie the threading cord to the new cable.
TIGHT (adj) (TIGHTER, TIGHTEST)	Not free	MAKE SURE THAT THE NUTS ARE TIGHT.	
	For other meanings, use:		
	SEAL (v)	MAKE SURE THAT THE FUEL TANK IS SEALED.	Make sure that the fuel tank is tight.
	TENSION (TN)	MAKE SURE THAT THERE IS TENSION IN THE LOCKWIRE.	Make sure that the lockwire is tight.
TIGHTEN (v), TIGHTENS, TIGHTENED, TIGHTENED	To cause to be tight	TIGHTEN THE SCREWS.	
TIGHTLY (adv)	In a tight manner	HOLD THE CYLINDER TIGHTLY.	
till (conj)	UNTIL (conj)	MOVE THE LEVER UNTIL IT TOUCHES THE STOP.	Move the lever till it hits the stop.

Word (part of speech)	Approved meaning/ ALTERNATIVES	APPROVED EXAMPLE	Not approved example
TILT (v), TILTS, TILTED, TILTED	To move at an angle or to put something at an angle between the vertical and the horizontal	TO ALIGN THE BOLT, TILT IT TO THE LEFT SIDE.	
		IF THE SHAFT TILTS, STOP THE TEST PROCEDURE.	
TIME (n)	1. A duration that you can measure	THE TIME BETWEEN CLICKS MUST BE 20 SECONDS.	
	2. An occurrence	DO STEP 2 THREE TIMES. DO NOT TIGHTEN THE NUTS AT THIS TIME.	
time (v)	TIME (n)	MEASURE THE TIME THAT IS NECESSARY FOR THE PRESSURE TO DECREASE.	Time the pressure decay.
tip (n)	POINT (n)	BEFORE YOU MAKE THE HOLE, MAKE SURE THAT THE POINT OF THE DRILL BIT IS SHARP.	Before you make the hole, make sure the tip of the drill bit is sharp.
tip (v)	TILT (v)	TO INSTALL THE INDICATOR, TILT IT REARWARD.	To install the indicator, tip it rearward.
TO (prep)	Function word that shows: - The direction of, point of arrival or time, connection, result, cause - That the word that follows is an infinitive	SET THE SWITCH TO "OFF".	
		THIS VALUE IS EQUIVALENT TO THE MAXIMUM.	
		ATTACH THE UNIONS TO THE VALVE.	
		GET ACCESS TO THE SELECTOR VALVES.	
		TRY TO LIFT THE COVER MANUALLY.	

Word (part of speech)	Approved meaning/ ALTERNATIVES	APPROVED EXAMPLE	Not approved example
TOGETHER (adv)	In one group or position	USE TAPE TO ATTACH THE CABLES TOGETHER.	
	For other meanings, use:		
	AT THE SAME TIME	DISASSEMBLE THE BYPASS VALVE AND THE DUMP VALVE AT THE SAME TIME.	Disassemble the bypass valve and the dump valve together.
TOLERANCE (n)	A permitted difference from a standard	THE TOLERANCE IS PLUS OR MINUS 0.2 mm.	
TOO (adv)	More than is necessary or correct	MAKE SURE THAT THE TEMPERATURE IS NOT TOO HIGH.	
TOOL (n)	An object used to make or do something	REMOVE ALL TOOLS AND EQUIPMENT.	
TOP (adj)	That is uppermost	A NUMBER IN THE TOP LEFT CORNER OF THE DISPLAY REFERS TO A RELATED PAGE.	
TOP (n)	The uppermost position	THE VERTICAL STABILIZER IS INSTALLED ON THE TOP OF THE FUSELAGE TAIL SECTION.	
top off (v)	FILL (v)	FILL THE HYDRAULIC SYSTEM AS NECESSARY.	Top off the hydraulic system as necessary.
	ADD (v)	ADD FLUID TO THE HYDRAULIC RESERVOIR AS NECESSARY.	Top off the hydraulic reservoir as necessary.
TORQUE (v), TORQUES, TORQUED, TORQUED	To tighten to a specified torque	TORQUE THE PLUG TO 6 Nm.	
TOTAL (adj)	All of	THE TOTAL QUANTITY OF FUEL IN THE CENTER TANK IS 5000 LB.	

Word (part of speech)	Approved meaning/ ALTERNATIVES	APPROVED EXAMPLE	Not approved example
TOTAL (n)	The full quantity	ADD THE VALUES. THE TOTAL MUST NOT BE MORE THAN 15 mm.	
TOUCH (v), TOUCHES, TOUCHED, TOUCHED	To be in contact	MAKE SURE THAT THE LOCK NUT TOUCHES THE SPRING.	
TOW (v), TOWS, TOWED, TOWED	To pull something along	MAKE SURE THAT YOU COMPLETE THE PROCEDURE BEFORE YOU TOW THE VEHICLE BACK TO ITS INITIAL POSITION.	
toward (prep)	TO (prep)	THE ARROW MUST POINT TO THE REAR.	The arrow must point toward the rear.
	DIRECTION (n)	MANUALLY PUSH THE RAM AIR TURBINE IN THE DIRECTION OF THE RETRACTED POSITION.	Manually push the Ram Air Turbine toward the retracted position.
toxic (adj)	POISONOUS (adj)	THIS COMPOUND IS POISONOUS.	This compound is toxic.
trace (n)	SIGN (n)	EXAMINE THE AREA AROUND THE DRAIN HOLES FOR SIGNS OF CORROSION.	Check the area surrounding the drain holes for traces of corrosion.
trace (v)	FIND (v)	FIND THE BROKEN WIRE.	Trace the broken wire.
track (v)	MONITOR (v)	MONITOR THE TEMPERATURE.	Track the temperature.
trained (adj)	APPROVED (adj)	ONLY APPROVED PERSONNEL CAN OPERATE THE BRAKES.	Only trained personnel can operate the brakes.

Word (part of speech)	Approved meaning/ ALTERNATIVES	APPROVED EXAMPLE	Not approved example
transfer (n)	MOVEMENT (n)	MAKE SURE THAT THE RATE OF MOVEMENT OF FUEL FROM THE WING TANKS TO THE CENTER TANK IS EQUAL.	Make sure the rate of fuel transfer from the wing tanks to the center tank is equal.
	SUPPLY (n)	MAKE SURE THAT THE RATE OF SUPPLY OF FUEL FROM THE WING TANKS TO THE CENTER TANK IS EQUAL.	Make sure the rate of fuel transfer from the wing tanks to the center tank is equal.
transfer (v)	MOVE (v)	MOVE THE COUPLINGS TO THE NEW UNIT.	The couplings must be transferred to the new unit.
	INSTALL (v)	INSTALL THE COUPLINGS ON THE NEW UNIT.	Transfer the couplings to the new unit.
transition (n)	CHANGE (n)	THE CHANGE FROM FULL-RATE OPERATION TO HALF-RATE OPERATION MUST OCCUR IMMEDIATELY.	The transition from full-rate operation to half-rate operation must occur immediately.
TRANSMIT (v), TRANSMITS, TRANSMITTED, TRANSMITTED	To send energy or a signal	MAKE SURE THAT THE TRANSPONDER TRANSMITS THE SIGNAL CORRECTLY.	
TRANSPARENT (adj)	Easy to see through	PUT A TRANSPARENT SHEET OF PLASTIC ON THE SURFACE.	
transport (v)	SEND (v)	SEND THE PART TO THE OVERHAUL SHOP.	Transport the part to the overhaul shop.
trapped (adj)	CAUGHT (adj)	MAKE SURE THAT THE CABLE IS NOT CAUGHT.	Make sure that the cable is not trapped.
TRAVEL (n)	The movement of an item	MAKE SURE THAT THE RANGE OF TRAVEL OF THE CONTROL SURFACES IS CLEAR.	

Word (part of speech)	Approved meaning/ ALTERNATIVES	APPROVED EXAMPLE	Not approved example
travel (v)	GO (v)	THE TURNBUCKLE GOES THROUGH A HOLE IN THE BULKHEAD.	The turnbuckle travels through a hole in the bulkhead.
	MOVE (v)	THE LEVER MOVES THROUGH A 30-DEG. ARC.	The lever travels through a 30-deg. arc.
treat (v)	APPLY (v)	APPLY WAX TO THE SURFACE.	Treat the surface with wax.
	TOUCH (v)	BE CAREFUL WHEN YOU TOUCH PRINTED CIRCUIT BOARDS.	Be careful when you treat printed circuit boards.
trigger (v)	CAUSE (v)	HEAT CAUSES THE PRESSURE TO INCREASE.	Heat triggers a pressure increase.
	START (v)	THIS SIGNAL STARTS THE OPERATION OF THE TEST SET.	This signal triggers test set operation.
trim (v)	CUT (v)	CUT THE PATCH TO THE CORRECT DIMENSION.	Trim the patch to fit.
trip (v)	OPEN (v)	OPEN THE CIRCUIT BREAKER.	Trip the circuit breaker.
	RELEASE (v)	THE HOLD-ON SWITCH RELEASES.	Hold-on switch trips.
	CAUSE TO FALL	AN IRREGULAR SURFACE CAN CAUSE YOU TO FALL.	You can trip and fall on an irregular surface.
triple (adj)	THREE (TN)	APPLY THREE LAYERS OF PRIMER.	Apply a triple coat of primer.
troubleshoot (v)	TROUBLESHOOTING (TN)	DO THE TROUBLESHOOTING PROCEDURE ON THE VHF NAVIGATION SYSTEM.	Troubleshoot the VHF navigation system.
true (adj)	CORRECT (adj)	MAKE SURE THAT THE VALUE SHOWN ON THE INDICATOR IS CORRECT.	Make sure that the indicator reading is true.
	AGREE (v)	MAKE SURE THAT THE INDICATION ON THE GAUGE AGREES WITH THE QUANTITY THAT IS IN THE TANK.	Make sure that the gauge shows the true quantity that is in the tank.

Word (part of speech)	Approved meaning/ ALTERNATIVES	APPROVED EXAMPLE	Not approved example
TRY (v), TRIES, TRIED, TRIED	To make an effort to do something	TRY TO MOVE THE CONTROL COLUMN.	
tuck (v)	PUT (v)	PUT THE HEADSTRAP INTO THE FACE PIECE.	Tuck the headstrap into the face piece.
TUNE (v), TUNES, TUNED, TUNED	To adjust equipment to the best performance	TUNE ADF 1 TO A KNOWN FREQUENCY.	
TURN (n)	One full cycle of movement around an axis	LET THE PRESSURE BECOME STABLE AFTER EACH TURN OF THE ADJUSTMENT SCREW.	
TURN (v), TURNS, TURNED, TURNED	To move or cause to move around an axis or a point	TURN THE CONTROL CLOCKWISE.	
		THE BEARING TURNS AT 1500 RPM.	
twice (adv)	TWO (TN)	DO THIS PROCEDURE TWO TIMES.	Do this procedure twice.
TWIST (v), TWISTS, TWISTED, TWISTED	1. To use a force that turns something and causes a distortion	DO NOT TWIST THE CABLES.	
	2. To turn or change shape as a result of torsion	IF THE CABLE TWISTS, DISCONNECT THE TWO CONNECTORS.	
TYPE (n)	A specified group	FIND THE TYPE AND DIMENSIONS OF THE DAMAGE.	
TYPICAL (adj)	That has the important qualities of a group	THIS INSTALLATION PROCEDURE IS TYPICAL FOR THIS TYPE OF FASTENER.	

Word (part of speech)	Approved meaning/ ALTERNATIVES	APPROVED EXAMPLE	Not approved example
unable (adj)	CANNOT (v)	IF YOU CANNOT TURN THE PULLEY, MAKE SURE THAT THE PIN IS REMOVED.	If you are unable to turn the pulley, make sure that the pin is removed.
unauthorized (adj)	NOT APPROVED	IF YOU ARE NOT APPROVED TO DO THIS WORK, DO NOT DO THIS ENGINE TEST.	If you are unauthorized, do not do this engine test.
uncap (v)	REMOVE (v) (A CAP [TN] OR CAPS [TN])	REMOVE THE CAPS FROM THE HOSES.	Uncap the hoses.
unclip (v)	REMOVE (v) (FROM A CLIP [TN] OR CLIPS [TN])	REMOVE THE VISOR FROM THE CLIP.	Unclip the visor.
uncoil (v)	UNWIND (v)	UNWIND THE CABLES CAREFULLY.	Uncoil the cables carefully.
uncontaminated (adj)	CLEAN (adj)	MAKE SURE THAT THE HYDRAULIC FLUID IS CLEAN.	Make sure that the hydraulic fluid is uncontaminated.
	CONTAMINATION (n)	MAKE SURE THAT THERE IS NO CONTAMINATION IN THE FUEL TANKS.	Make sure that the fuel tanks are uncontaminated.
uncouple (v)	DISCONNECT (v)	DISCONNECT THE TOW BAR FROM THE VEHICLE.	Uncouple the tow bar from the vehicle.
uncovered (adj)	COVER (TN)	DO NOT PUT A COVER ON THE CONTAINER.	Leave the container uncovered.
undamaged (adj)	NOT DAMAGED	MAKE SURE THAT THE SKIN IS NOT DAMAGED.	Make sure that the skin is undamaged.
UNDEMANDED (adj)	That occurs without an apparent cause	IF YOU GET AN UNDEMANDED MOVEMENT, DO A TEST OF THE SYSTEM.	

Word (part of speech)	Approved meaning/ ALTERNATIVES	APPROVED EXAMPLE	Not approved example
under (prep)	BELOW (prep)	INSTALL THE CABLE THROUGH THE GUIDE TUBE BELOW THE CABIN FLOOR.	Install the cable through the guide tube under the cabin floor.
	IN (prep)	THIS CAN OCCUR IN DIFFERENT CONDITIONS.	This can occur under different conditions.
	LESS THAN	MAKE SURE THAT THE PRESSURE IS LESS THAN 30 PSI.	Make sure the pressure is under 30 psi.
underneath (prep)	BELOW (prep)	PUT THE CONTAINER BELOW THE DRAIN VALVE.	Place the container underneath the drain valve.
underside (n)	BOTTOM (n)	EXAMINE THE FOUR ATTACHING HOLES ON THE BOTTOM.	Examine the four attaching holes on the underside.
undertake (v)	DO (v)	DO THIS TASK IN AN AREA THAT HAS GOOD AIRFLOW.	Undertake this task in an area that has good airflow.
	Use a specific action verb.	DO NOT START THE TEST WITHOUT PRECAUTIONS.	Do not undertake the test without precautions.
undo (v)	LOOSEN (v)	LOOSEN THE BOLTS THAT ATTACH THE UNIT.	Undo the bolts that attach the unit.
undue (adj)	UNWANTED (adj)	TO PREVENT UNWANTED FRICTION, POLISH THE SURFACE.	To prevent undue friction, polish the surface.
uneven (adj)	EQUAL (adj)	MAKE SURE THAT THE FUEL LOAD IS EQUAL ON EACH SIDE OF THE AIRCRAFT CENTERLINE.	Avoid uneven fuel load about the aircraft centerline.
	ROUGH (adj)	REPLACE THE SLIDE VALVE IF ITS SURFACE IS ROUGH.	Replace the slide valve if its surface is uneven.

Word (part of speech)	Approved meaning/ ALTERNATIVES	APPROVED EXAMPLE	Not approved example
UNFOLD (v), UNFOLDS, UNFOLDED, UNFOLDED	To open or to straighten or to cause to open or to straighten from a folded position or condition	UNFOLD THE SHEETING CAREFULLY.	
		MAKE SURE THAT THE INDICATOR LIGHT COMES ON WHEN THE BLADES UNFOLD.	
uniform (adj)	CONSTANT (adj)	CURE THE SEALANT AT A CONSTANT TEMPERATURE.	Cure the sealant at a uniform temperature.
	EQUAL (adj)	THE FUEL LOAD MUST BE EQUAL ON EACH SIDE OF THE AIRCRAFT CENTERLINE.	The fuel load must be uniform on each side of the aircraft centerline.
uniformly (adv)	EQUALLY (adv)	MAKE SURE THAT THE TWO PISTONS EXTEND EQUALLY.	Make sure that the two pistons extend uniformly.
unintentional (adj)	ACCIDENTAL (adj)	A GUARD ON THE CONTROL SWITCH OF THE PUMP PREVENTS ACCIDENTAL OPERATION.	A guard on the control switch of the pump prevents unintentional operation.
unintentionally (adv)	ACCIDENTALLY (adv)	IF YOU ACCIDENTALLY SPRAY RAIN REPELLENT FLUID ONTO A DRY WINDSHIELD, REMOVE THE FLUID WITH SOLVENT.	If rain repellent fluid is unintentionally sprayed onto a dry windshield, remove the fluid with solvent.
UNIT (n)	1. Equipment that does a function	YOU CAN REPLACE A LARGE NUMBER OF UNITS DURING ENGINE INSTALLATION.	
	2. A quantity that is a standard of measurement	THIS MANUAL USES SI UNITS.	
UNKNOWN (adj)	Not known	DO NOT USE FUEL WITH UNKNOWN PROPERTIES.	

Word (part of speech)	Approved meaning/ ALTERNATIVES	APPROVED EXAMPLE	Not approved example
unlatch (v)	RELEASE (v)	THE SWITCH MUST RELEASE.	Switch must unlatch.
	OPEN (v)	THE SWITCH MUST OPEN.	Switch must unlatch.
UNLESS (conj)	Except if	DO NOT REMOVE THE ADAPTER FROM THE STABILIZER UNLESS A REPAIR IS NECESSARY.	
unload (v)	RELEASE (v)	RELEASE THE SPRING TENSION.	Unload the spring tension.
	LOAD (TN)	REMOVE THE LOAD FROM THE SPRING BALANCE.	Unload the spring balance.
UNLOCK (v), UNLOCKS, UNLOCKED, UNLOCKED	To release or become released from a locked condition	MAKE SURE THAT THE CABIN DIFFERENTIAL PRESSURE IS ZERO BEFORE YOU UNLOCK THE DOOR.	
		IF THE LEVER UNLOCKS DURING THE TEST, SET THE LEVER TO THE LOCKED POSITION AGAIN.	
unnecessary (adj)	NOT NECESSARY	REMOVE THE GROUND EQUIPMENT THAT IS NOT NECESSARY.	Remove the unnecessary ground equipment.
unobstructed (adj)	CLOGGED (adj)	MAKE SURE THAT THE DRAIN OPENINGS ARE NOT CLOGGED.	Make sure that the drain openings are unobstructed.
	CLEAR (adj)	AN ATTENDANT MUST HAVE A CLEAR VIEW OF THE PERSON IN THE TANK.	An attendant must have an unobstructed view of the person in the tank.
unpack (v)	REMOVE (v)	REMOVE THE PARTS FROM THE SHIPPING CONTAINER.	Unpack the shipping container.
unpainted (adj)	BARE (adj)	APPLY PROTECTIVE TREATMENT ON THE BARE SURFACES.	Protect the unpainted surfaces.

Word (part of speech)	Approved meaning/ ALTERNATIVES	APPROVED EXAMPLE	Not approved example
unplug (v)	DISCONNECT (v)	DISCONNECT THE SOLDERING IRON.	Unplug the soldering iron.
	PLUG (TN)	REMOVE THE PLUG FROM THE FUEL LINE.	Unplug the fuel line.
unprotected (adj)	PROTECTION (n)	WHEN YOU REMOVE THE COVER, THE CONTACTS HAVE NO PROTECTION.	When you remove the cover, the contacts are unprotected.
unreel (v)	UNWIND (v)	UNWIND THE HOSE.	Unreel the hose.
unrestricted (adj)	FREE (adj)	THE BUSHINGS MUST BE FREE TO TURN.	Bushing rotation must be unrestricted.
unsafe (adj)	DANGEROUS (adj)	IN THESE CONDITIONS, IT IS DANGEROUS TO REMOVE THE CARTRIDGES FROM THE FIRE EXTINGUISHER BOTTLE.	In these conditions, it is unsafe to remove the cartridges from the fire extinguisher bottle.
UNSATISFAC-TORILY (adv)	In an unsatisfactory manner	IF THE ELEVATORS OPERATE UNSATISFACTORILY, MOVE THE CONTROL COLUMN.	
UNSATISFAC-TORY (adj)	Not satisfactory	IF THE CONDITION OF THE FILTERS IS UNSATISFACTORY, REPLACE THEM.	
unscrew (v)	REMOVE (v)	REMOVE THE SPACERS.	Unscrew the spacers.
	LOOSEN (v)	LOOSEN THE CAPTIVE BOLTS.	Unscrew the captive bolts.
	TURN (v)	TURN THE END FITTING TO ALIGN THE RIGGING HOLES.	Unscrew the end fitting to align the rigging holes.
UNSERVICEABLE (adj)	Not fit for its purpose	REPLACE THE UNSERVICEABLE BUSHINGS.	
unstable (adj)	NOT STABLE	IF THE SURFACE IS NOT STABLE, AN ACCIDENT CAN OCCUR.	If the surface is unstable, an accident can occur.

Word (part of speech)	Approved meaning/ ALTERNATIVES	APPROVED EXAMPLE	Not approved example
UNTIL (conj)	Up to the time that	MOVE THE LEVER UNTIL THE END OF THE LINK TOUCHES THE LOCK BAR.	
UNTIL (prep)	Up to the time of	KEEP THE FIRE BOTTLE ARMED UNTIL THE END OF THE TEST.	
unused (adj)	NEW (adj)	FILL THE TANK WITH NEW OIL.	Fill the tank with unused oil.
UNUSUAL (adj)	Not usual	UNUSUAL LOADS CAN CAUSE TREAD SEPARATION IN THE TIRES.	
UNUSUALLY (adv)	In an unusual manner	THE VALVE OPERATES WHEN THERE IS AN UNUSUALLY LARGE FLOW OF FLUID.	
UNWANTED (adj)	That must not be there	UNWANTED AIR GOES OUT THROUGH THE OVERFLOW PORT.	
		IF THE DRAIN HOLE IS BLOCKED, REMOVE THE UNWANTED MATERIAL.	
UNWIND (v), UNWINDS, UNWOUND, UNWOUND	To remove or become removed from around an object	UNWIND THE CABLE.	
		IF THE HOSE UNWINDS TOO QUICKLY FROM THE REEL, STOP THE TEST.	
UP (adj)	In a position above	MAKE SURE THAT THE ANCHOR IS UP.	
UP (adv)	To a position above	IF THE POINTER MOVES UP, DECREASE THE POWER.	
UP (prep)	To a position above	LET THE SLEEVE MOVE UP THE GUIDE TUBE.	

Word (part of speech)	Approved meaning/ ALTERNATIVES	APPROVED EXAMPLE	Not approved example
upon (prep)	ON (prep)	PUT THE OHMMETER ON A FLAT SURFACE.	Put the ohmmeter upon a flat surface.
	WHEN (conj)	WHEN YOU SET THE LEVER TO "UP", THE HYDRAULIC PRESSURE MUST DECREASE.	Upon UP selection, the hydraulic pressure must decrease.
upper (adj)	MAXIMUM (adj)	FILL THE CONTAINER TO THE MAXIMUM LEVEL.	Fill the container to the upper level.
	TOP (adj)	THE AIRFLOW THEN GOES ALONG THE TOP AND BOTTOM SURFACES OF THE SLATS.	The airflow then passes over the upper and lower surfaces of the slats.
uppermost (adj)	UP (adv)	INSTALL THE BOLTS WITH THEIR HEADS UP.	Install the bolts with their heads in uppermost position.
uppermost (adv)	UP (adv)	INSTALL THE BOLTS WITH THEIR HEADS UP.	Install the bolts with their heads uppermost.
UPSTREAM (adj)	In a direction opposite to the flow	DISCONNECT THE UPSTREAM CONNECTION OF THE VALVE.	
UPSTREAM (adv)	In a direction opposite to the flow	MOVE THE PROBE UPSTREAM AND MONITOR THE TEMPERATURE AGAIN.	
UPSTREAM OF (prep)	In a direction opposite to the flow	DISCONNECT THE CONNECTION UPSTREAM OF THE VALVE.	
up to (prep)	UNTIL (prep)	UNTIL THIS STEP, THE UNIT IGNORES ALL OUTPUT SIGNALS.	Up to this step the unit ignores all output signals.
	THRU (prep)	DO STEPS 1 THRU 4.	Do all steps up to and including step 4.
	MAXIMUM (n)	YOU CAN INSTALL A MAXIMUM OF TEN WASHERS.	You can install up to ten washers.

Word (part of speech)	Approved meaning/ ALTERNATIVES	APPROVED EXAMPLE	Not approved example
upward (adv)	UP (adv)	MOVE THE LEVER UP.	Move the lever upward.
urgently (adv)	IMMEDIATELY (adv)	GET MEDICAL AID IMMEDIATELY.	Seek medical aid urgently.
usage (n)	USE (v)	DO NOT USE INCORRECT THINNERS. THIS CAN CAUSE PROBLEMS.	The usage of incorrect thinners can cause problems.
use (n)	OPERATION (n)	WHEN THE MOTOR IS IN OPERATION, KEEP YOUR HANDS AWAY FROM THE BLADE.	Keep your hands away from the blade when the motor is in use.
	USE (v)	USE PROTECTIVE GLOVES DURING THIS TASK.	The use of protective gloves is mandatory during this task.
USE (v), USES, USED, USED	To make something do its specified function	USE THE TEST BOX TO DO A TEST ON THE SYSTEM.	
		USE A SPECIAL CONTAINER FOR STORAGE.	
		DO NOT USE HIGH PRESSURE DURING THE CLEANING PROCEDURE.	
using (v)	USE (v)	USE A 1/2 INCH WRENCH TO REMOVE THE COVER.	Remove the cover using a 1/2 inch wrench.
	WITH (prep)	REMOVE THE COVER WITH A 1/2 INCH WRENCH.	Remove the cover using a 1/2 inch wrench.
USUAL (adj)	That you use or that occurs most frequently	IF YOU DO NOT GET THE USUAL RESULTS, DO A SYSTEM TEST.	
USUALLY (adv)	In a usual manner	USUALLY, THE HYDRAULIC FLUID FLOWS INTO THE VALVE THROUGH PORT A AND OUT THROUGH PORT B.	

Word (part of speech)	Approved meaning/ ALTERNATIVES	APPROVED EXAMPLE	Not approved example
utilization (n)	USE (v)	READ THE INSTRUCTIONS THAT TELL YOU HOW TO USE THE OXYGEN MASK.	Read the oxygen mask instructions for its utilization.
utilize (v)	USE (v)	USE AN OIL CAN TO LUBRICATE THE TURNBUCKLE.	Utilize an oil can for turnbuckle lubrication.

Blank Page

Word (part of speech)	Approved meaning/ ALTERNATIVES	APPROVED EXAMPLE	Not approved example
valid (adj)	CORRECT (adj)	MAKE SURE THAT THE TEST RESULTS ARE CORRECT.	Make sure that the test results are valid.
	APPLICABLE (adj)	THIS PROCEDURE IS APPLICABLE ONLY IN DIGITAL MODE.	This procedure is only valid in digital mode.
	This word becomes part of a technical name when associated to an official document.	NDT PERSONNEL MUST HOLD A VALID CERTIFICATE OF COMPETENCE AT LEVEL 1, 2 OR 3.	
validity (n)	CORRECT (adj)	MAKE SURE THAT THE TEST RESULTS ARE CORRECT.	Check the validity of the test results.
	APPLICABLE (adj)	THIS STEP IS APPLICABLE ONLY WHEN YOU APPLY THE TOP LAYER OF PAINT.	The validity of this step is restricted to the application of the paint topcoat.
VALUE (n)	A quantity that is calculated or given	MAKE SURE THAT THE VALUES AGREE WITH THE TOLERANCES.	
variable (adj)	CHANGE (v)	IF THE HUMIDITY CHANGES FREQUENTLY, PUT A COVER ON THE UNIT.	If the humidity is variable, put a cover on the unit.
various (adj)	DIFFERENT (adj)	REFER TO TABLE 5 FOR THE DIFFERENT TORQUE VALUES THAT ARE APPLICABLE TO THE UNIT.	Refer to Table 5 for the various torque values that are applicable to the unit.

Word (part of speech)	Approved meaning/ ALTERNATIVES	APPROVED EXAMPLE	Not approved example
vary (v)	CHANGE (v)	CHANGE THE FREQUENCY AND RECORD THE RESULTS.	Vary the frequency and record the results.
vent (v)	RELEASE (v)	RELEASE THE PRESSURE.	Vent the pressure.
	FLOW (v)	LET THE FUEL FLOW FREELY OVERBOARD.	Let the fuel vent freely overboard.
	OPEN (adj)	MAKE SURE THAT THE TANK IS IN AN OPEN AREA THAT HAS A GOOD AIRFLOW.	Make sure that the tank is in an area that is vented to the atmosphere.
ventilate (v)	AIRFLOW (n)	MAKE SURE THAT THE AREA WHERE YOU WILL USE THIS SOLVENT HAS GOOD AIRFLOW.	Ventilate the area where this solvent is used.
ventilation (n)	AIRFLOW (n)	MAKE SURE THAT THE AIRFLOW IS SUFFICIENT.	Make sure that the ventilation is sufficient.
verbal (adj)	TELL (v)	TELL THE CHIEF ABOUT THE RESULTS.	Give a verbal message about the results to the chief.
	HEAR (v)	MAKE SURE THAT A PERSON WHO IS NOT IN THE TANK CAN ALWAYS HEAR YOU.	Make sure that you maintain verbal contact with an operator outside the tank.
verify (v)	MAKE SURE (v)	MAKE SURE THAT THE FITTINGS ARE TIGHT.	Verify the tightness of fittings.
VERTICAL (adj)	At 90 degrees to the horizon	MEASURE THE VERTICAL DISTANCE BETWEEN THE TWO POINTS ON THE GRAPH.	
VERTICALLY (adv)	At 90 degrees to the horizon	THE POINTER MOVES VERTICALLY ALONG THE SCALE.	
VERY (adv)	To a high degree	ADD THE OIL VERY SLOWLY.	
via (prep)	THROUGH (prep)	GET ACCESS THROUGH THE No. 6 BREAK-IN PANEL.	Get access via No. 6 break-in panel.

Word (part of speech)	Approved meaning/ ALTERNATIVES	APPROVED EXAMPLE	Not approved example
vibrate (v)	VIBRATION (n)	IF THERE IS VIBRATION IN THE UNIT, DISCONNECT THE POWER SUPPLY.	If the unit vibrates, cut the power.
VIBRATION (n)	Regular movement from a center position	PRESSURE SURGES AND VIBRATION CAN CAUSE WEAR ON THE HOSES.	
vicinity (n)	NEAR (prep)	MAKE SURE THAT SOLVENTS ARE NOT NEAR THE ENGINE EXHAUST.	Make sure that solvents are not in the vicinity of the engine exhaust.
	ADJACENT (adj)	DO AN INSPECTION OF THE DAMAGE AND OF THE ADJACENT AREA.	Do an inspection of the damage and of the immediate vicinity.
VIEW (n)	The ability to see something	MAKE SURE THAT YOU HAVE A SATISFACTORY VIEW OF ALL COMPONENTS.	
view (v)	SEE (v)	THE BOLT WILL BE AT THE 2 O'CLOCK POSITION WHEN SEEN FROM THE REAR.	The bolt will be at 2 o'clock when viewed from the rear.
	LOOK (v)	THE BOLT WILL BE AT THE 2 O'CLOCK POSITION WHEN YOU LOOK AT THE PUMP FROM THE REAR.	The bolt will be at 2 o'clock when viewed from the rear.
vigorous (adj)	FORCE (TN)	WHEN YOU CLEAN THE CARPET, DO NOT USE TOO MUCH FORCE ON THE BRUSH. TOO MUCH FORCE CAN CAUSE DAMAGE TO THE CARPET.	Vigorous cleaning with a brush can cause damage to the carpet.
vigorously (adv)	FORCE (TN)	WHEN YOU MOVE THE CONTROL WHEEL, DO NOT APPLY TOO MUCH FORCE.	Do not move the control wheel too vigorously.

Word (part of speech)	Approved meaning/ ALTERNATIVES	APPROVED EXAMPLE	Not approved example
visible (adj)	SEE (v)	MAKE SURE THAT YOU CAN SEE THE OIL LEVEL THROUGH THE SIGHT GAUGE.	Make certain that the oil level is visible through the sight gauge.
	VIEW (n)	WHEN THE INDICATOR COMES INTO VIEW, STOP THE PUMP.	When the indicator is visible, stop the pump.
VISUAL (adj)	With the eyes	REMOVE THE VISUAL INDICATOR FROM THE VALVE.	
VISUALLY (adv)	With the eyes	VISUALLY EXAMINE THE DUCT.	
vital (adj)	IMPORTANT (adj)	AFTER THE TEST, IT IS IMPORTANT TO SET THE INDICATOR BACK TO ZERO.	It is vital to set the indicator back to zero after the test.
	MANDATORY (adj)	AFTER THE TEST, IT IS MANDATORY TO SET THE INDICATOR BACK TO ZERO.	It is vital to set the indicator back to zero after the test.
	NECESSARY (adj)	AFTER THE TEST, IT IS NECESSARY TO SET THE INDICATOR BACK TO ZERO.	It is vital to set the indicator back to zero after the test.
VOICE (n)	The sound you make when you speak	SPEAK INTO THE MICROPHONE IN YOUR USUAL VOICE.	
volatile (adj)	Use a different construction. Tell the reader what the risk or effect is.	BE CAREFUL WITH THE SOLVENT NEAR FLAMES OR SPARKS. THIS SOLVENT CAN RELEASE FLAMMABLE FUMES.	Be careful with the solvent (this solvent is volatile and can be dangerous).
VOLUME (n)	1. The space that an object fills	MEASURE THE VOLUME OF THE OIL CAREFULLY.	
	2. How loud a sound is	TO ADJUST THE VOLUME, USE THE BUTTONS ON THE SIDE OF THE PHONE.	

Word (part of speech)	Approved meaning/ ALTERNATIVES	APPROVED EXAMPLE	Not approved example
WAIT (v), WAITS, WAITED, WAITED	To stop doing something while another thing occurs.	(1) IF THE STATUS OF THE FAULT IS "NOT CONFIRMED": (a) OPEN THE APPLICABLE CIRCUIT BREAKER. (b) WAIT FOR 4 MINUTES. (c) DO THE BITE TEST.	
WALK (v), WALKS, WALKED, WALKED	To move on foot from one location to a different location	DO NOT WALK ON THIS AREA.	
WANT (v), WANTS, WANTED, WANTED	To intend, to desire	RECORD THE NAME OF THE FILE THAT YOU WANT TO DOWNLOAD.	
		IF YOU WANT TO STOP THE PROCEDURE, RECORD THE NUMBER OF THE LAST COMPLETED STEP.	
WARM (adj)	Moderately hot	KEEP THE GASKETS IN A WARM DRY AREA.	
warm (v)	WARM (adj)	MAKE THE MIXTURE WARM BEFORE YOU USE IT.	Warm the mixture before use.
warn (v)	TELL (v)	TELL PERSONNEL TO STAY AWAY FROM THE CONVEYOR SYSTEM.	Warn personnel to stay away from the conveyor system.
	WARNING (TN)	IF THE INSTRUCTIONS GIVE A WARNING ABOUT DANGEROUS PRESSURE, DO NOT OPEN THE CONTAINER.	If the instructions warn of a dangerous pressure, do not open the container.
wash (v)	CLEAN (v)	CLEAN YOUR HANDS WITH SOAP.	Wash your hands using soap.

Word (part of speech)	Approved meaning/ ALTERNATIVES	APPROVED EXAMPLE	Not approved example
watch (v)	MONITOR (v)	MONITOR THE SPEED INDICATION CONTINUOUSLY UNTIL THE END OF THE TEST.	Watch the speed indication for the duration of the test.
	LOOK (v)	LOOK FOR AIR BUBBLES.	Watch for air bubbles.
watertight (adj)	SEAL (v)	MAKE SURE THAT THE TANK IS SEALED.	Check that the tank is watertight.
way (n)	PROCEDURE (n)	DO NOT USE OTHER PROCEDURES TO REPLACE THE UNIT.	Do not use other ways to replace the unit.
WE (pron)	The manufacturer or the company which releases the documentation Do not use this pronoun for other meanings.	WE DO NOT RECOMMEND OTHER ALTERNATIVES.	
WEAK (adj) (WEAKER, WEAKEST)	With small strength, power, or concentration	USE A WEAK CLEANING SOLUTION.	
weakness (n)	WEAK (adj)	THERE IS A WEAK POINT IN THE STRUT.	There is a point of weakness in the strut.
WEAR (v), WEARS, WORE, WORN	To become damaged by friction For other meanings, use:	THE CABLE CAN WEAR QUICKLY IN THIS POSITION.	
	USE (v)	DURING THE SOLDERING PROCEDURE, USE A PROTECTIVE SHIELD.	During the soldering procedure, wear a protective shield.
	PUT ON (v)	PUT ON PROTECTIVE CLOTHING WHEN YOU USE THE SOLVENT.	Wear protective clothing when you use the solvent.
WEATHER (n)	Conditions of the atmosphere: temperature, moisture, winds, and clouds	STEP 2 IS APPLICABLE ONLY IF THE WEATHER IS WET.	
weep (n)	LEAKAGE (n)	IF THERE IS FUEL LEAKAGE FROM THE COUPLINGS, REPAIR THEM.	If there are fuel weeps from the couplings, repair them.

Word (part of speech)	Approved meaning/ ALTERNATIVES	APPROVED EXAMPLE	Not approved example
WEIGH (v), WEIGHS, WEIGHED, WEIGHED	1. To measure the weight of something	WEIGH THE TWO PARTS OF THE COMPOUND.	
	2. To have a specified weight	THE UNIT WEIGHS 20 KG.	
WEIGHT (n)	The force caused when gravity acts on the mass of an object	THE BASIC WEIGHT OF THE UNIT DOES NOT INCLUDE ALL ITEMS.	
well (adv)	CORRECTLY (adv)	MAKE SURE THAT THE BEARING IS CORRECTLY INSTALLED.	Make sure that the bearing is well seated.
	GOOD (adj)	THERE MUST BE A GOOD AIRFLOW IN THE WORK AREA.	The work area must be well ventilated.
	FULLY (adv)	ALL LABORATORY EQUIPMENT MUST BE FULLY CLEANED.	All laboratory equipment must be well cleaned.
WET (adj) (WETTER, WETTEST)	That has liquid on it, in it, or absorbed into it	DRY THE WET SURFACES.	
WHEN (conj)	At the time that or during	WHEN THE PISTON MOVEMENT STOPS, MEASURE THE TRAVEL.	
whenever (conj)	WHEN (conj)	STOP THE TEST WHEN THE PRESSURE IS MORE THAN 7000 kPa.	Stop the test whenever the pressure is more than 7000 kPa.
WHERE (conj)	At, to, or in which location	CLEAN THE AREA WHERE YOU APPLIED THE SEALANT.	
whether (conj)	IF (conj)	THIS LIMIT IS APPLICABLE IF THE AIRCRAFT IS ON THE LANDING GEAR OR ON JACKS.	This limit applies whether the aircraft is on the landing gear or on jacks.
WHICH (pron)	The thing or things that	EXAMINE THE EMERGENCY FUEL PUMP, WHICH IS ON THE LEFT SIDE.	

Word (part of speech)	Approved meaning/ ALTERNATIVES	APPROVED EXAMPLE	Not approved example
whichever (pron)	THAT (pron)	USE THE ADHESIVE THAT IS AVAILABLE.	Use whichever adhesive is available.
WHILE (conj)	At the same time	MAKE SURE THAT A PERSON HOLDS THE ITEM, WHILE YOU DISCONNECT IT.	
whilst (conj)	WHILE (conj)	DO NOT OPEN THE HOUSING WHILE THE ELECTRICAL POWER IS CONNECTED.	Do not open the housing whilst the electrical power is connected.
WHO (pron)	The person or personnel that	PERSON B, WHO OPERATES THE REAR WINCH, MUST LOCK THE FIXTURE.	
whoever (pron)	WHO (pron)	ALL PERSONS WHO ARE IN THIS AREA MUST USE A RESPIRATOR.	Whoever is in this area must use a respirator.
whole (adj)	FULL (adj)	DO THE FULL PROCEDURE.	The whole procedure must be done.
	ALL (adj)	EXAMINE ALL OF THE SYSTEM TO FIND THE CAUSE OF INTERFERENCE.	Examine the whole system to find the cause of interference.
wholly (adv)	FULLY (adv)	THE SYSTEM IS FULLY AUTOMATIC.	The system is wholly automatic.
whose (pron)	Use a different construction.	THE GENERATORS SUPPLY A THREE-PHASE VOLTAGE. THE FREQUENCY OF THIS VOLTAGE INCREASES WHEN THE GENERATOR ROTOR SPEED INCREASES.	The generators produce a three-phase voltage whose frequency is proportional to the generator rotor speed.
WIDE (adj) (WIDER, WIDEST)	That has large width	REPAIR ALL CRACKS THAT ARE WIDER THAN 0.05 mm.	
WIDTH (n)	The smaller of two dimensions, the second-largest of three dimensions	MEASURE THE WIDTH OF THE CRACK AT THE WIDEST POINT.	

Word (part of speech)	Approved meaning/ ALTERNATIVES	APPROVED EXAMPLE	Not approved example
WILL (v) No other forms of this verb.	Helping verb that shows simple future tense	WARNINGS AND CAUTIONS IN THIS MANUAL WILL HELP YOU TO DO THE WORK SAFELY AND CORRECTLY.	
WIND (v), WINDS, WOUND, WOUND	To move around and around an object	WIND THE TAPE ON THE REEL.	
windy (adj)	WIND (TN)	WHEN THERE ARE STRONG WINDS, OBEY THESE SPECIAL PRECAUTIONS.	Take special precautions in very windy conditions.
wipe (v)	CLEAN (v)	BEFORE EACH CYCLE, CLEAN THE INDICATOR.	Wipe the indicator surfaces before each cycle.
wire (v)	ATTACH (v) (WITH WIRE [TN])	ATTACH THE HOSE TO THE FIXTURE WITH WIRE.	Wire the hose to the fixture.
wire-lock (v)	LOCK (v) (WITH LOCKWIRE [TN])	LOCK THE FOUR BOLTS TO EACH OTHER WITH LOCKWIRE.	Wire-lock the four bolts to each other.
	SAFETY (v) (WITH SAFETY WIRE [TN])	SAFETY THE FOUR BOLTS TO EACH OTHER WITH SAFETY WIRE.	Wire-lock the four bolts to each other.
WITH (prep)	Function word that shows association or relationship, help or sharing, a means or instrument	ALIGN THE MARK WITH THE LONGITUDINAL AXIS.	
		ATTACH THE FLANGE (10) WITH THE FOUR BOLTS (15).	
		REMOVE THE VALVE (2) WITH THE EXTRACTOR.	
withdraw (v)	REMOVE (v)	REMOVE THE INDICATOR FROM THE INSTRUMENT PANEL.	Withdraw the indicator from the instrument panel.

Word (part of speech)	Approved meaning/ ALTERNATIVES	APPROVED EXAMPLE	Not approved example
within (prep)	IN (prep)	A FILTER PREVENTS CONTAMINATION OF THE COMPONENTS IN THE SYSTEM.	A filter prevents contamination of the components within the system.
	IN...OR LESS	LIGHT L4 COMES ON IN 5 SECONDS OR LESS.	Light L4 illuminates within 5 seconds.
	IN LESS THAN	MAKE SURE THAT THE WARNING LIGHT COMES ON IN LESS THAN 3 SECONDS.	Make sure that the warning light illuminates within 3 seconds.
WITHOUT (prep)	Not with	SOME DAMAGE IS PERMITTED WITHOUT REPAIR. REFER TO TABLE 1 FOR THE APPLICABLE LIMITS.	
withstand (v)	RESISTANT (adj)	THIS MATERIAL IS RESISTANT TO FIRE.	This material can withstand fire.
WORK (n)	That which you do when you use physical strength, or mental power	DO THE WORK IN A CLEAN AREA.	
work (v)	WORK (n)	BE CAREFUL WHEN YOU DO WORK NEAR FUEL TANKS.	Be careful when you work near fuel tanks.
would (v)	CAN (v)	SOLVENTS THAT STAY ON THE PART CAN CAUSE CORROSION.	Solvents left on the part would corrode the part.
wrap (v)	PUT (v)	PUT THE PART IN OILPAPER.	Wrap the part in oilpaper.
	WIND (v)	WIND MASKING TAPE AROUND THE PART.	Wrap masking tape around the part.
WRITE (v), WRITES, WROTE, WRITTEN	To record data or information as words, letters, or symbols	WRITE THE TEST DATE ON THE CERTIFICATE.	

Word (part of speech)	Approved meaning/ ALTERNATIVES	APPROVED EXAMPLE	Not approved example
wrong (adj)	INCORRECT (adj)	IDENTIFY THE BELLCRANK AND SHAFT WITH MARKS. THIS WILL PREVENT AN INCORRECT INSTALLATION.	Mark the bellcrank and shaft to prevent wrong installation.
wrongly (adv)	INCORRECTLY (adv)	IF THE HOSES ARE INCORRECTLY INSTALLED, DAMAGE CAN OCCUR.	Damage can occur if hoses are wrongly installed.

Word (part of speech)	Approved meaning/ ALTERNATIVES	APPROVED EXAMPLE	Not approved example

Blank Page

Word (part of speech)	Approved meaning/ ALTERNATIVES	APPROVED EXAMPLE	Not approved example
YES (adv)	Function word that shows the positive answer to a question	DOES THE LIGHT COME ON? YES OR NO?	
yet (conj)	BUT (conj)	TIGHTEN THE NUTS, BUT KEEP THEM SUFFICIENTLY LOOSE TO REMOVE THEM WITH YOUR HAND.	The nuts should be tightened, yet loose enough to remove by hand.
	AT THIS TIME	DO NOT REMOVE THE FIXTURE COVER AT THIS TIME.	Do not remove the fixture cover yet.
YOU (pron)	The reader or the user	YOU CAN CONTINUE THE TEST.	
YOUR (adj)	Related to the reader or the user	IF YOU GET SOLVENT IN YOUR EYES, FLUSH THEM IMMEDIATELY WITH WATER.	
yourself (pron)	YOU (pron)	DO NOT USE YOUR HANDS TO PREVENT MOVEMENT OF THE LP COMPRESSOR. IF YOU DO, THE LP COMPRESSOR CAN CAUSE YOU INJURY AND CAN CAUSE DAMAGE TO THE EQUIPMENT.	Do not use your hands to prevent movement of the LP compressor. If you do, you can cause injury to yourself and damage to the equipment.
zero (v)	ZERO (TN)	ADJUST THE METER TO ZERO.	Zero the meter.

<Reference >

1. ASD-STE100 (issued by ASD, Brussels, Belgium)

2. ICAO definitions Annex

3. MECHANICAL/ELECTRICAL ENTRY COURSE(The BOEING Company)

4. Avionics for AMEs: Aircraft Maintenance Engineer Handbook
   (Ministry of Land, Infrastructure and Transport, 2015 Dec 31)

<편찬위원>

이상희 항공우주기술협회장

오경훈 항공우주기술협회 운영이사

한상경 항공우주기술협회 사무국장

하영태 호원대학교 항공정비공학과 교수

<감수진>

강영식 항공기술사

김형래 한국폴리텍대학 교수

도종봉 주)샤프에비에이션케이

박환관 경운대학교 항공기술교육원장

이학재 극동대학교 교수

김영준 인하공업전문대학 교수

나금풍 전)경북전문대학 교수

박상윤 경남도립남해대학 교수

이명원 정석항공과학고등학교 부원장

SIMPLIFIED TECHNICAL ENGLISH: Specification ASD-STE100

이 책의 한국어판 저작권은 사단법인 한국항공우주기술협회와 경문사에 있습니다.
저작권법에 의해 한국 내에서 보호를 받는 저작물이므로 무단전재와 복제를 금합니다.

저자와의
합의하에
인지를
생략합니다.

Aviation Technical English

항공기술영어 · ASD-STE100

지은이 ASD, AeroSpace and Defence Industrise Association of Europe
편찬위원 이상희 · 오경훈 · 한상경 · 하영태
펴낸이 조경희
펴낸곳 경문사
펴낸날 2022년 7월 18일 1판 1쇄
등 록 1979년 11월 9일 제1979-000023호
주 소 04057, 서울특별시 마포구 와우산로 174
전 화 (02)332-2004 팩스 (02)336-5193
이메일 kyungmoon@kyungmoon.com

값 25,000원

ISBN 979-11-6073-560-4

★ 경문사의 다양한 도서와 콘텐츠를 만나보세요!

홈페이지 www.kyungmoon.com	페이스북	facebook.com/kyungmoonsa
포스트 post.naver.com/kyungmoonbooks	블로그	blog.naver.com/kyungmoonbooks
북이오 buk.io/@pa9309	유튜브	https://www.youtube.com/channel/UClDC8x4xvA8eZlrVaD7QGoQ

경문사 출간 도서 중 수정판에 대한 **정오표**는 **홈페이지 자료실**에 있습니다.